Computational Approaches to Energy Materials

Computational Approaches to Energy Materials

Edited by

ARON WALSH

Department of Chemistry, University of Bath, UK

ALEXEY A. SOKOL

Department of Chemistry, University College London, UK

C. RICHARD A. CATLOW

Department of Chemistry, University College London, UK

WILEY

A John Wiley & Sons, Ltd., Publication

This edition first published 2013

Registered office
John Wiley & Sons Ltd, The Atrium, Southern Gate, Chichester, West Sussex, PO19 8SQ, United Kingdom

For details of our global editorial offices, for customer services and for information about how to apply for permission to reuse the copyright material in this book please see our website at www.wiley.com.

Library of Congress Cataloging-in-Publication Data

Computational approaches to energy materials / edited by Aron Walsh, Alexey A. Sokol, Richard Catlow.
1 online resource.
Includes bibliographical references and index.
Description based on print version record and CIP data provided by publisher; resource not viewed.
ISBN 978-1-118-55143-1 (MobiPocket) – ISBN 978-1-118-55144-8 (ePub) – ISBN 978-1-118-55145-5 (Adobe PDF) – ISBN 978-1-119-95093-6 (hardback) 1. Energy storage–Mathematical models. 2. Electron distribution–Mathematical models. 3. Energy conversion–Mathematical models. I. Walsh, Aron, editor of compilation. II. Sokol, Alexey A., editor of compilation. III. Catlow, C. R. A. (Charles Richard Arthur), 1947, editor of compilation.
TK2896
621.31–dc23

2013007643

A catalogue record for this book is available from the British Library.

ISBN: 9781119950936

Set in 11/13pt Times by Aptara Inc., New Delhi, India.

Printed and bound in Singapore by Markono Print Media Pte Ltd

Contents

About the Editors

Dr Aron Walsh is a Royal Society University Research Fellow in the Centre for Sustainable Chemical Technologies at the University of Bath. He obtained his BA (Mod) and PhD in computational chemistry from Trinity College Dublin. After receiving the Royal Irish Academy Young Chemist award, he worked for the US Department of Energy at the National Renewable Energy Laboratory and moved to the United Kingdom in 2009 as a Marie Curie Research Fellow at University College London. His research experience to date has followed a coherent path, applying a range of computational techniques to challenging problems in the areas of solid-state chemistry and physics, with a particular emphasis on the description of defect processes in semiconductors. He has authored over 100 peer-reviewed publications.

Dr Alexey A. Sokol, a senior research associate in the Department of Chemistry at University College London (UCL), has worked on the development and applications of computational methods to solid-state physics, chemistry and materials science for over 20 years. His early work concerned disordered materials and interaction of high-energy radiation with semiconductor devices. His PhD project at the Royal Institution of Great Britain and UCL concentrated on the development of a theory of defects in zeolites as centers of chemical activity. On completing this project, Dr Sokol took part in the EU ESPRIT project QUASI, where he has developed a solid-state embedded cluster QM/MM technique implemented in the computational chemistry environment software ChemShell. More recently, these methods have been successfully applied to nanosystems, starting with small clusters and then being extended to larger nanoparticles, wires, tubes and thin films. His current work focuses on excited states of bulk defects and nanostructures.

Professor C. Richard A. Catlow, Dean of Mathematical and Physical Sciences at University College London and Fellow of the Royal Society, has worked for over 30 years in the field of computational and experimental studies of complex inorganic materials. His group has pioneered a wide range of applications of computational techniques in solid-state chemistry to systems and problems including microporous and oxide catalysts, ionic conductors, electronic ceramics and silicate minerals. This applications program has been supported by technique and code development, including recent work on embedded cluster methodologies for application to the study of catalytic reactions. The computational work has been firmly linked with experimental studies, using both neutron scattering and synchrotron radiation techniques, where the Royal Institution group has also made notable contributions to development as well as application studies. Professor Catlow's research has led to over 800 publications, and in 2004 he was elected to Fellowship of the Royal Society for "pioneering the development and application of computer modeling in solid state and materials chemistry."

List of Contributors

Silvana Botti, Laboratoire des Solides Irradiés and ETSF, École Polytechnique, CNRS, CEA-DSM, Palaiseau, France; LPMCN, CNRS, Université Lyon 1, Villeurbanne, France

C. Richard A. Catlow, Department of Chemistry, University College London, London, UK

Dorothy Duffy, Department of Physics and Astronomy, University College London, London, UK

Craig A.J. Fisher, Nanostructures Research Laboratory, Japan Fine Ceramics Center, Nagoya, Japan

Emiel J.M. Hensen, Department of Chemical Engineering and Chemistry, Eindhoven University of Technology, Eindhoven, The Netherlands

M. Saiful Islam, Department of Chemistry, University of Bath, Bath, UK

Anderson Janotti, Materials Department, University of California, Santa Barbara, USA

Yong-Hyun Kim, Graduate School of Nanoscience and Technology, KAIST, Daejeon, South Korea

Emmanouil Kioupakis, Materials Department, University of California, Santa Barbara, USA; Department of Materials Science and Engineering, University of Michigan, Ann Arbor, USA

Eugene A. Kotomin, Max Planck Institute for Solid State Research, Stuttgart, Germany; Institute for Solid State Physics, University of Latvia, Riga, Latvia

Maija M. Kuklja, Materials Science and Engineering Department, University of Maryland, College Park, USA

Viet-Duc Le, Graduate School of Nanoscience and Technology, KAIST, Daejeon, South Korea

Joachim Maier, Max Planck Institute for Solid State Research, Stuttgart, Germany

Rapela R. Maphanga, Materials Modelling Centre, University of Limpopo, Sovenga, South Africa

Yuri A. Mastrikov, Institute for Solid State Physics, University of Latvia, Riga, Latvia; Materials Science and Engineering Department, University of Maryland, College Park, USA

Rotraut Merkle, Max Planck Institute for Solid State Research, Stuttgart, Germany

Phuti E. Ngoepe, Materials Modelling Centre, University of Limpopo, Sovenga, South Africa

Evgeny A. Pidko, Department of Chemical Engineering and Chemistry, Eindhoven University of Technology, Eindhoven, The Netherlands

Patrick Rinke, Materials Department, University of California, Santa Barbara, USA; Fritz-Haber-Institut der Max-Planck-Gesellschaft, Berlin, Germany

Rutger A. van Santen, Department of Chemical Engineering and Chemistry, Eindhoven University of Technology, Eindhoven, The Netherlands

Dean C. Sayle, Defence College of Management and Technology, Cranfield University, Shrivenham, UK

Alexey A. Sokol, Department of Chemistry, University College London, London, UK

Chris G. Van de Walle, Materials Department, University of California, Santa Barbara, USA

Julien Vidal, Physics Department, King's College London, London, UK

Aron Walsh, Department of Chemistry, University of Bath, Bath, UK

Qimin Yan, Materials Department, University of California, Santa Barbara, USA

Preface

The importance of efficient and sustainable energy technologies has grown enormously over the past decade, driven primarily by concerns over global warming, diminishing fossil-fuel reserves, the need for energy security and increasing consumer demand for portable electronics.

Stricter legislation regarding carbon dioxide emissions from road transport vehicles, combined with increasing fuel prices, will no doubt continue to encourage the introduction of even greater numbers of hybrid electric and fully electric vehicles. Next-generation energy technologies such as these can only come about through the development and optimization of high-performance materials. This, of course, cannot be achieved efficiently without a thorough understanding of the fundamental science of complex solids, an understanding that underpins applied research in this multidisciplinary field.

Computational methods are now an integral and indispensable part of the materials characterization and development process, as experimental techniques are often used at their fundamental limits (i.e., at the atomic and subatomic scales). Today, modeling of structures and properties of materials at this fundamental level is not only useful for confirming experimental results and enabling their correct interpretation but also increasingly being used as a predictive tool that can guide experimental research efforts.

Following an overview of the principles of atomistic modeling, our book focuses on the materials used for clean energy generation and storage, specifically on the development of new materials for thin-film solar cells, radiation-resistant materials for nuclear power and ion-conducting materials for batteries. The challenges involved in using hydrogen as an energy carrier necessitate research into new classes of materials for hydrogen storage. The latter naturally couples to the key issues of alternative energy storage and utilization in fuel cells, batteries and solid-state lighting. We conclude with a general account of the search for stable nanostructures, as nanostructured materials are of key importance in contemporary energy technologies and pose significant challenges for computational structure and property prediction.

Acknowledgments

Computational materials science represents a vibrant and rapidly expanding subject area. The techniques described in this book have been built from developments spanning across the last century, and we acknowledge all of the scientists who have contributed to their advancement and application, especially those who have not been explicitly referenced in this text.

We thank all authors for their contributions, Déborah Demathieu for her assistance, and the team at Wiley for their help throughout the production process.

1

Computational Techniques

C. Richard A. Catlow[1], *Alexey A. Sokol*[1], *and Aron Walsh*[2]
[1]*Department of Chemistry, University College London, London, UK*
[2]*Department of Chemistry, University of Bath, Bath, UK*

1.1 Introduction

In this chapter, we introduce fundamental approaches and ideas, which will be exploited in the rest of the book. These can be divided into two main streams: one dealing with the motion of atoms or ions described at a simplified level of theory and another focusing on electrons. The modeling framework, which covers both streams, is outlined next.

1.2 Atomistic Simulations

1.2.1 Basic Concepts

Methods based on interatomic potentials have a major and continuing role in molecular and materials simulation. The concept of the potential is simple: the energy (E) of the system is written as either an analytical or possibly a numerical function of the nuclear coordinates, r_i, of particles $i = 1$ to n:

$$E = E(r_1, r_2, r_3, \ldots, r_n).$$

The function will normally be written as a sum of terms that depend on the coordinates of two, three or more atoms, although in many potential models,

Computational Approaches to Energy Materials, First Edition. Edited by Aron Walsh, Alexey A. Sokol and C. Richard A. Catlow.

especially those for ionic and semi-ionic solids, only two body terms are considered; for the latter class of material, the electrostatic term is normally separated, that is,

$$E = E_{\text{coul}} + E_{\text{SR}},$$

where the Coulomb energy, E_{coul}, is obtained by summing over all the interactions between the atomic charges, which are a parameter of the model and must be assigned. The short-range energy, E_{SR}, represents the remainder of the interactions including Pauli repulsion and covalent and dispersive attractive terms. Simple analytical functions are commonly used for E_{SR}, including inverse power (r^{-m}) and exponential terms ($\exp(-r/r_0)$. Detailed discussions can be found in the seminal book of Born and Huang [1], and more recent literature [2,3].

In modeling metallic systems, rather different approaches must be adopted; in particular, the effects of the conduction band electrons on atomic interactions must be included—a difficult problem for which there is no simple solution. Nevertheless, a number of effective and useful potential models have been developed for metallic systems, which may be based on the "embedded atom concept." Details and examples are given in [4].

Accurate models especially for ionic solids must include a representation of polarization. It has long been known that in solid-state modeling simple point dipole models have serious inadequacies, leading to excessive polarization, as they omit any representation of the damping of polarization by the resulting increase in short-range repulsion with neighboring ions. This problem was simply and elegantly solved by the development over 50 years ago of the shell model by Dick and Overhauser [5]. This crude but effective model describes an atom or ion in terms of a "core," which contains all the mass and represents the nucleus and core electrons, and a "shell" (of charge, Y), which is massless and represents the polarizable valence shell electrons; the core and the shell are coupled by an harmonic spring (of constant, k), and the development of a dipole moment is modeled by the displacement of the shell relative to the core. The charge of the shell (Y) and the value of the spring constant (k) are parameters of the model; and of course, the sum of core and shell charges must equal the total atomic charge. Moreover, the shell model parameters can be related to the polarizability (α) by the simple relationship:

$$\alpha = \frac{Y^2}{k}.$$

Elaborations such as the "breathing shell" model have been developed, but the basic shell model remains the most widely used treatment of polarizability in materials simulation.

A potential model will normally therefore consist of (i) a set of atomic charges, where appropriate, (ii) analytical (or occasionally numerical) functions, containing variable parameters, and (iii) a representation of polarizability for short-range interactions, which will require specification of the parameters Y and k when the shell model is used. In Section 1.2.2, we review the methods used to set the variable parameters and then we return to some of the more common potential models.

1.2.2 Parameterization

Once the choice of the form of the potential model has been made, the crucial next step is to parameterize the model, that is, fix the variable parameters, so that the model describes the specific material (or materials) under investigation. Here, there are two broad strategies, which may in some cases be used in concert:

1. *Empirical fitting*: involves variation of the parameters in order to reproduce, as accurately as possible, experimental data of the material. Standard procedures are available for calculating a wide range of properties using potential models (see, e.g., [3]). These are usually coupled to a least-squares minimization procedure to achieve the best fit of calculated to experimental data. Commonly used data include cohesive or lattice energies, crystal structures, elastic and dielectric properties and where available lattice dynamical data. The procedure is simple in concept and highly automated in principle, but in practice it may prove difficult and lengthy and require extensive user intervention and direction to achieve the optimum parameter set. And, of course, it requires that suitable and accurate experimental data be available.
2. *Fitting to energy surfaces*: requires no empirical data, but rather uses energy surfaces calculated by electronic structure methods, with parameters in the potential model being varied to ensure that the surface calculated using the potential model matches as closely as possible that determined by the electronic structure technique. The energy surface is constructed by varying the structural parameters of the material or molecule in a systematic manner, followed by a least squares fitting of the potential parameters. The approach is again in principle straightforward but of course requires an accurate energy surface to which to fit the potential parameters.

Both approaches are widely used and as noted they may be used together, and indeed a potential derived by the latter approach should always be tested in regards to the extent to which it reproduces any available experimental data. More generally, in evaluating a potential model, it is necessary to examine carefully its mode of derivation. When empirical methods are used, the range and accuracy of the data will be crucial; when parameters have been derived from calculated energy surfaces at a higher level of theory, the quality of electronic structure technique will determine the accuracy of the parameterized model.

1.2.3 Parameter Sets

A wide range of parameter sets are available for different classes of material and many can be found in online databases [6]. For oxides, which are extensively used in energy materials, the Born model parameter set derived by Catlow and Lewis [7] may often provide a useful starting point as these parameters have the merit of simplicity and transferability between different materials, which may be an important factor in assessing the suitability of a potential model for applications, in which several materials are investigated and compared. Other significant considerations when deciding on the suitability of a model are accuracy—that is, the extent to which the model reproduces known crystal properties—and stability—an important consideration as models may perform well around the equilibrium configuration of a crystal but have instabilities for other configurations that may be sampled in dynamical simulations or simulations of defective crystals. More generally the assessment and choice of a potential model is crucially important and needs careful and detailed consideration.

1.2.4 Implementation

Having developed or chosen a suitable model for calculating energies and forces as a function of nuclear coordinates, they may be implemented in a wide range of powerful simulation tools (e.g., CP2K, DL-POLY, GULP, GROMACS, KLMC, LAMMPS, METADISE), based on three main concepts:

1. *Minimization*: A conceptually simple approach, in which the aim is to locate the energy minimum configuration of the system modeled, with the energy calculated using an interatomic potential model or by an electronic structure technique. The complexity of energy landscapes may, however, make the identification of the global minimum far from straightforward, and a range of both sophisticated search and minimization algorithms have been developed. Minimization is perhaps at its most effective when refining approximately known structures, although developments in search procedures for energy landscapes have given the techniques an increasingly predictive value [8, 9]. Minimization may be applied to any type of atomic assembly including crystals, molecules and adsorbed species. The approach has been applied with particular effect to defects in solids where the method, originally pioneered by Mott [10], effectively minimizes the energy of a region of crystal surrounding the defect with more approximate quasi-continuum treatments of the more distant regions of the lattice. Energy minimization may also be extended to free energy minimization when entropies can be calculated by, for example, the vibrational partition function in a crystalline solid [11]. The technique has been further developed to study transition states, or more generally, minimum energy pathways as in the popular nudged-elastic band (NEB) approach. Overall, despite its basic simplicity and obvious limitations in omitting any explicit representations of dynamic effects, minimization is a robust and powerful approach and should often be the first approach of a simulation study.
2. *Molecular Dynamics (MD)*: Here, again the basic idea is simple: the time evolution of a system at the molecular level is followed by solving the classical equations of motion—essential Newtonian mechanics for molecules. In practice, the equations of motion are solved using a numerical, iterative procedure, employing repeated applications of a time step, during which both atomic positions and velocities are updated using the known velocities and forces, respectively. The

time step must, of course, be shorter than the characteristic time associated with any important process (e.g., the period of an atomic or molecular vibration) and values of typically 1 fs are chosen. A simulation proceeds by initially setting the positions and velocities of the particles to be included in the simulation; the latter are chosen with target temperature in mind. During the initial "equilibration" stage of the simulation, the system reaches thermal equilibrium, and it is normally necessary to scale the velocities a number of times in order to retain the simulation at the target temperature. Once the system is in equilibrium at the specified temperature, the production stage commences and the data from the simulation (positions and velocities) are stored for subsequent analysis.

MD simulations yield a wealth of information—structural properties via radial distribution functions, information on dynamical properties via a range of correlation factors and diffusion coefficients via the variation of particle mean square displacements with time. Although originally developed to model properties of liquid systems, they have found extensive applications in solids, especially those with high ionic mobility, that is, fast ion conductors, which find important applications in both batteries and fuel cells. The technique does, however, have substantial limitations, most notably the limited amount of "real time" that can be explored with a simulation and the limitations on the size of the simulation cell. The two are, of course, related; but with contemporary simulations with realistic models, implemented on high performance computing (HPC) platforms, simulations are generally confined to the nanosecond time range, with simulation boxes containing several thousand particles. Of course, longer time scales and larger simulation boxes may be achieved given very substantial resources, but in general the phenomena of interest must be accessible within the limitations placed on the technique by the time and size constraints. One major consequence is that standard MD cannot effectively model rare events, which we may define as those involving energies considerably in excess of k_BT, as such events will be insufficiently sampled (or not sampled at all) during the course of a simulation. There are a number of recent developments including the increasingly used metadynamics that alleviate this problem, but it does remain intrinsic to the technique. As with minimization techniques, MD can be applied to molecules, clusters and systems with 2D or 3D periodicity. The technique is powerful and flexible, yielding quantitative information can often give valuable insight via graphical displays of configurations and migration mechanisms.

3. *Monte Carlo (MC) simulations*: A stochastic method alternative to MD for the sampling of large and complex configurational spaces using random numbers. This method is particularly useful to assess higher energy configurations or take the system of interest over a high potential energy barrier, which would be inaccessible in realistic MD time scales. Another advantage of MC is the possibility to explore remote regions of configurational space. In modeling materials structure and thermodynamic properties, the Metropolis algorithm is commonly applied to bias the search toward thermally accessible configurations, which is controlled

by the simulation temperature. The approach can be extended to include variation in the number of particles based on the chemical potential in Grand Canonical MC. In contrast to direct minimization, both MC and MD techniques are suitable to study nonequilibrium, rapidly evolving systems and processes. While MC methods, unlike MD, do not give information on the time evolution of a system, an adaptation of the technique—Kinetic Monte Carlo (KMC)—can allow time dependent processes to be modeled by sampling events, such as atomic migration processes, rather than particle positions.

1.3 Electronic Structure Techniques

The atomistic simulation techniques previously discussed are concerned with describing the energy landscape of individual atoms or ions, where classical mechanics can be usefully employed as the first successful approximation. Electrons are much lighter particles with essential quantum behavior. The goal of quantum chemical approaches, or more broadly, the electronic structure techniques, is to provide the description of electrons. The distribution of electrons in turn determines the structural, optical and magnetic structure of molecules and solids. As electrons are charged particles with spin 1/2 in atomic units, they interact with each other coulombically, and their motion is correlated in different manner depending on mutual orientation of their spins. Quantum mechanics of many-electron systems describes these as effects of electron exchange and correlation, which should be taken into account for a proper quantitative description. Further in materials, the behavior of large numbers of electrons is described using quantum statistical methods suitable for fermions (i.e., Fermi-Dirac statistics), where one typically deals with the Fermi gas or liquid models.

The most advanced quantum chemical methods are based on the (approximate) solution of the Schrödinger equation for the many-electron wavefunction, while density functional theory (DFT) is developed around the electron density, which is a one-electron property. The result for the majority of contemporary electronic structure techniques is the ground-state electron distribution and associated one-electron properties. A distinction is generally made between first-principles (or *ab initio*) methods that contain no external parameterization beyond fundamental physical constants, and semi-empirical methods, which typically replace more computationally demanding terms with parameters obtained from a fit to a large data set of experimental measurements or calculations at a higher level of theory.

Beyond the one-electron approaches, the full many-body system can be treated, or excited states can be calculated, building up from the ground-state single-particle description. The methods discussed in this section are well developed and have been implemented in a wide range of academic and commercial software packages; a number of such codes are listed in Table 1.1. While any chemical system can, in principle, be calculated using these methods, care must be taken, especially

Table 1.1 *A selection of actively developed electronic structure codes suitable for modeling energy materials*

Code	Method	Basis set	Web site
ABINIT	DFT	Plane waves	http://www.abinit.org
ADF/BAND	DFT	Slater local orbitals	http://www.scm.com
CASTEP	DFT	Plane waves	http://www.castep.org
CONQUEST	DFT/HF	Numerical functions	http://hamlin.phys.ucl.ac.uk
CPMD	DFT/HF	Plane waves	http://www.cpmd.org
CP2K	DFT/HF	Gaussian orbitals/plane waves	http://cp2k.berlios.de
CRYSTAL	DFT/HF	Gaussian local orbitals	http://www.crystal.unito.it
DFTB+	DFTB	Tight binding atomic orbitals	http://www.dftb-plus.info
$DMOL^3$	DFT	Numerical functions	http://accelrys.com
ELK	DFT/HF	Augmented plane waves	http://elk.sourceforge.net
EXCITING	DFT	Augmented plane waves	http://exciting-code.org
FLEUR	DFT	Augmented plane waves	http://www.flapw.de
FHI-AIMS	DFT/HF	Numerical functions	https://aimsclub.fh-berlin.mpg.de
GAUSSIAN	DFT/HF	Gaussian local orbitals	http://www.gaussian.com
GPAW	DFT	Numerical functions	https://wiki.fysik.dtu.dk/gpaw
ONETEP	DFT	Wannier functions	http://www2.tcm.phy.cam.ac.uk/onetep
PETOT	DFT	Plane waves	https://hpcrd.lbl.gov/~linwang/
SIESTA	DFT	Numerical functions	http://www.icmab.es/siesta
SPHINX	DFT/DFTB	Plane waves	http://www.mpie.de/index.php?id=sxlib
QUANTUM-ESPRESSO	DFT/HF	Plane waves	http://www.quantum-espresso.org
VASP	DFT/HF	Plane waves	http://cms.mpi.univie.ac.at/vasp
WIEN2K	DFT/HF	Augmented plane waves	http://www.wien2k.at
Opium		Pseudopotential generator	http://opium.sourceforge.net
BSE		Gaussian basis set/effective core potential database	https://bse.pnl.gov/bse

DFT, density functional theory; HF, Hartree–Fock; DFTB, density functional tight-binding.
For updated descriptions see http://www.psi-k.org/codes.shtml

for open-shell or highly correlated d and f shell systems to ensure that physically meaningful results are obtained.

For any numerical electronic structure approach, a tractable representation of the one-electron wavefunction (or orbital) is required, which is usually obtained from a linear combination of simple functions referred to as the basis set. These can take many forms, and are one distinguishing factor between computer codes (see Table 1.1). Introduction of a basis set reduces the problem of solving simultaneous differential equations to that of linear algebraic equations with coefficients determined by matrix elements (integrals in real space) between such basis functions. Due to the ease of computing electron integrals, Gaussian functions, $x^k y^l z^m e^{-\alpha r^2}$, are the most widely used basis for molecular calculations, while for solids, plane waves of the form $e^{i\mathbf{kr}}$ act as a more suitable basis (following the Bloch theorem for a periodic potential as discussed below) and offer high accuracy and efficiency.

As the more weakly bound valence electrons make the primary contributions to chemical bonding, a common approximation is to replace the core states by an analytical function that results in the same effective potential for the valence electrons, but at a greatly reduced computational cost. These are known as pseudopotentials or effective core potentials, and exist in many flavors, for example, accurate norm-conserving pseudopotentials [12] or computationally efficient ultrasoft pseudopotentials developed by Vanderbilt [13]. This approach to saving the computational effort has close similarities with a more recent method of projector-augmented wave potentials developed by Blöchl [14], which relies on an explicit but simplified representation of core electronic states in the inner atomic regions. Many codes now come with their own optimized set of potentials, but there are also databases of transferable potentials and pseudopotential generators available for the entire periodic table.

1.3.1 Wavefunction Methods

The time-independent, nonrelativistic Schrödinger equation can be expressed concisely as

$$\hat{H}\Psi = E\Psi,$$

where Ψ is the wavefunction, E is the energy and $\hat{H}$ is the Hamiltonian, the sum of the kinetic and potential energy operators. The kinetic operator can be divided into the contributions from the nuclei and the electrons, and the potential as the sum of the nuclei–nuclei interactions, the electron–electron interactions and the nuclei–electron interactions:

$$\hat{H} = \hat{K}_{\text{nuclei}} + \hat{K}_{\text{electrons}} + \hat{V}_{\text{nuclei}} + \hat{V}_{\text{electrons}} + \hat{V}_{\text{nuclei-electrons}}.$$

Solution of the eigenvalue equation results in the eigenfunctions (state wavefunctions) and the eigenvalues (state energies).

To make solution of the Schrödinger equation practical for many-atom and many-electron systems, we can separate nuclear and electronic degrees of freedom. To this end, we rearrange the Hamiltonian as follows:

$$\hat{H} = [\hat{K}_{\text{nuclei}} + \hat{V}_{\text{nuclei}}] + \hat{K}_{\text{electrons}} + \hat{V}_{\text{electrons}} + \hat{V}_{\text{nuclei-electrons}}.$$

Differences in inertia between the nuclei and electrons mean that the nuclei are almost static compared to the electrons and that their kinetic energy and the nuclear–nuclear interactions can be calculated separately. This forms the basis of the Born–Oppenheimer approximation and is valid for most chemical systems of interest, except for the motion of very light atoms (i.e., hydrogen) or where correlations between the ionic and electronic motion are important (e.g., vibrations in solids or the interactions of fast ions with a solid).

1.3.1.1 Hartree–Fock Theory

The Hartree–Fock (HF) method is widely used in quantum chemistry. The predicted equilibrium interatomic distances and bond angles for the majority of molecules are typically within a small percentage of experimental measurements. Many codes now allow for HF calculations of periodic systems, but these are more expensive than simple DFT-based approaches described below, and have inherent difficulties in the description of metallic systems, and hence are less widely used.

In 1928, Hartree introduced a self-consistent field method to evaluate the approximate wavefunctions and energies in many-electron systems [15]. The total electronic wavefunction for N electrons is constructed as the product of the individual one-electron orbitals:

$$\Psi_0 = \Psi_0(1)\,\Psi_0(2)\ldots\Psi_0(N).$$

Individual electrons are assumed to move as independent particles in the mean field (potential) due to all electrons, hence the probability to find all electrons in a certain configuration can be calculated as a product of probabilities of finding individual electrons. Solving the approximate Schrödinger equation for each one-electron orbital, in a mean field of all electrons, results in a new wavefunction, thus causing a change in the electron distribution and therefore in the potential. The procedure would typically start from a trial set of orbitals (initial guess) to generate a mean-field (Hartree) potential and should be iterated self-consistently until the wavefunctions stop changing (or remain within a certain numerical threshold).

Following the variational principle, the energy calculated with any trial (approximate) wavefunction (Ψ_T) is higher than the true energy obtained using the exact wavefunction (Ψ_0):

$$[<\Psi_T|\hat{H}|\Psi_T>= E_T] \geq [E_0 =< \Psi_0\left|\hat{H}\right|\Psi_0>].$$

This fact is sometimes used as the basis for an alternative solution to the self-consistent field procedure outlined above, where the energy is minimized with respect to the free coefficients that define the trial wavefunction.

Within the original Hartree approximation, the effect of the electron–exchange interaction is neglected. Fock recognized that the Hartree wavefunction was not antisymmetric with respect to electron interchange [16]. In 1930, he demonstrated that the Hartree approach could be made antisymmetric by appropriately adding and subtracting all possible electron combinations, which was later redefined by Slater as the determinant of the matrix formed by the system of N electron orbitals [17]. Electron exchange is a nonlocal property as "Pauli repulsion" exists between electrons finite distances apart, and forms the basis of the modern HF theory. In spite of its name, the exchange is a stabilizing interaction as it keeps electrons of the same spin further apart, thus effectively reducing electron–electron Coulomb repulsion. In practical applications, one obtains a large set of orbitals, only a part of which of the lowest energy is occupied by core and valence electrons, while the higher energy virtual orbitals (conduction states in solids) normally remain empty.

The HF approach captures a significant amount of the physics of many-electron systems, but it will never result in an exact wavefunction for an interacting system of electrons. From the definition of the variational principle, the energy of an HF wavefunction will always be greater than the energy of the true wavefunction. Löwdin described this measure of the error as the correlation energy. In the mean-field approach, each electron moves independently of all others in the system, except for having a Coulombic repulsion to the average positions of all electrons. As the motion of electrons is in fact correlated, they tend to "avoid" each other even more than the theory would suggest. This dynamic correlation is the main source of error.

1.3.1.2 Post-Hartree–Fock Approaches

There are a number of methods that go beyond the HF approximation in order to recover the electron correlation that is absent in the original theory. At the heart of such approaches is the use of one-electron excited states (unoccupied orbitals) along with the ground state (occupied orbitals) in constructing more representative many-electron wavefunctions, thus leading to more accurate energies and other fundamental properties. As usual, increased accuracy comes at much greater computational expense, so the applicability of these methods to complex systems remains limited. Importantly, these approaches are typically valid only for localized states, which are not routinely accessible from solid-state calculations. The situation has started changing with the advance of new methods using localized Wannier functions (available in, e.g., CRYSTAL, VASP, and FHI-AIMS).

One such method is known as configuration interaction (CI), and involves the calculation of excited states to create a more accurate electronic wavefunction. Many-electron excited states are obtained by swapping occupied and unoccupied states in Slater determinants. Due to the large number of possible excited states, only a subset are considered to make the problem tractable; for example, the addition of single and double excitations to the ground-state wavefunction produces the CISD method. In the multiconfiguration self-consistent field method, and related approaches such as complete active space self-consistent field (CASSCF) method, the one-electron HF molecular orbitals themselves are re-optimized when determining the new electronic wavefunction. Sometimes, such methods, which rely on the initial HF solution of particular symmetry, or reference, may become insufficient, and the need for multireference wavefunctions becomes particularly important for systems with low-lying and/or degenerate excited states.

An alternative approach is given by many-body perturbation theory, where the ground-state HF wavefunction is subject to a perturbation with respect to the correlation potential. The most widely used method is second-order Møller–Plesset (MP2) theory, while calculations up to sixth order (MP6) are feasible on small molecules. Of the two approaches, CI is typically the more accurate, but scales poorly with increasing system sizes; for example, HF conventionally scales as $O(N^4)$, MP2 as $O(N^5)$ and CISD as $O(N^6)$, where N is the number of basis functions used to construct the wavefunction. A modern alternative to the Møller–Plesset series is given by the coupled-cluster approaches, which include partial sums over infinite series of different excitations of certain types. While still being very computationally expensive, they provide a highly accurate treatment for both ground and excited states of small molecules. An alternative way of handling the electron correlation problem without the heavy computational burden of post-HF methods is through DFT, which is discussed in Section 1.3.2.

1.3.1.3 Semi-empirical Wavefunction Methods

For large or complex systems, first-principles methods can become too expensive to apply directly. However, further approximations can be made in order to make calculations more affordable. One of the earliest approximate methods is Hückel theory, which describes the π orbitals of aromatic hydrocarbons. Only nearest-neighbor interactions are considered, and all remaining electron integrals are parameterized. This approach is closely related to the tight-binding approximation employed in solid-state theory and, more generally, model Hamiltonian methods. Despite their simplicity, both the original Hückel theory and the extended Hückel method, developed by Hoffmann to treat π and σ bonds, have been highly successful approaches in the field of organic chemistry.

An extended family of more general semi-empirical approaches can be created by reducing the number of two-electron integrals present in the HF approach. These include zero differential overlap, intermediate neglect of differential overlap and so on. Parameterization of integrals can be performed over a large experimental database to reproduce first-principles results. The most advanced of these are the AM1, MNDO and PM3 (PM6) approaches, which are parameterized over a wide range of molecular systems to reproduce structural, thermodynamic and electronic properties. Similar to empirical potentials, the validity of these models for systems or properties that were not in the initial database is not guaranteed, but nonetheless they can produce valuable insight into the processes that are beyond the length and time scales accessible to higher level approaches, such as protein folding and biosolvation.

1.3.2 Density Functional Theory

DFT describes a chemical system in terms of its real-space electron density (n), rather than the $3N$-dimensional wavefunction associated with solution of the Schrödinger equation. The theory originates from the Thomas–Fermi model (1927), which provided a simple expression for the kinetic energy based on the uniform electron gas model and combined it with a classical description of nuclear–electron and electron–electron interactions. Moreover, the Hartree approximation introduces the mean potential that depends only on the electron density, and likewise the exchange potential models have been advanced for atoms and electron gas by Dirac, Slater, and Gáspár. Building upon these ideas, in 1964 Hohenberg and Kohn [18] reported two important theorems. Firstly, they proved that every observable of a stationary quantum mechanical (QM) system (including energy) could be calculated exactly from the ground-state electron density. Secondly, they showed that the total energy calculated for any trial density could not be lower than the true ground-state energy, and thus could be optimized via the variational principle. There is therefore a direct analogy between HF and DFT methods, except that the latter implicitly includes electron correlation, as it is an exact method by definition.

Within the self-consistent Kohn–Sham approach [19], *independent* (HF-like) one-electron particles are used to construct an *interacting* electron density:

$$n(r) = \sum_{i=1}^{N} |\Psi_i(r)|^2.$$

In practice, the success of DFT-based approaches relies on the description of exchange–correlation effects, which comes down to the choice of the exchange–correlation functional used to describe the interacting system of electrons.

1.3.2.1 Exchange–Correlation Functionals

While an exchange–correlation functional (E_{XC}) that results in the exact ground-state energy in principle exists, its form is hitherto unknown and only approximate forms are available. However, one reason that DFT is so appealing is that even relatively simple approximations to E_{XC} can give reasonable results for equilibrium atomic and electronic structure. The majority of density functionals are based upon the uniform electron gas model, for which accurate energies are known from quantum Monte Carlo simulations (here, MC techniques described above are used to solve the many-electron Schrödinger equation).

The simplest way to derive an E_{XC} for a real system of variable electron density based on contributions from the uniform electron gas is the local density approximation (LDA), where E_{XC} depends only on the value of the density at a position r in the system of interest. The density calculated at this point is then referenced to the E_{XC} contributions that a uniform electron gas of equivalent density would have. This process is repeated for, and integrated over, each point in space. Such an approach is expected to work well for systems, in which the electron density does not vary rapidly, such as a metal; however, practice shows that the LDA performs reasonably, at least at a semi-quantitative level across the whole range of chemical compounds and processes. Concerning errors in this approach, calculations based on the LDA generally result in exaggerated binding energies and underestimated bond lengths. The semi-local generalized gradient approximation (GGA) is dependent on both the value of the density at a position r and its variation close to r (the first derivative, or gradient of density). This accounts better for the nonhomogeneity of the true electron density. In contrast to the LDA, the GGA has a tendency to overestimate bond lengths.

Perdew has presented the hierarchy of density functionals metaphorically as Jacob's ladder, which rests upon the "Hartree world" without exchange or correlation and reaches toward the utopia of chemical accuracy [20]. At the bottom, the local potential experienced by each electron at each point is determined by the charge density at the same point. In heaven, the potential at each point is determined by the global electron distribution. To climb the ladder we use auxiliary means in the form of the gradient expansion of the charge density and different constructs using Kohn–Sham orbitals. Above the rungs of the LDA and GGA are the more complex meta-GGA functionals, which depend on second derivatives, or Laplacian, of the charge density along with the Kohn–Sham orbital kinetic energy density:

$$\tau(r) = \frac{1}{2}\sum_{i=1}^{N} |\nabla\Psi_i(r)|^2.$$

Hybrid exact-exchange functionals, which incorporate an element of nonlocal HF-like electron exchange, can be considered as one of the examples of the next hyper-GGA rung, which makes use of the Kohn–Sham orbital exchange energy. Generally, hyper-GGA functionals can include dependence on the orbital exchange both in exchange and correlation functionals. These methods are now under development and still experimental. While both orbital kinetic and exchange energy densities are local with respect to orbitals, their dependence on the charge density is essentially nonlocal. Finally, unoccupied Kohn–Sham orbitals can be used on the fifth rung, which explicitly includes some correlation effects as they are considered by various post-HF theories, for example in MP2 and the random phase approximations.

The application of hybrid functionals has become increasingly popular in the last decade, for both solid-state and molecular systems: the error cancellation between HF and DFT can generally provide better structural, electronic and thermodynamic properties than either of the pure methods. While the amount of exact-exchange and/or screening of that exchange is not necessarily universal, there has been notable success for a wide range of systems using standard functionals [21,22]. A number of modern functionals are listed in Table 1.2.

DFT has proved a successful approach applied to a wide range of systems, from insulators to metals, and from the molecular to the solid state. However, the biggest challenge for electron density-based methods can be found in the low- and high-density limits—weak interactions (van der Waals forces) and highly correlated (congested) d and f electron systems.

Table 1.2 A selection of commonly used DFT exchange–correlation functionals

Functional	Type	References
AM05	GGA	R. Armiento and A.E. Mattsson, Phys. Rev. B7, 085108 (2005)
BJ06	Meta GGA	A.D. Becke and E.R. Johnson, J. Chem. Phys. 124, 221101 (2006)
B3LYP	Hybrid GGA/LDA	P.J. Stephens *et al.*, J. Chem. Phys. 98, 11623 (1994)
B97	Hybrid GGA	A.D. Becke, J. Chem. Phys. 107, 8554 (1997)
B97-3	Hybrid GGA	T.W. Keal and D.J. Tozer, J. Chem. Phys. 123, 121103 (2005)
HSE06	Hybrid GGA	A.V. Krukau *et al.*, J. Chem. Phys. 125, 224106 (2006)
M05	Hybrid Meta GGA	Y. Zhao *et al.*, J. Chem. Phys. 123, 161103 (2005)
PBE	GGA	J.P. Perdew *et al.*, Phys. Rev. Lett. 77, 3865 (1996)
PBE0	Hybrid GGA	C. Adamo and V. Barone, J. Chem. Phys. 110, 6158 (1999)
PBEsol	GGA	J.P. Perdew *et al.*, Phys. Rev. Lett. 100, 136406 (2008)
PW91	GGA	J.P. Perdew *et al.*, Phys. Rev. B 46, 6671 (1992)
PW92	LDA	J.P. Perdew and Y. Wang, Phys. Rev. B 45, 13244 (1992)
PZ81	LDA	J.P. Perdew and A. Zunger, Phys. Rev. B 23, 5048 (1981)
TPSS	Meta GGA	J. Tao *et al.*, Phys. Rev. Lett. 91, 146401 (2003)
vdW-DF	GGA + dispersion	M. Dion *et al.*, Phys. Rev. Lett. 92, 246401 (2004)

While nonbonding interactions can be neglected for strongly bound covalent or ionic materials, for molecular solids or physisorbed states, which may be present in gas storage or catalytic systems, their contributions can become critical. A number of exchange–correlation functionals have been developed to account for these terms. One approach is to incorporate an empirical r^{-6} correction as found in the Lennard-Jones potential (termed DFT-D) [23], while an alternative approach derives the dispersion forces from a partitioning of the charge density itself [24]. Such methods have had recent success in the description of phase transitions in metal–organic frameworks [25], as well as the conformational landscape of polypeptides [26].

For highly correlated systems, where electrons occupy localized orbitals, standard LDA or GGA functionals have a tendency to delocalize the electronic states, which affects, for example, the description of magnetism in $3d$ systems. While hybrid functionals are starting to be applied to these systems, a more popular and approximate method formulated for treating d and f shell systems is DFT + U, where U is an orbital-dependent external potential [27]. One of the most prolific applications of DFT + U in recent times has been in the treatment of chemical reductions in the catalytic materials TiO_2 and CeO_2, where standard LDA and GGA functionals have been shown to result in an unphysical description of the excess electron distribution.

1.3.2.2 Semi-empirical Density Functional Approaches

Similar to semi-empirical wavefunction-based methods, it is possible to reduce the cost of DFT through further approximations. The most successful approach in this area, which can facilitate the simulation of thousands of atoms, is density functional tight-binding (DFTB) theory. In solid-state physics, the tight-binding method based on linear combinations of atomic orbitals, where the interaction potentials are determined empirically or intuitively, has been successful when applied to a range of solids, as reviewed extensively by Harrison [28]. DFTB extends this method to the Kohn–Sham formalism using a minimal localized atomic basis set, and where the level of self-consistency can be controlled, giving rise to different orders of theory. It is a chemically intuitive approach that allows for the calculation of large-system sizes and complex response functions. A detailed discussion of these techniques can be found in the book of Martin [29].

1.3.3 Excited States

Many of the processes relating to energy materials involve excited states, most commonly the absorption or emission of light by materials, which are addressed in Chapters 2 and 8, respectively. In the calculation of excited states, the concepts of excitonic and quasi-particle states are most important, and commonly confused. It is possible to have excitations in which the number of particles is conserved (e.g., optical absorption creating an N^*-excited state), or in which the number of particles changes (e.g., characterized by the $N-1$ ionization potential or $N+1$ electron affinity). For charge-conserving excitations, the dielectric response can be calculated starting from the single-particle density at various levels of theory—a zeroth-order response is based purely on the DFT or HF orbital energies, while higher level theories include electron–hole (excitonic) interactions explicitly (e.g., time-dependent DFT or HF) [30].

Alternatively, one could start with a description of quasi-particle states (e.g., the $N-1/N+1$ excitation spectra) obtained from many-body perturbation theory (e.g., the *GW* method) [31] and include excitonic effects using explicit two-particle calculations (e.g., the Bethe–Salpeter method) [32]. These approaches are currently the state of the art for calculating the optical properties of solids [33], discussed in more detail in the context of solar cell materials in Chapter 2 and solid-state lighting in Chapter 8.

One drawback of one-electron theories is that the band gap (the energy separation between filled and empty states) of semiconducting and insulating materials is typically incorrect by a significant amount: for LDA/GGA-DFT it is too small, while for HF it is too large. For comparison with experimental measurements, a rigid shift (or scissors operator) is generally applied, or a hybrid DFT functional is constructed that results in the correct eigenvalue spectrum. Neither approach is entirely satisfactory, and this is an area where many-body perturbation theory is being rapidly developed for, and applied to [34–36]. The first-principles quantitative prediction of band gaps in nonmetallic materials remains one of the major challenges in contemporary computational materials science.

1.4 Multiscale Approaches

The techniques outlined above are generally applicable directly to small molecules or homogeneous crystalline solids with small- to medium-sized unit cells in equilibrium. However, often we are interested in large molecules, inhomogeneous solids, complex environments or geometrical arrangements, systems that are far away from equilibrium or have particularly long evolution times. Moreover, when a direct treatment by *ab initio* methods is not viable, and we want to use semi-empirical or interatomic potential-based techniques, suitable parameters will not always be available or easily derivable. Then in many cases, multiscale approaches could be the method of choice. A typical multiscale approach would combine in one model two or three levels of description applied in different regions and/or time intervals.

1.4.1 Hybrid QM/MM Embedding Techniques

Hybrid techniques can be applied to systems with large numbers of atoms, where there is a clearly defined site of interest comprising a small group of atoms, or a cluster with an active site or point defect. The method aims at describing this cluster at the highest affordable level of theory, post-Hartree–Fock or hybrid DFT at present, while the remaining atoms are treated with a parameterized interatomic potential, or molecular mechanical (MM) approach. But other schemes combining different MM or QM methods have also been advanced and widely used. Further, these approaches closely relate to solvation models employed in studies of liquid phase, for example, the behavior of molecular complexes in dielectric solutions.

Hybrid techniques could be broadly divided into two classes [37, 38]: in one, *subtractive* or *multilayered*, more than one level of theory would be applied within the same region of interest [39], whereas in another, *additive* or *multiregion*, each atomic group is treated with one specific approach [40]. As the fundamental extensive measure used in evaluation of system structure and properties is its energy (which could be, for example, potential, internal, including vibrational contributions, or free energy), the terms refer to the way the energy and forces are calculated. Suitable techniques have been developed for both the length and time scales.

Depending on the physicochemical nature of the system of interest, system partitioning may involve (i) only steric constraints on the embedded cluster, which constitutes *mechanical* embedding, (ii) long-range Coulomb interactions included in the Hamiltonian used to describe the electron subsystem of the embedded region, which defines the *electrostatic* embedding, and (iii) short-range Coulomb and exchange–correlation interactions of the electron system of the embedded region with the environment, which could be included into the electronic Hamiltonian as an *embedding potential* [12, 41], a frozen (or relaxed generalized hybrid [42]) orbital or a charge (or spin) density [43].

A distinct problem in hybrid embedding is the termination of the inner QM region. For molecular and ionic systems, the boundary between QM and MM regions can be chosen along the minimum density path, so that all ions or molecules in the QM region remain intact. Then a suitable form of embedding would result in a correct electron localization within the QM region. For extended covalent and metallic systems this is not possible, but additional effort is required. In the

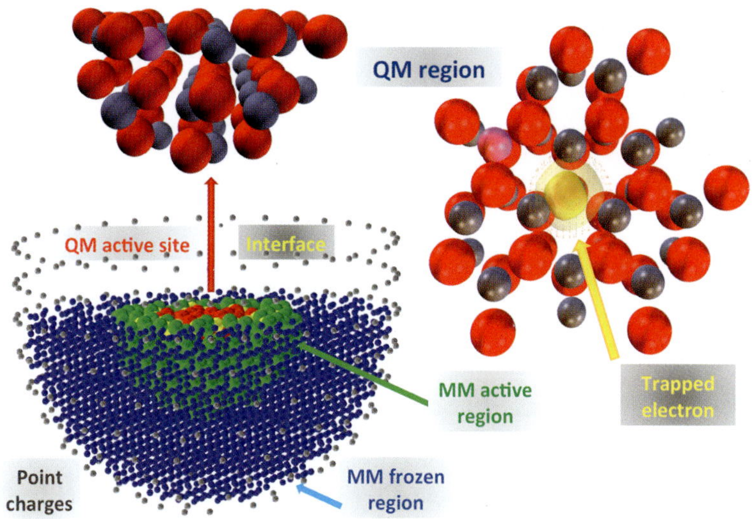

Figure 1.1 *A hybrid QM/MM model of a surface-active site. Atoms or ions are shown as spheres while a trapped electron as spin isodensity contours. Auxiliary point charges placed around the large MM cluster emulate the accurate Madelung field in the active region of the model.*

case of broken covalent bonds, especially of the sigma character, a QM region necessarily terminates in dangling bonds that need to be saturated. The simplest way to deal with it is by saturation (passivation) of dangling bonds by hydrogen, halide or a molecular group (e.g., methyl) [44]. The metallic systems in turn are characterized by strongly delocalized electrons; to account for their behavior, typically an electron localizing procedure is employed, in which the electrons in the QM cluster are treated as a perturbation of the ideal crystal [4, 45].

As an example of a typical hybrid QM/MM embedded cluster approach, used in the studies of heterogeneous systems, we highlight a ChemShell implementation of the solid-state embedding procedure [46], illustrated in Figure 1.1. Here, simple effective core potentials are centered on the cationic sites around the QM cluster forming an interface region, whose role is to contain the electrons within from spilling over the positive charges in the MM region. Similar models have been widely used in the studies of numerous technologically important materials used in catalysis, energy technologies and electronics [41, 44, 47–53]. The QM treatment used in these types of schemes varies from a straightforward application of a suitable QM method, which is available off the shelf, to sophisticated methods using electron localising procedures to generate a customized embedding potential.

1.4.2 Beyond Atomistic Models

We are often interested in processes that occur at much larger length and longer time scales than atomistic models are appropriate for, where the systems are huge, inhomogeneous and processes are far from equilibrium. At these scales different

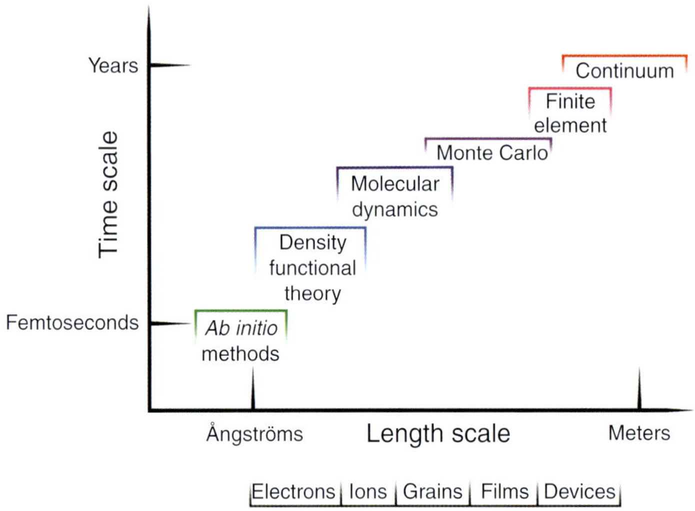

Figure 1.2 *Diagrammatic representation of the computational approaches employed at different temporal and spatial scales.*

laws govern the motion of matter; sometimes they can be reduced to the underlying quantum mechanics of atoms and electrons, but they can also require new approaches to the understanding of meso- and macrophysics and chemistry of materials. Appropriate methods have been developed, and are widely available, but lie outside the scope of this book. Nevertheless, we give some pointers as to where atomistic simulation techniques fit in this wider picture.

In Figure 1.2, we illustrate the hierarchy of the simulation methods employed to meet the challenges that arise in modeling and design of advanced functional materials, in particular those used in energy technologies.

The typical problems that we wish to tackle at a larger than atomic scale would include modeling macroscopic response to mechanical, thermal, chemical, electromagnetic or high-energy radiation impact on bulk or thin-film materials used for components of devices. One example is the mechanical resistance of a material to deformation and plasticity, which requires an understanding of phase stability, domain or grain structure and evolution, dislocation behavior or microstructure in general. Another, perhaps even more challenging aspect is the growth and self-assembly kinetics of nanoparticulate and nanocomposite materials, and indeed charged particle dynamics in porous systems used as reactors or membranes [38]. In all such situations, we move to the description of rigid or deformable groups of atoms, sometimes comprising quite large entities such as grains, nanoparticles, strands of polymers, or more generally whole macromolecules or residues (*coarse graining*). Thus, the collective motion of an atomic group is modeled explicitly, whereas its internal degrees of freedom and interactions are represented by averaged quantities. As the atomistic detail is lost, vast arrays of data pertinent to individual electrons or atoms are no longer required, while the necessary time step in dynamical or Monte Carlo stochastic simulations can be increased by orders of magnitude. Both interactions between the new objects and their motion can be described to differing levels of realism according to the simulation approach. For example, simple isotropic Lennard-Jones potentials are often applied in the context of lattice Boltzmann or molecular dynamics, but realistic size and shape, anisotropic potentials are also employed.

The methods to treat inorganic solids, which we are concerned most in this book, strongly overlap with those developed for liquids and soft matter and could be generally related to fluid dynamics. For the description of macroscopic phenomena, including mass and heat transport or kinetics, for which continuum (or continuous media) methods are employed, a number of accurate modern methods have been developed. Solving the corresponding equations of kinetics or hydrodynamics, usually on simulation grids, presents the next step of coarse-graining, now from meso- to macroscale. Finally, we note that "bottom-up" approaches try to make use of electronic techniques or atomistic simulations to parameterize mesoscopic models, the data from, which are used to derive relevant macroscopic parameters. The opposite "top-down" approaches, in which empirical macroscopic data are used to parameterize meso- and microscopic models, are also widespread.

1.5 Boundary Conditions

Electronic structure and atomistic simulations require us to solve relevant "master" equations for particular boundary conditions, which reflect the system dimensionality and complexity. Small clusters of 10–100 atoms would typically be treated as a molecule in the gas phase, that is, in vacuum, whereas simple bulk solids are considered using periodic Born–von Kármán boundary conditions in three dimensions. However, if we wish to reach toward a fully converged basis set for a particular cluster(s), then a plane-wave electronic structure code could be used, and thus artificial periodicity is imposed on this cluster. To reduce the unwanted effects of the periodicity, that is, cluster–cluster interactions, the simulation cell should be chosen to be as large as possible. In turn, systems, extended in one dimension, are often described as periodic in this dimension, even if in reality such a periodicity is not present as is the case for polymers. However, 1D periodic models are valid for the description of nanowires and nanotubes and still serve as effective models for dislocations and polymeric strands. Systems extended in two dimensions are typically material surfaces or interfaces. In practical calculations, 3D boundary conditions are employed, with the third dimension consisting of a finite slab and a large vacuum layer to reduce surface–surface interactions.

Remarkably, however different in nature, the classical vibrational motion of atoms in the harmonic approximation in condensed matter and quantum electronic motion are described by similar homogeneous wave equations, which have qualitatively similar wave solutions, both obeying the Bloch theorem. As a consequence, motion of atoms and electrons can be described as a plane wave propagating in the direction of periodicity modulated by a fully periodic function. When there are two or three periodic directions as in the case of surfaces or bulk crystals, the corresponding waves spread along any vector in which the system is periodic. A wave number of any such wave can always be chosen as a real number between zero and a certain maximum determined by the system period, with all allowed values thus forming a segment (1D), a polygon (2D) or a polyhedron (3D) and referred to as the first Brillouin zone (BZ). The vibrations (phonons) or one-electron orbitals with the period of the unit cell correspond to the first BZ center (Γ point) and form the basis of the entire phonon or electronic structure of the periodic system. For each point in the first BZ, there are only a finite number of solutions of the wave equations: $3N$ vibrations or M orbitals, where N is the number of atoms in the periodic unit cell and M is the number of basis functions. This defines $3N$ vibrational or M electronic energy bands across the first BZ, respectively. Importantly, wave solutions in directions where the system is finite are confined, which results in phonons propagating normal to surfaces being reflected back in the bulk direction and the wavefunctions that describe bound electrons decaying exponentially into free space.

Vibrational or electronic states, based on the plane-wave solutions described above (so-called Bloch states), are fully delocalized in the periodic directions of extended systems, which is often inconvenient if we wish to consider localized processes. For example, to include electronic correlation effects via many-body perturbation theory, CI or coupled cluster techniques would require permutation of delocalized states that lead to severe mathematical problems, generally requiring to sum over divergent series. To avoid this problem, as Bloch states with all allowed wave numbers form a complete basis set, they can be transformed—any linear combination of such states would form another proper solution of the wave equation. Localized solutions could be obtained using a Wannier procedure and then the most relevant localized states could be selected to calculate a desired property of interest, as for example implemented in the code CRYSCOR [54].

The periodic boundary conditions for extended systems are of course suitable for the true periodic systems, which are characterized by sharp X-ray, neutron or electron diffraction patterns. A large class of disordered materials, including glasses, metal and semiconductor alloys on one hand and soft matter on another, do not fit this description. Two types of approaches to modeling such systems have been developed: (i) the atomic or electronic correlation is assumed to decay in real space rapidly, and the system is represented by a large periodic simulation box, usually chosen as a cubic unit cell, with zone-center solutions used to represent the entire extended system; and (ii) effective medium (commonly in the form of mean field or coherent potential) approximations are employed, in which a small representative fragment of the extended system is embedded in a continuous medium characterized by some dispersion, which can in turn be obtained from counterpart crystalline systems, self-consistently or from empirical data. Both approaches work best for metallic or covalent systems, but experience problems when dealing with ionic systems, characterized by long-range correlation tails.

Ionic or semi-ionic/semi-covalent Coulombic systems in general require a careful treatment in order to avoid artefacts in our simulations. The problem occurs as we try to describe a finite object such as a crystallite as infinite. When charging a molecule or a particle, we allow the long-range Coulomb tail to decay completely in real space, which results in a finite Coulomb energy (excluding of course self-interaction). By charging an infinite object and using a periodic boundary condition, we create an infinite charge, which leads to infinite Coulomb energy density that is unsustainable within a normal physical system. However, often we wish to study local charged centers or nanotubes or surfaces, which are in reality locally compensated. In Section 1.6, we describe how this problem is dealt with when modeling charged point defects.

Even where there is no net charge, the local electrostatic field also has to be considered. For material surfaces, this issue is of particular importance. Depending on the terminating planes, for a binary ionic system, this can lead to dipolar (type III), quadripolar (type II) or nonpolar (type I) surfaces, following the notation developed by Tasker [55]. The presence of an uncompensated surface dipole leads to an infinitely large voltage between opposite surfaces, which necessarily results in a dielectric breakdown. This is the most fundamental reason why models of polar surfaces should always be carefully reconstructed; however, this may not always be evident from experimental analysis.

1.6 Point-Defect Simulations

The role of point defects in determining the properties of condensed matter has been recognized from the onset of solid-state studies in the nineteenth century. These include a combination of lattice site vacancies, interstitials and antisites that may contribute to the optical, electrical, thermal and mechanical response of the material. One of the principal objectives of atomistic simulations is to derive an accurate and coherent approach to the prediction of defect structure, energetics and properties. In Sections 1.6.1 and 1.6.2, we outline two of the most widely employed methods.

1.6.1 Mott–Littleton Approach

The Mott–Littleton approach was originally developed in 1938 as a method to calculate the energy of a charged point defect in alkali salts [10]. The method has since been developed, initially at the Harwell Laboratory, to provide an accurate and flexible description of defects in solids using computer technologies [41].

A charged defect in this approach is treated at the limit of infinite dilution as a local perturbation, the energy of which is minimized to determine the defect equilibrium energy, structure and physical properties. As the total energy of the perfect infinite system is infinite, when using periodic boundary conditions we normally deal with the energy density. This, however, not a useful quantity when dealing with an infinite system containing just one defect, as the change in the mean energy density is negligible. Considering the difference in total energies of the system, with a defect and that of the ideal system, thus proves to be a reasonable course of action. Using the method of interatomic potentials, calculation of such differences is a straightforward procedure as the total energy is summed up over a series of on-site, pairwise, three-body, four-body and so on interaction energies. All contributions from a particular site around the defect center can then be evaluated for the two systems and subtracted from one another to yield a contribution of interest to the defect energy.

On creation of a charged defect, the total energy of an infinite system changes in the first place by the Madelung energy of the introduced charge, qV_{Madelung}. Following the original derivation of Mott and Littleton, the defect energy can be summed up over contributions from the defect site itself (vacant, substitutional or interstitial) and its nearest and, perhaps, next nearest neighbors. To include longer range, polarization effects, a simplified continuous medium approximation of the dielectric material of a given permittivity, ϵ^0, could be employed, for which one can use the Jost formula, for example $-q^2/2R\left(1 - 1/\epsilon^0\right)$, where R is the radius of a sphere containing the defect site with all the neighboring atoms treated explicitly. This approximation is essentially employed in the majority of hybrid QM/MM methods, described in Section 1.4 [46]. Although this approach is robust and provides quite a reasonable estimate of the defect energy, there is still some arbitrariness in how the radius of the dielectric cavity containing defect is chosen or the scalar character of the dielectric constant, which fails for many technologically important materials with a large degree of anisotropy (one celebrated example is TiO_2 rutile for which $\epsilon^0_{33}/\epsilon^0_{11} \approx 2.3$ at 0 K).

In modern implementations [3], no recourse is made to continuous media approximations; all lattice sums are calculated explicitly until specified criteria of convergence are reached. The system is split into an inner region 1, contained within a sphere of a given radius, and the system remainder referred to as region 2. Within region 1, all ions are treated explicitly and their interactions with each other are calculated exactly (at a given level of theory). In region 2, ions are considered to be perturbed by forces generated by a defect in region 1, but remain in harmonic wells around their perfect crystalline sites and in equilibrium with the defect. The unit cells to which local groups of atoms belong would deform elastically in response to a nearly uniform stress generated by the defect region at a large separation distance. Moreover, parameters of the wells (or curvature) are approximated to be the same as in a perfect crystal. In other words, a response of the crystal with a defect to an external field (electrostatic, short-range or elastic in nature) is equated to the response of the perfect crystal.

In the linear response (or harmonic) approximation for any given defect configuration in region 1, displacements of atoms from their perfect positions in region 2 can be calculated in just one step. Then forces acting on displaced atoms in region 2 can be re-evaluated and new displacements calculated with the procedure applied iteratively until it converges. Following this procedure, the total defect energy is calculated as the sum of (i) the defect energy of atoms in region 1, (ii) the defect energy of atoms in region 1 due to the interaction with atoms in region 2 and (iii) the energy of displacement of atoms in region 2 from their crystalline ideal positions. The latter term is the polarization energy of the system remainder, which provides a much more accurate atomistic expression, thus replacing the simple Jost formula given earlier.

The Mott–Littleton approach has been very successful but still has certain limitations. Firstly, it has slow convergence with respect to region size, where quite reasonable estimates could be obtained on a modern single processor/core machine within seconds, but accurate energetics, converged to 1 meV for example (see Figure 1.3), may require calculations involving explicit sums over many thousands of atoms taking of the order of a hundred processor hours.

Secondly, the method itself may be quite sensitive to the initial (guess) defect configuration and get stuck in an artificial local minimum overstabilized by strong relaxations in region 2a, especially for relatively small region 1 sizes. Thirdly, molecular dynamics and Monte Carlo simulations cannot be routinely performed in this approach although would be possible in principle. The main obstacle is the nonvariational character of the defect energy for nonequilibrated region 1–2

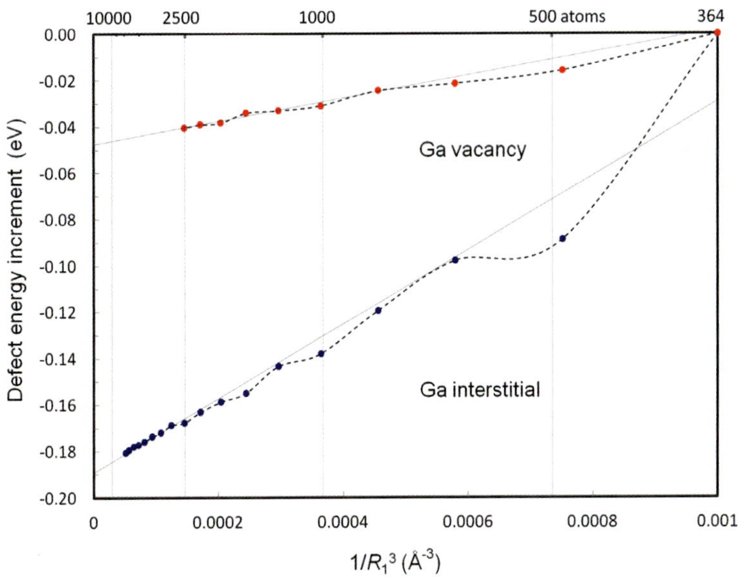

Figure 1.3 *Convergence in defect energy of fundamental Ga defects in the hexagonal phase of GaN with region 1 size. The graphs show approximately inverse cubic asymptotes for large values of* R_1.

structures. Finally, a fully *ab initio* counterpart of Mott–Littleton simulations has not been formulated yet, and possibly is mathematically intractable. As a remedy, hybrid QM/MM procedures have been employed, which use simplified *M-L* approaches to calculate defect polarization energy. These problems are mostly avoided in periodic supercell techniques described in Section 1.6.2.

1.6.2 Periodic Supercell Approach

An alternative to the embedded cluster techniques for the calculation of lattice defects is to maintain the 3D boundary conditions of the host material, where the defect is placed at the center of an expanded unit cell, the supercell approach. While in the past, the typical supercell expansion consisted of the order of 64 atoms, developments in high-performance computer architecture mean that calculations involving hundreds or thousands of atoms are now feasible, in particular when combined with linear-scaling electronic structure techniques.

The benefit of the supercell approach is that it requires no methodological advancement and hence works with all existing condensed matter codes. The drawback is that the isolated point defect of interest is treated as a periodic array – for both the total energy and electronic structure, the size of the supercell must be carefully checked for convergence. The issue is most serious for charged defect centers, as the Coulombic interaction is long-ranged, and the error remains significant even for relatively large unit cells, as illustrated in Figure 1.4 for the case of an oxygen vacancy in In_2O_3. As introduced by Leslie and Gillan [56], the electrostatic correction to the total energy is governed by the supercell length (L), the Madelung

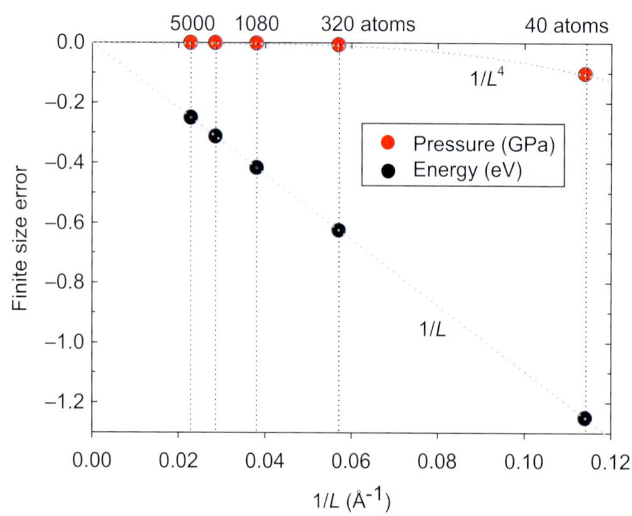

Figure 1.4 *Errors in total energy and pressure caused by finite-size effects for a doubly charged oxygen vacancy defect in In_2O_3, where L represents the length of the supercell expansion.*

constant of the material (α_m), the charge of the defect (q) and the dielectric constant (ϵ^0):

$$\Delta E = -\frac{\alpha_m q^2}{2\epsilon^0 L}.$$

The corresponding correction to the pressure is

$$\Delta p = -\frac{\alpha_m q^2}{6\epsilon^0 L^4}.$$

Makov and Payne extended this formalism to include multipole corrections [57]. Such correction methods are now widely used in the community, with a variety of coefficients [58]. Freysolt *et al.* recently presented a straightforward implementation that automates the correction for calculations based on electronic structure techniques [59]; however, it does rely on the input of the dielectric tensor of the material, which may be calculated separately or inferred from experiments.

1.7 Summary

Having introduced the fundamentals of atomistic simulation techniques, their utility for a range of energy-related technologies is illustrated in the following eight chapters. For example, excited-state methods are further developed in Chapters 2 and 8 in relation to solar cells and solid-state lighting, while molecular dynamics approaches are extensively used in Chapters 3 and 4 covering nuclear and energy storage materials. A suggested reading list (Further Reading) is included at the end of the book for any reader interested in the finer details, and especially to those wishing to develop or implement their own techniques in the future.

References

1. Born, M. and Huang, K. (1956) *Dynamical Theory of Crystal Lattices*, Oxford University Press, Oxford.
2. Bush, T.S., Gale, J.D., Catlow, C.R.A. and Battle, P.D. (1994) Self-consistent interatomic potentials for the simulation of binary and ternary oxides, *J. Mater. Chem.* **4**(6), 831–837.
3. Gale, J.D. and Rohl, A.L. (2003) The general utility lattice program, *Mol. Simul.* **29**(5), 291–341.
4. Daw, M.S. and Baskes, M.I. (1983) Semiempirical, Quantum Mechanical Calculation of Hydrogen Embrittlement in Metals, *Phys. Rev. Lett.* **50**, 1285–1288.
5. Dick, B.G. and Overhauser, A.W. (1958) Theory of the dielectric constants of alkali halide crystals, *Phys. Rev.* **112**(1), 90.

6. KLMC Potential Database (2011) *Database of Published Interatomic Potential Parameters*, http://www.ucl.ac.uk/klmc/Potentials/ (accessed 7 November 2012).
7. Lewis, G.V. and Catlow, C.R.A. (1985) Potential models for ionic oxides, *J. Phys. C Solid State Phys.* **18**(6), 1149–1161.
8. Oganov, A.R. (2011) *Modern Methods of Crystal Structure Prediction*, Wiley-VCH Verlag GmbH.
9. Woodley, S.M. and Catlow, R. (2008) Crystal structure prediction from first principles, *Nat. Mater.* **7**(12), 937–946.
10. Mott, N.F. and Littleton, M.J. (1938) Conduction in polar crystals. I. Electrolytic conduction in solid salts, *Trans. Faraday. Soc.* **34**, 485.
11. Walsh, A., Sokol, A.A. and Catlow, C.R.A. (2011) Free energy of defect formation: thermodynamics of anion Frenkel pairs in indium oxide, *Phys. Rev. B* **83**(22), 224105.
12. Bachelet, G.B., Hamann, D.R. and Schlüter, M. (1982) Pseudopotentials that work: from H to Pu, *Phys. Rev. B* **26**(8), 4199–4228.
13. Vanderbilt, D. (1990) Soft self-consistent pseudopotentials in a generalized eigenvalue formalism, *Phys. Rev. B* **41**(11), 7892.
14. Blöchl, P.E. (1994) Projector augmented-wave method, *Phys. Rev. B* **50**(24), 17953.
15. Hartree, D.R. (1928) The wave mechanics of an atom with a non-coulomb central field. Part I. Theory and methods, *Math. Proc. Camb. Phil. Soc.* **24**(01), 89–110.
16. Fock, V. (1930) Näherungsmethode zur Lösung des quantenmechanischen Mehrkörperproblems, *Z. Phys. A Hadron. Nucl.* **61**(1), 126–148.
17. Slater, J.C. (1963) The electronic structure of atoms—the Hartree-Fock method and correlation, *Rev. Mod. Phys.* **35**(3), 484.
18. Hohenberg, P. and Kohn, W. (1964) Inhomogeneous electron gas, *Phys. Rev.* **136**(3B), B864.
19. Kohn, W. and Sham, L.J. (1965) Self-consistent equations including exchange and correlation effects, *Phys. Rev.* **140**(4A), A1133.
20. Perdew, J.P., Ruzsinszky, A., Constantin, L. *et al.* (2009) Some fundamental issues in ground-state density functional theory: a guide for the perplexed, *J. Chem. Theory Comput.* **5**, 902.
21. Marsman, M., Paier, J., Stroppa, A. and Kresse, G. (2008) Hybrid functionals applied to extended systems, *J. Phy. Condens. Matter* **20**(6), 064201.
22. Heyd, J. and Scuseria, G.E. (2004) Efficient hybrid density functional calculations in solids: assessment of the Heyd–Scuseria–Ernzerhof screened Coulomb hybrid functional, *J. Chem. Phys.* **121**(3), 1187–1192.
23. Grimme, S. (2004) Accurate description of van der Waals complexes by density functional theory including empirical corrections, *J. Comput. Chem.* **25**(12), 1463–1473.
24. Tkatchenko, A. and Scheffler, M. (2009) Accurate molecular Van Der Waals interactions from ground-state electron density and free-atom reference data, *Phys. Rev. Lett.* **102**(7), 073005.
25. Walker, A.M., Civalleri, B., Slater, B. *et al.* (2010) Flexibility in a metal-organic framework material controlled by weak dispersion forces: the bistability of MIL-53(Al). *Angew. Chem. Int. Ed.* **49**(41), 7501–7503.
26. Tkatchenko, A., Rossi, M., Blum, V. *et al.* (2011) Unraveling the stability of polypeptide helices: critical role of van der Waals interactions, *Phys. Rev. Lett.* **106**(11), 118102.

27. Dudarev, S.L., Savrasov, S.Y., Humphreys, C.J. and Sutton, A.P. (1998) Electron-energy-loss spectra and the structural stability of nickel oxide: an LSDA + U study, *Phys. Rev. B* **57**(3), 1505.
28. Harrison, W.A. (1980) *Electronic Structure and the Properties of Solids*, W H Freeman & Co.
29. Martin, R.M. (2004) *Electronic Structure*, Cambridge University Press, Cambridge.
30. Runge, E. and Gross, E.K.U. (1984) Density-functional theory for time-dependent systems, *Phys. Rev. Lett.* **52**(12), 997.
31. Hedin, L. (1965) New method for calculating the one-particle Green's function with application to the electron-gas problem, *Phys. Rev.* **139**, A796.
32. Salpeter, E.E. and Bethe, H.A. (1951) A relativistic equation for bound-state problems, *Phys. Rev.* **84**(6), 1232.
33. Ramos, L.E., Paier, J., Kresse, G. and Bechstedt, F. (2008) Optical spectra of Si nanocrystallites: Bethe-Salpeter approach versus time-dependent density-functional theory. *Phys. Rev. B* **78**(19), 195423.
34. Jiang, H., Gomez-Abal, R.I., Rinke, P. and Scheffler, M. (2009) Localized and itinerant states in lanthanide oxides united by GW@LDA + U, *Phys. Rev. Lett.* **102**, 126403.
35. Rinke, P., Janotti, A., Scheffler, M. and Van de Walle, C.G. (2009) Defect formation energies without the band-gap problem: combining density-functional theory and the GW approach for the silicon self-interstitial, *Phys. Rev. Lett.* **102**, 026402.
36. Vidal, J., Trai, F., Bruneval, F. *et al.* (2010) Effects of electronic and lattice polarization on the band structure of delafossite transparent conductive oxides, *Phys. Rev. Lett.* **104**(13), 136401.
37. Sherwood, P., de Vries, A.H., Guest, M.F. *et al.* (2003) Quasi: a general purpose implementation of the QM/MM approach and its application to problems in catalysis, *J. Mol. Struct. Theochem* **632**(1–3), 1–28.
38. Jorn, R. and Voth, G.A. (2012) Mesoscale simulation of proton transport in proton exchange membranes, *J. Phys. Chem. C* **116**(19), 10476–10489.
39. Svelle, S., Tuma, C., Rozanska, X. *et al.* (2009) Quantum chemical modeling of zeolite-catalyzed methylation reactions: toward chemical accuracy for barriers, *J. Am. Chem. Soc.* **131**(2), 816–825.
40. Harding, J.H., Harker, A.H., Keegstra, P.B. *et al.* (1985) Hartree-Fock cluster computations of defect and perfect ionic crystal properties, *Physica B + C* **131**(1–3), 151–156.
41. Lidiard, A.B. (1989) The Mott-Littleton method: an introductory survey. *J. Chem. Soc. Faraday Trans. 2* **85**(5), 341–349.
42. Gao, J., Amara, P., Alhambra, C. and Field, M.J. (1998) A generalized hybrid orbital (GHO) method for the treatment of boundary atoms in combined QM/MM calculations, *J. Phys. Chem. A* **102**(24), 4714–4721.
43. Wesolowski, T.A. and Warshel, A. (1993) Frozen density functional approach for ab initio calculations of solvated molecules, *J. Phys. Chem.* **97**(30), 8050–8053.
44. Sherwood, P., de Vries, A.H., Collins, S.J. *et al.* (1997) Computer simulation of zeolite structure and reactivity using embedded cluster methods, *Faraday Discuss.* **106**, 79–92.
45. Voloshina, E., Gaston, N. and Paulus, B. (2007) Embedding procedure for ab initio correlation calculations in group II metals, *J. Chem. Phys.* **126**(13), 134115.

46. Sokol, A.A., Bromley, S.T., French, S.A. *et al.* (2004) Hybrid QM/MM embedding approach for the treatment of localized surface states in ionic materials, *Int. J. Quant. Chem.* **99**(5), 695–712.
47. Nygren, M.A., Pettersson, L.G.M., Barandiarán, Z. and Seijo, L. *et al.* (1994) Bonding between CO and the MgO(001) surface: a modified picture, *J. Chem. Phys.* **100**(3), 2010–2018.
48. Seijo, L. and Barandiarán, Z. (1992) Self-consistent embedded clusters: building block equations for localized orthogonal orbitals, *J. Math. Chem.* **10**(1), 41–56.
49. Shluger, A.L., Sushko, P.V. and Kantorovich, L.N. (1999) Spectroscopy of low-coordinated surface sites: theoretical study of MgO, *Phys. Rev. B* **59**(3), 2417.
50. Grimes, R.W., Catlow, C.R.A. and Stoneham, A.M. (1989) A comparison of defect energies in MgO using Mott-Littleton and quantum mechanical procedures, *J. Phys. Condens. Matter* **1**(40), 7367.
51. Sokol, A.A., French, S.A., Bromley, S.T. *et al.* (2007) Point defects in ZnO, *Faraday Discuss.* **134**, 267–282.
52. Galea, N.M., Scanlon, D.O., Martin, P. and Sherwood, P. (2009) Testing interatomic potentials for QM/MM embedded-cluster calculations on ceria surfaces, *e-J. Surf. Sci. Nanotech.* **7**, 413–420.
53. Stoneham, A.M., Gavartin, J., Shluger, A.L. *et al.* (2007) Trapping, self-trapping and the polaron family, *J. Phys. Condens. Matter* **19**(25), 255208.
54. Pisani, C., Schütz, M., Casassa, S. *et al.* (2012) Cryscor: a program for the post-Hartree-Fock treatment of periodic systems, *Phys. Chem. Chem. Phys.* **14**(21), 7615–7628.
55. Tasker, P.W. (1979) The stability of ionic crystal surfaces, *J. Phys. C Solid State Phys.* **12**(22), 4977.
56. Leslie, M. and Gillan, M.J. (1985) The energy and elastic dipole tensor of defects in ionic crystals calculated by the supercell method, *J. Phys. C Solid State Phys.* **18**(5), 973.
57. Makov, G. and Payne, M.C. (1995) Periodic boundary conditions in ab initio calculations, *Phys. Rev. B* **51**(7), 4014–4022.
58. Lany, S. and Zunger, A. (2008) Assessment of correction methods for the band-gap problem and for finite-size effects in supercell defect calculations: case studies for ZnO and GaAs, *Phys. Rev. B* **78**(23), 235104.
59. Freysoldt, C., Neugebauer, J. and Van de Walle, C.G. (2009) Fully ab initio finite-size corrections for charged-defect supercell calculations, *Phys. Rev. Lett.* **102**(1), 016402.

2

Energy Generation: Solar Energy

Silvana Botti[1,2] *and Julien Vidal*[3]
[1]*Laboratoire des Solides Irradiés and ETSF, École Polytechnique, CNRS, CEA-DSM, Palaiseau, France*
[2]*LPMCN, CNRS, Université Lyon 1, Villeurbanne, France*
[3]*Physics Department, King's College London, London, UK*

2.1 Thin-Film Photovoltaics

The energy received by the Earth from the Sun (with a rate of 174 PW in the upper atmosphere) is by far larger than the current world energy consumption. In view of the present crisis of the dominant energy resources and the raising concern about environment, solar power emerges as a concrete alternative for a sustainable and environment-friendly growth of the world energy supply [1].

The discovery of the photovoltaic (PV) effect by Alexandre Edmond Becquerel in 1839 provided a way to directly convert solar radiation into electricity. Although the thermodynamic limit for the efficiency of a PV device is about 85.0%, according to the Carnot law, the first solar cells, made of gold-coated selenium, were only around 1% efficient. The first modern PV cells, with an energy conversion efficiency of about 6%, were built only in the 1950s at Bell Laboratories by Chapin *et al.* [2] using silicon. Chapin and coworkers invented the basic design for solar devices that would be used in the following 60 years. At the heart of the cell, there is a p–n junction, obtained by placing in contact a p-doped and an n-doped semiconductor. When two materials with opposite type of doping are put together, an electric field is built at the interface, in the so-called space-charge region. This electric field drives the separation of the positively charged hole and the negatively charged electron

Computational Approaches to Energy Materials, First Edition. Edited by Aron Walsh, Alexey A. Sokol and C. Richard A. Catlow.

that are generated when a photon is absorbed. The charges further diffuse to the electrodes and finally create a current by recombining in the external circuit [3].

The PV technology of the first generation relied almost exclusively on one material, silicon. Indeed, silicon was the only semiconductor whose growth was mastered to the extent required for the precise control of doping and impurity concentration [4]. Since the 1950s, the increasing performance of crystalline silicon solar cells has benefited from the technological progresses in semiconductor industry. Despite reaching record efficiencies of 25% [5] and despite its domination on the solar cell market, the PV technology based on crystalline silicon struggles to break the symbolic \$1/W barrier. The reduction of costs is mainly limited by the requirement of growing a thick layer of highly pure silicon. In fact, due to the indirect gap of silicon and the consequent poor absorption of light at the onset, a crystalline layer of 100 μm is necessary in order to generate enough electron–hole pairs. Moreover, a high purity is required to increase the diffusion length of electrons and holes in the quasi-neutral region, in order to assure that the charges reach the electrical contacts before recombining. For years this issue has seriously impaired the viability of PV solar power, making it hard to compete with traditional energy resources [6].

A second-generation technology based on thin-film absorbers emerged already 30 years ago, with the main goal of decreasing the production cost of solar panels [7]. At the heart of the device, there is still a p–n junction that is responsible for the separation of charge carriers. In contrast with crystalline absorbing layers of hundreds of micrometers, thin-film absorbers are just 2–3 μm thick. The thickness of the absorber layer can be reduced only if extremely efficient light-harvesting materials are employed: that is why materials such as CdTe [8] and Cu (In, Ga)(Se, $S)_2$ [9], with remarkably large absorption coefficients, are ideal absorbers in thin-film PVs. The use of thin films implies significant saving on the quantity of material required. Besides, it brings a simplification in the growth processes and allows for the introduction of new substrates. Thanks to that, the \$1/W barrier for grid parity becomes accessible even if thin-film solar cells still display lower conversion efficiencies than their crystalline silicon counterparts.

In comparison to the first solar cells produced in the 1950s, the design of thin-film devices underwent some essential changes. Figure 2.1 displays the scheme of a thin-film solar cell—at least four different materials with specific electrical and optical properties are stacked. The single material properties that determine the PV efficiency, and that one would like to control, are band gaps, doping levels, and absorption coefficients. In addition, band offsets at the interfaces, which depend on the structure of the interface, on doping levels, and on band-edge alignments, have a strong influence on the device performance. These electronic properties are nowadays accessible and predictable via *ab initio* calculations.

In this chapter we present the state-of-the-art *ab initio* approaches able to determine electronic states and optical properties of complex compounds used in thin-film solar cells. Such approaches are based on the *GW* approximation within

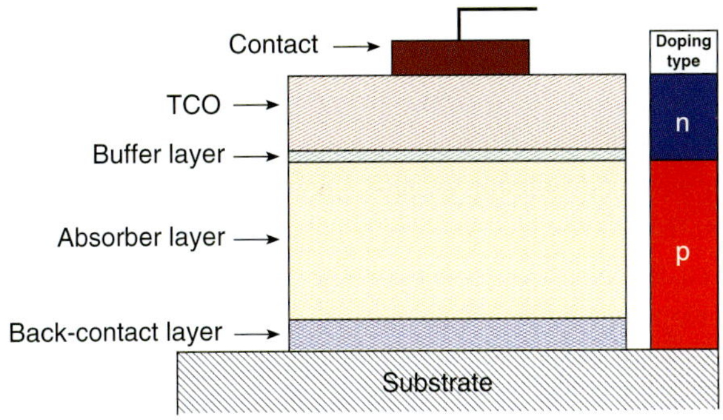

Figure 2.1 *Schematic description of a thin-film solar cell. TCO, transparent conducting oxide.*

many-body perturbation theory (MBPT). This is a rapidly expanding field of investigation and it is not possible to include here all recent developments. We decided therefore to focus on two applications—Cu-based absorbers and Cu-based transparent conducting oxides—and to use these examples to discuss achievements and limitations of the most accurate theories.

For the past years, there has been a substantial effort to study the properties of Cu-based compounds for PVs, not only to understand them from a fundamental point of view, but also to optimize their properties and design solar cells with better performance. However, there is still a clear lack of consistent theoretical and computational studies that can explain and describe electronic excitations in this class of materials. As a consequence, whenever silicon is not used as an absorber, a major obstacle for a fast development of highly efficient devices is the lack of knowledge of basic material properties, in particular, related to defect physics. In fact, silicon and other III-V semiconductors have also a broad use outside the world PVs (e.g., in electronics, lasers, and LEDs), and therefore a much larger effort has been devoted to the understanding of their basic semiconductor properties with beneficial consequences for PV technologies.

Cu-based chalcopyrite semiconductors of the form $Cu(Ga,In)(S,Se)_2$ (CIGS) have already gained a reputation as the leading candidates for second-generation PVs, with the highest efficiency among thin-film solar cells (20.1%) [9, 10]. This is possible thanks to the peculiar optical and structural properties of CIGS, which possess an extraordinary stability under operating conditions [11], contrary to their binary analog Cu_2S [12]. In fact, these materials conserve, to a very high degree, their electronic properties in a large nonstoichiometric range and are remarkably insensitive to radiation damage or impurities. This appears to be a consequence of self-healing mechanisms that compensate for the creation of defects [13, 14]. Such stability, together with direct gaps and high absorption coefficients at the absorption edge, makes them ideal PV materials.

Another class of chalcogenides shares with chalcopyrite CIGS their excellent electronic properties. It is the family of quaternary $Cu_2ZnSn(S,Se)_4$ (CZTS) with kesterite or stannite structure. CZTS have an extra advantage—they are only composed of elements that are naturally abundant and nontoxic. Similarly to CIGS, the alloys $Cu_2ZnSnS_xSe_{1-x}$ have optimal gaps for PV conversion, according to the Shockley–Queisser limit, and high absorption coefficients. Their use as absorbers in thin-film solar cells is getting established with a growing energy conversion efficiency (almost 10% [15, 16] for lab cells). However, the understanding of the properties of the different phases is still rather superficial.

The quest for improving the efficiency of energy conversion does not concern only the absorbing layer. Another source of loss of efficiency is the front contact of the device, made of transparent conducting oxides (TCOs). These are insulating oxides that, for a certain range of doping, become conductive while retaining transparency in the visible spectrum. The most common examples are electron (n-)doped SnO_2, In_2O_3, and ZnO. Hole (p-)doping of wide-gap semiconductors was, for a long time, very hard to obtain [17, 18]. It is therefore not surprising that the discovery of p-doping in $CuAlO_2$ thin films with a carrier mobility of about $10\,cm^2/(Vs)$ attracted great interest [19, 20]. TCOs are key elements for different applications, including flat screens, which implies the need for a diversification of the offer in terms of chemical elements used. We will analyze here the electronic and optical properties of Cu-based transparent conductive materials belonging to the delafossite family, such as $CuAlO_2$, $CuGaO_2$ [21], and $CuInO_2$ [22]. The latter compound is particularly interesting as it exhibits bipolar (n- and p-type) conductivity by doping with appropriate impurities and tuning the film-deposition conditions [22]. This opens the way to the development of transparent p–n junctions, and therefore fully transparent optoelectronic devices.

2.2 First-Principles Methods for Electronic Excitations

A reliable description from first principles of electronic excitations in complex materials starts from identifying the best theoretical approaches, which can simplify ideally both the form of the many-body Hamiltonian and the analysis of the physical quantities involved, while conserving the desired accuracy of results.

As introduced in Chapter 1, density functional theory (DFT) [23, 24] was until recently the only *ab initio* tool capable to calculate the electronic states of materials with a large number of electrons in the unit cell. However, standard DFT is designed to access ground-state properties and not excited states. This restriction is not a question of available approximations. In fact, even if one could calculate the exact Kohn–Sham (KS) eigenvalues, their differences would not necessarily be close to measured excitation energies. Neither, by definition, they stand for electron addition or removal energies [25]. Hence, the fact that the KS gap is in general reported to be too small with respect to measured gaps does not tell us *a priori* anything about the quality of a chosen approximation for the exchange–correlation potential appearing in the KS equations.

The case of materials with localized *d* or *f* states is particularly delicate. In fact, the most common approximations for the exchange–correlation functional (e.g., the local density approximation, LDA [24] and generalized gradient approximations, GGAs [26]), which are known to be reliable for the ground-state properties of *sp* semiconductors, may become inadequate for the system with *d* or *f* electrons. In the case of the Cu-based compounds that we are interested in, this problem can be traced to a poor description of the hybridization of the *d* states of the transition metal with the *p* states of S, Se, and O close to the Fermi level, leading to remarkably large errors in the evaluation of bonding lengths. As a consequence, also ground-state properties, such as the equilibrium atomic structure or phonon frequencies, can be unsatisfactorily described by standard density functionals.

Another strong reason to go beyond DFT and its standard approximations remains the need to calculate excited states for applications in the domain of PVs (i.e., to evaluate quasi-particle energies, absorption spectra, excitonic binding energies, and so on). A possible solution consists in studying particle propagations and fluctuations in the system. This gives correlation functions that can then be related to response functions yielding, for example, the linear response for optical absorption. These correlation functions are one- or two-body Green's functions [27]. The one-body Green's function (that can essentially be understood as a time-dependent particle and hole density matrix) has phase fluctuations (i.e., poles in frequency space) given by electron addition and removal energies (quasi-particle energies), measured in photoemission or inverse photoemission experiments. The particle-hole part of the two-particle Green's function has instead poles at the energies of neutral excitations. MBPT yields a framework where one can find suitable approximations for those Green's functions. In particular, the *GW* approximation, introduced in 1965 by Lars Hedin [28], has been extremely successful in describing quasi-particle energies for metals, semiconductors, and insulators, in the bulk as well as at surfaces and in confined structures. Restricted self-consistent procedures within the *GW* approximation [29–31] have been recently proposed and used with success for transition metal compounds, where perturbative *GW* fails. Using MBPT, the price to pay for a physically intuitive and in general quite reliable description is the high computational cost because now Green's functions appear instead of the electronic density. In view of that, more efficient (and less reliable) approaches, often dependent on some empirical parameters, such as DFT + *U* [32], DFT-corrected methods [33, 34], and hybrid functionals [35–37], have been extensively explored for the past years to calculate electronic band structures with a reduced computational effort.

Concerning neutral excitations, the Bethe–Salpeter equation (BSE) within MBPT is the state-of-the-art approximation [27, 38–40]. Another possible path to access neutral excitations, closer to the spirit of DFT, consists in actually exposing the system to a time-dependent external potential and calculating the evolution of the density in time. This route has become accessible thanks to the extension of DFT to its time-dependent formulation (TDDFT) [41–45], put on rigorous basis by the Runge–Gross theorem [43]. By analogy to the case of classical mechanics, one can intuitively understand that in TDDFT the quantum mechanical "trajectory" of the system under the influence of a time-dependent external potential is found by searching for the extrema of an action, instead of the minimization of a total energy, as done for the ground state. One obtains hence the time-dependent KS equations as a generalization of the static case, and from these, the response functions describing the neutral excitations of a system [46]. At this point, the nontrivial difficulty resides in finding appropriate approximations for the time-dependent exchange–correlation potential $v_{xc}[n](r, t)$. Note that the functional dependence is on the density in the whole space and at all past times. In particular, simple adiabatic approximations based on static functionals proved to work particularly bad for solids [47].

Thanks to many recent developments of improved linear response kernels derived from MBPT [48–55], TDDFT is today a real alternative to MBPT for the calculation of electronic spectra, even for solids, especially when the number of atoms in the unit cells is large and the solution of the BSE becomes unfeasible.

In this section we give a brief review of these theoretical approaches.

2.2.1 Hedin's Equations and the *GW* Approximation

The time-ordered one-particle Green's function is defined as

$$G(1, 2) = -i \langle \Psi_0 | T \left[\hat{\psi}(1) \hat{\psi}^\dagger(2) \right] | \Psi_0 \rangle, \tag{2.1}$$

where $|\Psi_0\rangle$ is the many-body N-particle ground state, $\hat{\psi}(1)[\hat{\psi}^\dagger(1)]$ is the annihilation (creation) operator of an electron in the Heisenberg picture, T is the time-ordering operator, and $1 = \{r_1, t_1, \sigma_1\}$ stands for the set of the real space and time coordinates plus the spin degree of freedom. The Green's function G has poles at the frequencies of electron addition and removal energies.

The time evolution of G corresponds to the propagation of an electron (or a hole) and obeys

$$\left[i \frac{\partial}{\partial t_1} + \frac{\nabla^2}{2} - V_{\text{ext}} \right] G(1, 2) = \delta(1, 2) - i \int \mathrm{d}3 v(1, 3) G_2(1, 3, 2, 3^+), \tag{2.2}$$

where v is the bare Coulomb potential, V_{ext} is an external potential, and the two-particle Green's function G_2 describes the creation and annihilation of pairs of particles. The equation of motion (2.2) tells us how a moving charged particle polarizes the system, creating electron–hole pairs along its way. That is the reason why the two-particle quantity G_2 is involved in (2.2). In order to derive the equations of motion for G_2, G_3, and higher order Green's functions must be considered analogously. The hierarchy of equations containing Green's functions of increasing order can be evaluated using diagrammatic expansions based on Wick's theorem. Alternatively, one can use Schwinger's functional derivatives to obtain a set of equations for G that can be further approximated. Here, we follow the second line to give a brief summary of the *GW* formalism, developed in its present form by Lars Hedin [28].

Even though G is still unknown, (2.2) tells us that its evolution involves the polarization of the system (i.e., G_2). The physical idea at the heart of Schwinger's formalism is that polarization can be created as a response to a time-dependent perturbing external potential U. In the spirit of the linear-response theory, the time-dependent part of the potential will be made to vanish at the end of the derivation, so that $U \rightarrow V_{\text{ext}}(r)$ at $\pm\infty$. To apply this idea, the one-particle term of the Hamiltonian in (2.2), namely, $h_0 = -\nabla^2/2 + V_{\text{ext}}$, becomes $h_0 = -\nabla^2/2 + U$.

It is possible to show that the variation of G with respect to U gives

$$\frac{\delta G(1,2)}{\delta U(3)} = G(1,2)G(3,3^+) - G_2(1,3,2,3^+). \tag{2.3}$$

Using (2.3), the equation of motion (2.2) reads

$$\begin{aligned}&\left[i\frac{\partial}{\partial t_1} - h_0(1) + i\int \mathrm{d}3 v(1,3)G(3,3^+)\right]G(1,2)\\ &\quad = \delta(1,2) + i\int \mathrm{d}3 v(1^+,3)\frac{\delta G(1,2)}{\delta U(3)},\end{aligned} \tag{2.4}$$

where the two-particle Green's function does not appear anymore. As $-iG(3,3^+) = \rho(3)$, where ρ is the electronic density, the term $-i\int \mathrm{d}3 v(1,3)G(3,3^+)$ is nothing more than the Hartree potential $V_{\mathrm{H}}(1)$. This means that the effects of the perturbation U on the system can be split into a classical contribution (i.e., the Hartree term) and the contribution that encompasses quantum effects. The latter term can be rewritten by defining the self-energy Σ:

$$i\int \mathrm{d}3 v(1^+,3)\frac{\delta G(1,2)}{\delta U(3)} = \int \mathrm{d}3\, \Sigma(1,3)G(3,2). \tag{2.5}$$

The self-energy therefore plays the role of a nonlocal and dynamical potential that accounts for all the effects of exchange and correlation in the system. Physically, it represents the effective potential that an extra fermion in the system feels for the polarization that its propagation induces due to exchange effects. Using (2.5) in (2.2) one obtains

$$\left[i\frac{\partial}{\partial t_1} - h_0(1) - V_{\mathrm{H}}(1)\right]G(1,2) = \delta(1,2) + \int \mathrm{d}3\Sigma(1,3)G(3,2). \tag{2.6}$$

Setting $\Sigma = 0$ in (2.6), it is possible to define the Hartree Green's function G_{H}, which is the resolvent of the Hartree Hamiltonian calculated with the exact electronic density ρ:

$$\left[i\frac{\partial}{\partial t_1} - h_0(1) - V_{\mathrm{H}}(1)\right]G_{\mathrm{H}}(1,2) = \delta(1,2). \tag{2.7}$$

Combining (2.6) and (2.7), one finds the Dyson equation for G:

$$G(1,2) = G_{\mathrm{H}}(1,2) + \int \mathrm{d}34 G_{\mathrm{H}}(1,3)\Sigma(3,4)G(4,2). \tag{2.8}$$

The self-energy is responsible for the renormalization effect that yields the difference between the propagation of an independent (Hartree) particle and the propagation of an interacting fermionic particle.

From (2.5) one can then extract with a little algebra a definition of Σ of practical use:

$$\Sigma(1,2) = i \int \mathrm{d}34 G(1,4)\Gamma(4,2;5)\frac{\delta V(5)}{\delta U(3)} v(3,1^{+}), \tag{2.9}$$

where $V(1) = U(1) + V_{\mathrm{H}}(1)$ is the total classical potential, and where the irreducible vertex function has been introduced:

$$\Gamma(1,2;3) = -\frac{\delta G^{-1}(1,2)}{\delta V(3)} = \delta(1,2)\delta(1,3) + \frac{\delta \Sigma(1,2)}{\delta V(3)}. \tag{2.10}$$

The derivative $\delta\Sigma/\delta V$ is then usually replaced by the chain rule $(\delta\Sigma/\delta G)(\delta G/\delta V)$, and (2.10) is transformed into an integral equation:

$$\Gamma(1,2;3) = \delta(1,2)\delta(1,3) + \int \mathrm{d}4567 \frac{\delta \Sigma(1,2)}{\delta G(4,5)} G(4,6)\, G(7,5)\Gamma(6,7;3). \tag{2.11}$$

Following an analogous procedure, one also obtains the equations

$$\Sigma(1,2) = i \int \mathrm{d}34 G(1,4)\Gamma(4,2;3)W(3,1+), \tag{2.12}$$

$$W(1,2) = v(1,2) + \int \mathrm{d}34 v(1,3)P(3,4)W(4,2), \tag{2.13}$$

$$P(1,2) = -i \int \mathrm{d}34 G(2,3)G(4,2)\Gamma(3,4;1), \tag{2.14}$$

in terms of the time-ordered polarization operator $P(1, 2)$ and the dynamical screened interaction $W(1,2) = \delta V(1)/\delta U(3) v(3,2) = \epsilon^{-1}(1,3)v(3,2)$. Equations (2.8), (2.11), (2.12), (2.13), and (2.14) form the closed set of five equations known as Hedin's equations. They depend on one another and should be solved iteratively following the scheme of Figure 2.2.

The "famous" GW approximation is obtained by disregarding the second term on the right-hand side of (2.11) and inserting Γ in (2.12):

$$\Gamma(1,2;3) = \delta(1,2)\delta(1,3), \tag{2.15}$$

$$\Sigma(1,2) = iG(1,2)W(2,1). \tag{2.16}$$

Going beyond the GW approximation requires the inclusion of vertex corrections, by solving (2.11) with its four-point kernel dominated by $\delta\Sigma(1,2)/\delta G(3,4)$.

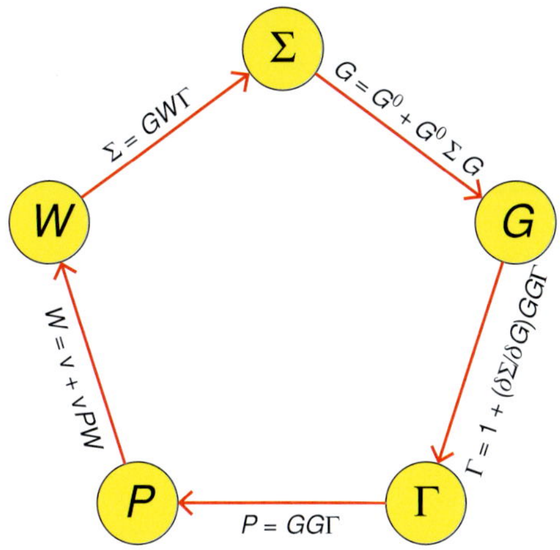

Figure 2.2 *Hedin's pentagon connecting the Green's function G, the self-energy Σ, the vertex Γ, the polarizability P, and the screened Coulomb interaction W.*

Indeed, (2.11) is the main obstacle to the calculation of polarizabilities and self-energies beyond the random phase approximation (RPA) and the *GW* approximation.

The main plus of the *GW* approach is that polarization effects, which screen the propagation of an extra particle, are explicitly taken into account through a dynamically screened self-energy. Instead, in Hartree–Fock approach there is no polarization and no relaxation of the system and the self-energy is static. Even if vertex corrections are neglected and the self-energy is evaluated within the *GW* approximation, the Dyson equation (2.8) is still a self-consistent equation in G and should be solved iteratively. In this sense the starting point is arbitrary. In fact, it is possible to start with Green's functions of any noninteracting system and iterate self-consistently (keeping the vertex $\Gamma = 1$).

Even if Hedin obtained the *GW* approximation by iterating Hedin's equations starting from $\Sigma = 0$ (i.e., from the Hartree Green's function), modern calculations starting from the works of Strinati *et al.* [56], Hybertsen and Louie [57, 58] and Godby *et al.* [25, 59] use rather a "best G, best W" approach [60]. This means that the idea of a strict iterative solution of Hedin's equations is dismissed, and the *GW* self-energy is constructed instead using the best mean-field results that are available. The simplest choice is in general to use the KS eigenvalues and eigenfunctions coming from a DFT calculation within LDA or GGA. We refer to this procedure as perturbative *GW*.

It is clear that a perturbative approach is justified when the departure wavefunctions are already close to the quasi-particle ones. This is indeed the case for many systems, and therefore perturbative *GW* has been extremely successful in describing electron addition and removal energies for metals, semiconductors, and insulators [61]. However, perturbative GW has also proved to be insufficient for many transition metal compounds [14, 29, 31, 62–65]. The problem in many cases is that the LDA wavefunctions are not localized enough, leading even to an artificial metallic character in materials that are insulators. Perturbative GW may already improve upon this situation by opening the gap, but usually by an insufficient amount.

Several methods have appeared to solve this problem. They follow two main lines:

1. *Replacing the simple LDA with a better starting point and then performing a single-shot GW calculation*: Several attempts to improve the starting point with respect to LDA can be found in the literature, like using GGA [66], exact exchange [67], LDA/GGA + U [68, 69], or hybrid functionals [70].
2. *Performing (restricted) self-consistent (sc) GW* [29–31]: The technique proposed by Faleev *et al.* [30] is called quasi-particle scGW and has the advantage of being independent of the starting point (obviously at the price of a larger computational complexity).

There is an alternative self-consistent procedure that yields wavefunctions extremely close to those obtained in a quasi-particle scGW calculation at a reduced computational cost. This method is scCOHSEX, as explained by Bruneval *et al.* [29]. The dynamical effects that are absent in COHSEX calculations can then be accounted for by performing a final perturbative GW step starting from the scCOHSEX result. scCOHSEX + G_0W_0 has already been applied to many transition metal compounds, yielding excellent results for the band gaps and the quasi-particle band structure [14, 29, 62–64]. Note that whenever compared, quasi-particle scGW and scCOHSEX + G_0W_0 gave very similar band gaps and wavefunctions.

2.2.2 Hybrid Functionals

Even in its perturbative form, GW is by all measures an expensive technique, with a very unfavorable scaling with the number of atoms in the unit cell. It is therefore unpractical for the study of band structures of large systems and prohibitive to obtain total energies even for simple systems. Much of the computational effort in GW comes from the dynamically screened Coulomb interaction W. It is therefore interesting to explore to which extent dynamical effects are mandatory, or whether nonlocality is the dominating characteristic. The move from local KS potentials to nonlocal functionals has first been pushed forward in quantum chemistry, where today the so-called hybrid functionals are very popular. These functionals mix a fraction α of Fock exchange with a combination of LDA and GGA functionals. The application of hybrid functionals to the solid state had a slower start [35, 36]. The situation changed recently, helped by the wider availability of computer codes that support hybrids [71, 72] and the steady increase of computational power covering the additional cost with respect to a local potential. Besides yielding good structural properties [73], hybrids have proved to correct to a large extent the band-gap problem [35, 74]. Another landmark came with the introduction of screened hybrids [75, 76]. These functionals lead to faster calculations and improved band gaps, especially for small band-gap systems. Furthermore, by screening the Coulomb interaction at large distances, they also give access to metals.

The intuition lying behind a hybrid functional is rather clear. While LDA or GGA calculations strongly underestimate the gap, Hartree–Fock calculations overestimate it typically by more than a factor of two. By changing the mixing α from 0 to 1, one has a continuous change between local KS and Hartree–Fock, and an essentially linear variation between the respective gaps. Therefore, to obtain the experimental gap, one simply has to use the appropriate mixing parameter. This value can be determined from a fit to a series of systems, and is often set to around $\alpha \sim 0.2 - 0.3$. This choice gives very good results for a large class of systems, but it usually fails when the gap is very large or very small. This fact is easy to understand if we move away from a generalized KS picture [37] and consider the hybrid as an approximation to the self-energy Σ. In the *GW* approximation, the latter can be written as

$$\sum(r, r'; \omega) = \sum{}_{\mathrm{sX}}(r, r') + \sum{}_{\mathrm{rest}}(r, r'; \omega), \tag{2.17}$$

with $\sum_{\mathrm{sX}}(r, r')$ being the statically screened-exchange (sX) term, and $\sum_{\mathrm{rest}}(r, r'; \omega)$ containing the static Coulomb hole and dynamical contributions. If the screening in the sX term is replaced by an effective static dielectric constant $\epsilon_\infty = 1/\alpha$, and $\sum_{\mathrm{rest}}$ is modeled by the static and local part of the hybrid functional [77], the quasi-particle equation has the same form as the generalized KS equation solved for hybrid functionals. Note that this is similar to the sX LDA method [37, 78, 79], where Σ is replaced by a Thomas–Fermi-screened exchange part and the LDA xc functional. From these arguments we can conclude that the physical value for the mixing parameter α is related to the inverse of the dielectric constant of the material at hand [14, 80, 81].

At present, hybrid functional calculations are more and more common in theoretical studies for PVs as they offer a very good compromise between accuracy and efficiency. In particular, band gaps of the order of 1–2 eV are included in the ideal window [81] for reliable band-gap calculations using screened hybrids, such as HSE06. However, a major problem still remains, related to the use of one single mixing parameter for all materials, resulting in significant errors of traditional hybrid functionals for large and small gap materials. Nevertheless, hybrid functionals represent for the time being the only possible way to study electronic excitations in systems with large unit cells (such as supercells containing hundreds of atoms to describe defects), which explain the interest for finding improved functionals [81].

2.2.3 Bethe–Salpeter Equation

Band structures and photoemission experiments are based on a description in terms of a one-particle effective Hamiltonian for one-particle excitations (i.e., quasi-particles). In case of absorption experiments, instead, a one-particle description is no longer adequate, since one has to deal with the simultaneous creation of a quasi-electron and a quasi-hole that interact in the system. In this section, we therefore introduce the effective two-particle Hamiltonian capable of dealing with these neutral excitations.

The BSE uses the intuitive scheme of MBPT, which makes the task to identify efficient approximations easier. The key quantity is the four-point reducible polarizability L, which can be expressed in terms of two-particle Green's function G_2 describing the propagation of two particles (for absorption, the relevant part describes the propagation of an electron and a hole):

$$L(1,2,3,4) = L_0(1,2,3,4) - G_2(1,2,3,4), \tag{2.18}$$

where L_0 is the disconnected part consisting of two one-particle Green's functions G:

$$L_0(1,2,3,4) = iG(1,3)G(4,2). \tag{2.19}$$

The function L satisfies a Dyson-like screening equation, known as the Bethe–Salpeter equation:

$$\begin{aligned} L(1,2,3,4) = L_0(1,2,3,4) + \int \mathrm{d}5678\, L_0(1,2,5,6)[v(5,7)\delta(5,6)\delta(7,8) \\ + \Xi(5,6,7,8)]L(7,8,3,4). \end{aligned} \tag{2.20}$$

In order to solve (2.20), one has to approximate the many-body interaction kernel Ξ. The standard approximation consists in using the GW self-energy of (2.16). In this way the solution of the BSE corresponds to the inclusion of vertex corrections in P through a second iteration of Hedin's pentagon of equations. When the BS kernel is approximated to first order in the screened Coulomb interaction W, the BSE reads

$$L = L_0 + L_0({}^4v - {}^4W)L. \tag{2.21}$$

In (2.20), we have defined the four-point extension of the Coulomb potential ${}^4v(1,2,3,4) = \delta(1,2)\delta(3,4)v(1,3)$, whereas ${}^4W = \delta(1,3)\delta(2,4)W(1,2)$ is the four-point extension of the screened Coulomb potential. The first term, the unscreened electron–hole exchange, is repulsive, while the second term, the screened electron–hole Coulomb interaction, is attractive. Note that the δ-functions connect different indices in the two cases: this is because the former stems from variations of the Hartree potential, whereas the latter is due to variations of an exchange-like self-energy contribution. Because of the way the indices are connected in this second case, the BSE cannot be written in a two-point form. This issue represents the real bottleneck in practical calculations. Instead, the measurable response function is obtained via a two-point contraction of L, namely,

$$P^{\text{red}}(1,2) = -L(1,1,2,2). \tag{2.22}$$

From the reduced polarizability P^{red} that is understood to be time- (or contour-) ordered, the causal response function χ can be inferred; the relation between the time-ordered and causal responses is in fact $\chi(\omega > 0) = P^{\text{red}}(\omega > 0)$. Another useful quantity is the independent quasi-particle polarizability, $P_0(1,2) = -L_0(1,1,2,2)$. Note that this is not equal to $\chi_{\text{KS}} = -\text{i}G_{\text{KS}}G_{\text{KS}}$. The expression looks similar, but the KS states and eigenvalues in χ_{KS} are now replaced by their MBPT counterparts, which are, at this level of approximation, determined in the quasi-particle GW approximation [28]. To be precise, practical calculations use GW eigenvalues and KS wavefunctions to build G.

For simple semiconductors, dynamical effects in the electron–hole screening in W and in the Green's function G tend to cancel; therefore, they are usually neglected in both terms. This means that instead of $W(1,2)$ one uses a statically screened instantaneous interaction and for G the Green's function obtained by a perturbative or scGW calculation. In practice, the BSE (2.20) is solved by diagonalizing a two-particle excitonic Hamiltonian, providing information about the excitonic eigenstates and eigenvalues. The BSE yields excellent results for the optical properties of solids with both continuum and bound excitons, in bulk and nanostructured systems [39].

2.2.4 Model Kernels for TDDFT

Whenever the solution of the BSE is unfeasible, it is however possible to identify efficient ways to calculate absorption spectra, going beyond the simple sum over independent transitions between KS states. An appealing idea is to use the time-dependent (TD) generalization of DFT, TDDFT [43, 82], whose central quantity is the electronic density that responds to a time-dependent external field. TDDFT retains all the advantages of the DFT formalism in terms of computational efficiency. However, the lack of good approximations for the time-dependent exchange–correlation (xc) functional, especially in the case of solids [39, 47], is its principal limitation.

The exchange–correlation kernel f_{XC} of TDDFT, which is the variation of the exchange–correlation potential with respect to the electronic density, is most often approximated by (i) $f_{\text{XC}} = 0$, leading to the RPA, or (ii) $f_{\text{XC}} = \delta v_{\text{XC}}^{\text{LDA}}/\delta\rho$, which gives the adiabatic local density approximation (ALDA) [45, 82]. Note that the RPA differs from an independent-particle calculation, even if the xc kernel is zero, because it accounts for variations of the Hartree potential upon excitation. This is a classical contribution, which in the optical spectra corresponds to local field or depolarization effects, particularly important when charge inhomogeneities are pronounced.

Concerning the calculation of absorption spectra of semiconductors, neither the RPA nor the ALDA is sufficient to yield results in quantitative (and sometimes not even in qualitative) agreement with experiments. In particular, the lack of a long-range ($1/r$) decay of the ALDA linear response kernel f_{xc} can cause large errors in the calculation of absorption spectra of extended systems [39, 47, 83]. Nevertheless, it has been shown [47, 51] that in some cases an agreement with experiments can be recovered by retaining only the long-range contribution (LRC) that the exact f_{xc} should have in the asymptotic limit. In the following we will see an example of resulting model kernels. Botti *et al.* [49, 84] account for continuum excitonic effects, giving very good optical properties for small-medium gap semiconductors, with a computational burden comparable to a DFT approach.

One should remember however that CIGS materials possess bound excitons [85–88], and that the TDDFT–LRC approach has proved to work better for systems with continuum excitons. That is why the only fully reliable approach in this case is the solution of the computationally demanding BSE or the use of a more sophisticated model kernel, which accurately describes the electron–hole pair dynamics [48, 50, 52, 53]. Nevertheless, the simple model kernels often remain the best compromise between accuracy and efficiency whenever one wants to deal with large supercells to study interfaces or defects.

2.3 Examples of Applications

The first-principles approaches, presented in Section 2.2, have recently been applied for the first time to complex materials of technological interest. In this chapter, we select three illustrative applications of self-consistent GW in the field of PV energy conversion, where the use of accurate state-of-the-art methods has allowed to clarify controversial results in literature.

Our first example concerns copper indium/gallium diselenide/disulfide (CIGS) compounds. These materials are doped by native defects. These defects play an important role in the definition of the exceptional stability under operating conditions of the electronic properties of these compounds. We discuss more in detail here the effect of Cu vacancies.

The second example is kesterite and stannite $Cu_2ZnSn(S,Se)_4$ compounds. This family of quaternary chalcogenide is emerging as an alternative to CIGS materials and shares with the parent family extraordinary structural and electronic properties.

The third example is the family of $Cu(Al,In,Ga)O_2$ delafossite transparent conductive oxides, which possess a chemical composition similar to CIGS compounds. A long-standing debate on the direct and indirect fundamental gaps of these materials has intrigued experimentalists and theoreticians for years.

Concerning the calculations discussed in this section, the set of methods used includes DFT in the standard LDA and GGA, LDA/GGA + U, hybrid functionals and GW approaches. The reported standard DFT and GW calculations were performed within the plane-wave scheme implemented in ABINIT [89], using norm-conserving pseudopotentials [90] and including semi-core states in the valence. Hybrid functional and LDA/GGA + U calculations were performed with the Vienna *ab initio* simulation package (VASP) [71, 72, 91, 92]. TDDFT and BSE spectra were obtained using the code YAMBO [93]. More details on calculations can be obtained in the cited references.

2.3.1 Cu-Based Thin-Film Absorbers

Ternary chalcopyrites crystals are obtained from the zincblende structure by replacing the Zn cations alternatively with Cu and In or Ga. $CuIn(S,Se)_2$ compounds are commonly referred to as CIS, and $CuGa(S,Se)_2$ compounds are known as CGS. The two families together are the so-called CIGS materials. In ternary chalcopyrites each anion (Se or S) is coordinated by two In or Ga and two Cu atoms, while each cation is tetrahedrally coordinated by four anions. The existence of two different cations results in two different bonding lengths $R_{(\text{in,Ga})-(\text{S,Se})}$ and $R_{\text{Cu}-(\text{S,Se})}$, leading to two structural anomalies [94]: (i) The tetragonal cell exhibits a distortion, defined by the parameter $\eta = c/2a \neq 1$, the ratio between the lattice constants a and c. (ii) The ideal zincblende site for the anion is perturbed, yielding a deformation of the anion tetrahedron, which is measured by the anion displacement parameter

$$u = \frac{1}{4} + \left[R^2_{\text{Cu}-(\text{S,Se})} - R^2_{(\text{In,Ga})-(\text{S,Se})}\right]/a^2 \neq \frac{1}{4}. \tag{2.23}$$

Both structural anomalies are small, but not negligible. The precise experimental determination of the anion displacement u is more difficult than the measurement of the lattice constants a and c due to the inhomogeneity of the samples. Indeed, the dispersion of the measured values is very small for the lattice parameters (usually $<1\%$), while it is significant (as large as 7%) for the anion displacement [94, 95].

Merino *et al.* [95] reported a relatively narrow spread in the band gap ($\approx 10\%$) as a function of the dispersion of the anion displacement u. This finding is coherent with the observation of an extraordinary stability of CIGS absorbers under operating conditions. Unfortunately, experiments alone are not conclusive due to the lack of information on the composition and uniformity of the samples. In fact, in real samples there is a dispersion of values of u. It is therefore interesting to calculate the evolution of the gaps as a function of u. This was first addressed theoretically using DFT [96], in some cases including empirical corrections to get closer to the experimental gap [94, 97]. An uncommonly large variation was found, in apparent contradiction with the experimentally proved insensitivity of the CIGS gap to any kind of damage or perturbation.

In the following we will focus on CIS materials, in particular using $CuInS_2$ as a test case. Similar results were found for $CuInSe_2$, $CuGaSe_2$, and $CuGaS_2$ [14, 98]. A first question to answer is if the DFT data can be trusted, as they were obtained from KS eigenvalues using LDA or GGA functionals. Indeed, these approximations are known to perform badly in CIGS materials. For example, in CIS the KS band gap is vanishing, in contrast to the experimental values of 1.54 eV and 1.05 eV for $CuInS_2$ and $CuInSe_2$ [99], respectively. Moreover, *ad hoc* corrections, such as the scissor operator, often used to adjust the DFT band gap to the experimental value, are too simplistic to analyze the dependence of the band gap on u.

In spite of the underestimation of the band-gap, LDA and GGA usually yield good structural parameters for semiconductors and insulators. Unfortunately, for CIS, the theoretical range of anion displacements obtained within these approximations ($0.215 < u < 0.220$) lies outside the experimental range. It is clear, thus, that to understand the paradox of the band-gap dependence on the unit cell deformations in CIGS chalcopyrites one has to go beyond standard DFT.

We have already seen that scGW and hybrid functional are the most pertinent approaches beyond standard DFT. These approaches were applied by Vidal *et al.* [14] to study the interplay between structural and electronic properties in CIS compounds. Here, we analyze the main findings of that work in comparison with previous experimental and theoretical results.

If the atomic arrangement in the unit cell is relaxed using LDA or GGA functionals the anion displacement u is systematically underestimated by 5–10%, in agreement with previous results [96, 100, 101]. Moreover, always within LDA/GGA, the KS bottom conduction band overlaps the KS top valence band, yielding negative band gaps [102]. On the other hand, values of u in agreement with experiments can be obtained using the Heyd–Scuseria–Ernzerhof (HSE06) [75] hybrid functional [14]. In fact, the structural relaxation within HSE06 yields $u = 0.226$ ($u = 0.229$) for the ideal monocrystal of $CuInS_2$ ($CuInSe_2$). HSE06 gives slightly underestimated band gaps for CIGS. Band gaps in excellent agreement with experiments are obtained from scGW once the experimental geometry (or the HSE06 geometry) is fixed. Unfortunately, scGW cannot be used to perform structural relaxations.

In order to study the variations of the gap upon structural distortion, Vidal *et al.* [14] varied u in the interval $0.2 < u < 0.25$, which encompasses both experimental and theoretical ranges, and evaluated the quasi-particle band structures using and comparing different *ab initio* schemes. They also verified that sensible variations of the lattice parameters a and c have negligible effects on the gap.

The dependence on u of the band gap of $CuInS_2$ is shown in Figure 2.3 [14]. We can see that the KS–LDA curve (magenta with filled circles) has the same slope as the theoretical curve (dotted line) obtained by Jiang and Feng [97] using the LDA corrected by a scissor operator. By applying the perturbative G_0W_0 approach on top of DFT (blue line with crosses), bands are reordered, reversing the sign of the band gap for $u > 0.215$. In any case, G_0W_0 gaps remain quite small, and the slope of the curve does not change for most of the u range when compared to the KS results. The slope of the G_0W_0 calculation increases when $u = 0.25$, becoming comparable to the slope of the curves obtained applying a self-consistent approach. This is the only case for which the KS gap is already positive and no band reordering occurs.

The scCOHSEX procedure (yellow lines with stars) leads to a very different result—the gap enlarges (to values slightly larger than the experimental ones), and

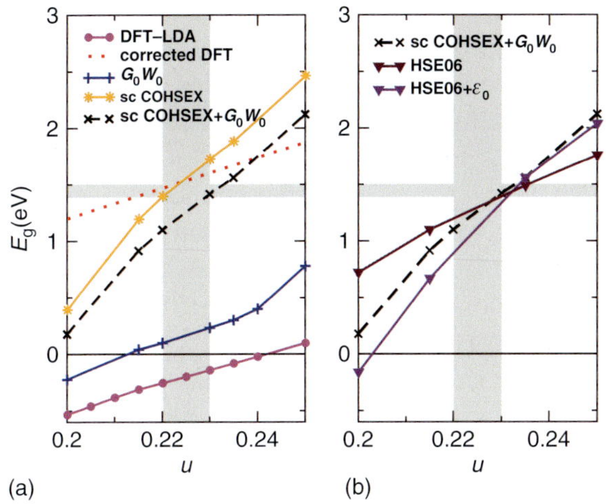

Figure 2.3 *Photoemission band gap versus the anion displacement u for $CuInS_2$. The vertical shaded areas give the spread of experimental data for u, while the horizontal shaded areas give the dispersion of measured band gaps. (a) Calculations using DFT–LDA, G_0W_0, scCOHSEX, and scCOHSEX + G_0W_0. The dotted line refers to the work of Jiang and Feng [97]. (b) Comparison of scCOHSEX + G_0W_0 and hybrid functional calculations, using HSE06 and a modified HSE06. (From [14].)*

the slope of the curve significantly increases. This is due to the progressive modification in the self-consistent iterations of the unsatisfactory LDA starting point. When the scCOHSEX eigenstates are used as a starting point for a perturbative G_0W_0 calculation (black dashed lines), the band gap gets reduced by a constant value of about 0.3 eV, without affecting the slope of the dependence on u. Note that the scCOHSEX + G_0W_0 band gap is in agreement with experimental data for the anion displacement measured for the monocrystal [103].

Something more can be learnt from HSE06 calculations for the gaps (maroon line with filled triangles) that yield an intermediate slope between DFT–LDA and scCOHSEX + G_0W_0. The effect of the HSE06 functional is determined by the fixed amount of Hartree–Fock (HF) exchange included:

$$E_{\rm xc}^{\rm HSE06} = E_{\rm xc}^{\rm GGA} + \frac{1}{4}E_{\rm x}^{\rm HF,sr} - \frac{1}{4}E_{\rm x}^{\rm GGA,sr}, \tag{2.24}$$

where the exchange term $\frac{1}{4}E_{\rm HF,sr}$ can be seen as an approximated contribution to the self-energy, whose Coulomb interaction is screened by the mixing parameter $\frac{1}{4}$ and the short-range (sr) screening factor, while $\frac{1}{4}E_{\rm x}^{\rm GGA,sr}$ is a screened GGA exchange. In contrast, in a GW calculation the exchange part of the self-energy is screened by the inverse dielectric constant, as a first approximation using the static ε_∞; the latter varies for different values of u. To shed light on this point, Vidal *et al.* [14] performed a series of HSE06 calculations using $\frac{1}{\varepsilon_\infty}$, computed for each value of u, to replace the mixing parameter in front of the screened Hartree–Fock exchange. The resulting curve (violet line with open triangles) has the same slope as the scCOHSEX curve, pointing to the fact that the electronic screening is the essential ingredient that controls the band gap. This idea was further exploited by Marques *et al.* [81] to obtain a nonfixed mixing parameter for hybrid functionals, sensitive to screening.

Note that sc*GW* cannot be applied at the moment to supercells containing *d* electrons if the unit cell is larger than 20–30 atoms. As long as larger cells are out of reach for *GW* calculations, hybrid functionals, in particular with improved approximations for the mixing parameter, promise to be a computationally efficient alternative with a satisfactory degree of accuracy.

In summary, all theoretical results, independently of the level of sophistication, reveal variations of the band gap well outside the experimental range of values. In fact, considering as an upper limit for Δu, the range where both theoretical and experimental values are included ($\Delta u \lesssim 0.02$), we deduce that the gap variation ΔE_g due to the anion displacement alone would be $\Delta E_g = 32.2 \times \Delta u \approx 0.65$eV. This suggests that the puzzle of electronic stability of CIGS compounds cannot be solved simply by replacing LDA for a more accurate quasi-particle calculation.

A hint of the possible solution of this apparent paradox of band-gap stability can be found by also considering the effect of high concentrations of defects [11]. Here, we consider as an example the effect due to Cu vacancies V_{Cu}, shallow acceptor defects that are known to be present in all samples, consistent with the observation of p-type conductivity. In particular, V_{Cu} is thought to have very low formation energies [104] compared to other intrinsic defects. One should not forget that other defects are known to be present in large concentrations, in particular $\mathrm{Cu_{In}}$ and $\mathrm{In_{Cu}}$ antisites, which participate in the formation of complexes of defects with very low formation energy [104]. Indeed, a complete model should include the effects of all most common defects.

Already at the KS level, the presence of Cu vacancies V_{Cu} opens up the band gap [104, 105], with a larger effect for increasing concentrations of V_{Cu}, once again in contradiction with the observed stability of experimental band gaps against stoichiometry deviations. This trend is confirmed by G_0W_0 calculations for supercells [14] corresponding to concentrations of V_{Cu} in the usual experimental range of off-stoichiometry [95]. The concentration of V_{Cu} is related to the formation energy ΔE_{f} through a Boltzmann distribution, whose relevant temperature and chemical potentials are set by the growth conditions ($T \approx 500$–$600\,^{\circ}$C, copper poor samples) at which defect equilibrium occurs. Moreover, this equilibrium is assumed to be quenched during the rapid cool down of the samples. However, previous calculations of formation energies were done at the level of DFT and suffered from the unsatisfactory description of the localized states contributing to the VBM. Lany and Zunger [33] proposed to use LDA + U to correct the formation energy as follows:

$$\Delta E_{\mathrm{f}} = \Delta E_{\mathrm{f}}^{\mathrm{LDA}} - \Delta E_{v}^{\mathrm{LDA}+U}. \tag{2.25}$$

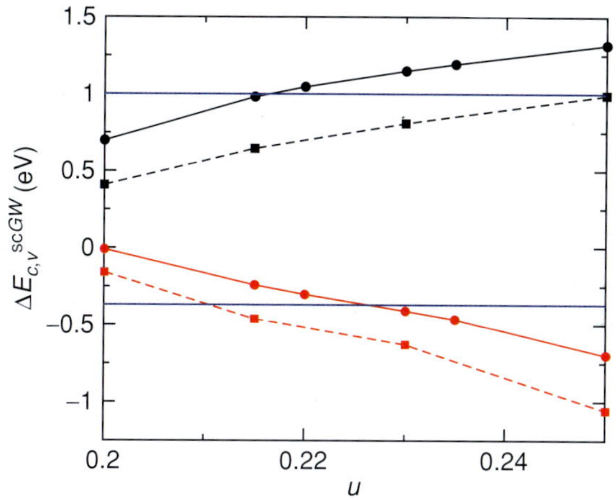

Figure 2.4 The shifts of the conduction band minimum (CBM) and valence band maximum (VBM) $\Delta E_{c,v}$ of $CuInS_2$ with respect to the LDA band edge versus the anion displacement u. We compare results for scCOHSEX with self-consistency only in energy (dashed curve) and in both wavefunctions and energies (continuous curve). The blue horizontal lines represent $\Delta E_{c,v}$ given by DFT + U [106], where U is set to 6 eV. (Modified from [14].)

Vidal *et al.* [14] used this idea, evaluating the VBM shift ΔE_v within sc*GW*. In this case, ΔE_v does not correct a shortcoming of KS theory, but improves on the LDA functional, as the exact exchange–correlation functional would give the exact VBM already within DFT. In Figure 2.4, one can find the band edge corrections $\Delta E_{c,v}$ with respect to the KS VBM, calculated using sc*GW* and shown for different levels of self-consistency. The VBM shift has a dispersion of around 1 eV in the considered range of *u*. This dispersion is totally absent in DFT (+ *U*) calculations [106] (horizontal blue line). Note that the shifts are significantly different according to whether both wavefunctions and eigenvalues or only eigenvalues are updated in the self-consistent procedure. Figure 2.4 proves as well that the variations of the band gap with *u* are due to both variations of the CBM and the VBM.

Following the logic of the feedback loop scheme of Figure 2.5, one can realize that if a distortion of the lattice occurred for any reason (e.g., extended defect and axial strain), it would actually have little influence on the gap value because of two compensating effects.

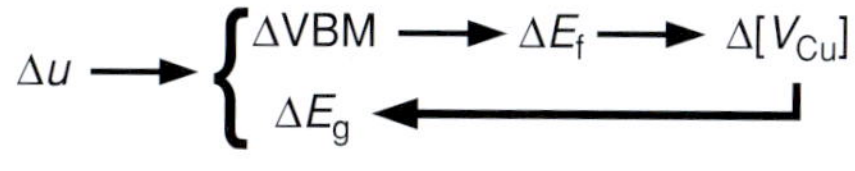

Figure 2.5 Schematic representation of the feedback loop that stabilizes the band gap. (From [14].)

Let assume that the band gap is a functional of u and $[V_{Cu}]$ so that $E_g = E_g(u, [V_{Cu}])$. Any variation of the band gap ΔE_g can be expressed as

$$\Delta E_g = \frac{\partial E_g}{\partial u}\Delta u + \frac{\partial E_g}{\partial [V_{Cu}]}\Delta[V_{Cu}], \tag{2.26}$$

where u is the anion displacement and $[V_{Cu}]$ is the concentration of Cu vacancies. The two variables u and $[V_{Cu}]$ are in fact not independent but linked through the variation of the VBM induced by variations of u. In the case of a shallow acceptor, its formation energy should be corrected by the correct position of the VBM (see equation (2.25)) as

$$\Delta E_f(u) = \Delta E_f^{DFT} - \Delta E_v^{scGW}(u). \tag{2.27}$$

The derivatives entering in equation (2.26) were calculated by fitting the curves in Figure 2.6.

Shallow acceptors are supposed to behave like the VBM due to their proximity to it. This has been the common strategy used for the past 20 years in order to

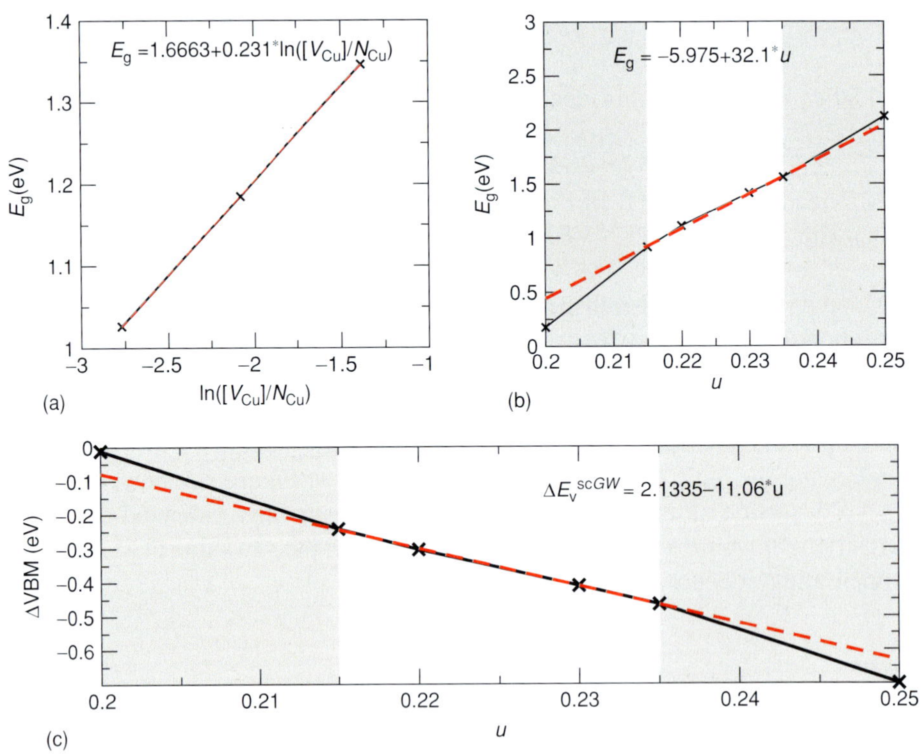

Figure 2.6 (a) Value of the band gap of $CuInS_2$ calculated in G_0W_0 versus the concentration of Cu vacancies. (b) Value of the band gap of $CuInS_2$ calculated in scCOHSEX+G_0W_0 versus the anion displacement. (c) ΔE_v^{scGW} of $CuInS_2$ versus the anion displacement. (From [14].)

correct formation energies [33]. Then, it is possible to relate the concentration of V_{Cu} to the formation energy through a Boltzmann distribution as

$$[V_{Cu}] = N_{Cu} e^{\frac{-\Delta E_f}{k_B T}}, \tag{2.28}$$

with T being the growth temperature for which the diffusion of defects is quenched ($T \approx 500$–$600\,°C$). Moreover, high concentrations of V_{Cu} can significantly enlarge the band gap. Thus, any variation of u can change both the band gap and the formation energy of V_{Cu}, and as a consequence vary the concentration of V_{Cu}, which will change the band gap accordingly. Overall, the effects of u and V_{Cu} cancel each other.

The same dependence of the gap on u was found for CGS compounds, except for the fact that the GGA values of u are inside the experimental window, due to a compensation of errors in the calculation of bond length [98]. Analogous results are also found for quaternary CZTS compounds, whose electronic properties are extremely similar to the ones of the parent CIGS family. Let us analyze this statement more in detail. This will give us the possibility to have a closer look at the band structures obtained using the different levels of approximation.

In literature there are only few recent studies that address the structural [107–110], electronic [108, 110, 111], and defect properties [112, 113] of CZTS. The zincblende-derived kesterite structure ($I\bar{4}$) is recognized to be the most stable. The energy difference per atom with respect to the stannite structure ($I\bar{4}2m$) is, however, only of few millielectron volts per atom [108, 110, 114], proving that kesterite and stannite phases can coexist in experimental samples, and explaining the reported disordered structures [115]. Similarly to chalcopyrites CIGS, quaternary kesterite and stannite CZTS are obtained from the zincblende structure by replacing the Zn cations in such a way that each anion (Se or S) is coordinated by one Zn, one Sn, and two Cu atoms. The existence of three distinct cations results in three different cation–anion bond lengths, which induce a displacement of the anion from its ideal zincblende site. That distortion is measured by the anion displacement parameters (u_x, u_y, u_z), that is, the relative coordinates of the anion in the conventional body-centered tetragonal cell. The dispersion of experimental data for u is known to be large in CIGS compounds [94, 95]. Concerning CZTS, fewer structural measurements can be found in literature and in most cases only a and c are reported [116–120]. Concerning the measured band gaps [121–127], early studies suggested a value of about 1.4–1.6 eV both for S and Se compounds. This was in disagreement with DFT calculations [108, 110], using semi-local or hybrid functionals, which obtained systematically a smaller gap for Se compounds. More recent measurements [121] delineate a gap of about 1 eV for $Cu_2ZnSnSe_4$, explaining the previous overestimation with the presence of ZnSe in the sample.

In Figure 2.7, we display band structures for kesterite Cu_2ZnSnS_4, obtained using different theoretical schemes at the experimental geometry. In Figure 2.7a, the KS LDA band structure is compared with the sc*GW* bands. We observe that

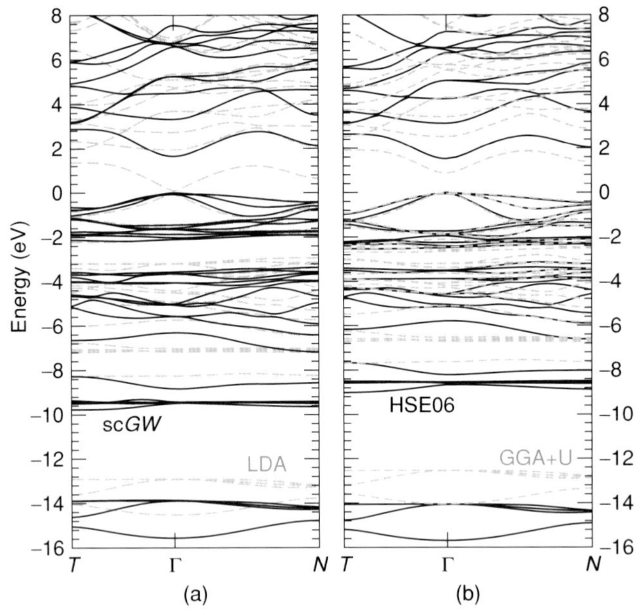

Figure 2.7 *Band structures for kesterite Cu_2ZnSnS_4 calculated using (a) DFT–LDA (dashed line) and scGW (continuous line), and (b) GGA + U (dashed line) and HSE06 (continuous line). Top valence bands are set to zero. (From [107].)*

scGW corrections upshift almost rigidly the lowest conduction states. Concerning the valence:

1. The dispersion of the S p–Cu d antibonding states at the top of the valence remains fairly unaltered, even if the overlap of LDA and quasi-particle wavefunctions shows remarkable variations in this region.
2. The bandwidth of the S p–Cu d bonding states (located between −3.5 and −6.7 eV) slightly increases.
3. The bands associated to the (Zn,Sn)–S bond (between −8 and −10 eV) are inverted and downshifted by about 2 eV with respect to LDA.
4. Also, the S s states are moved down by about 2 eV.

In Figure 2.7b, we show the same bands as obtained from GGA + U and HSE06 calculations. In this case HSE06 bands are remarkably similar to scGW bands. This is not particularly surprising as the Hartree–Fock mixing of HSE06 is particularly suited for materials with gaps of about 1–2 eV [74, 81]. As expected, GGA + U shifts down the states with Cu d character, thereby opening the gap to a reasonable value. However, it is evident from Figure 2.7 that the overall description of the band dispersions is quite inaccurate. Note that the analysis of the band structure of a CIGS compound or another CZTS compound would have led to the same conclusions [98, 107].

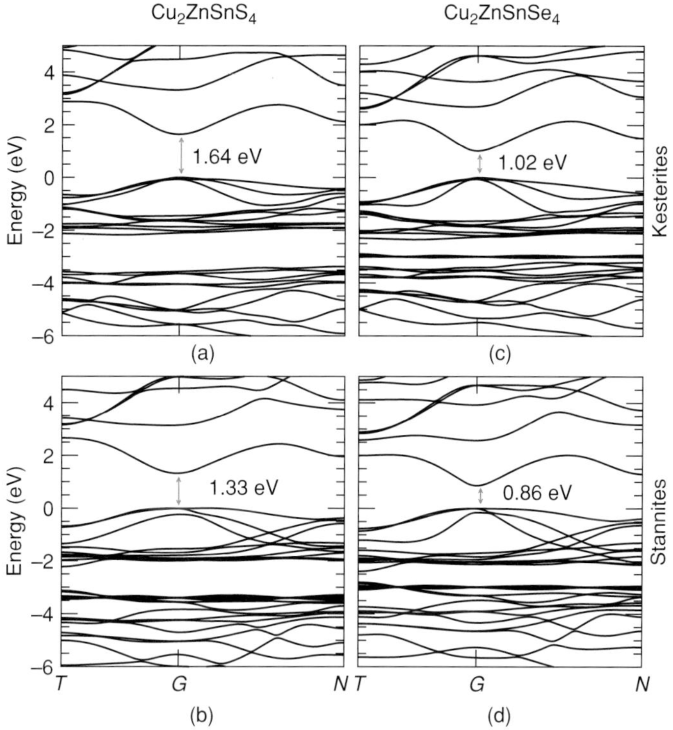

Figure 2.8 Band structures from scGW for (a) kesterite Cu_2ZnSnS_4, (b) stannite Cu_2ZnSnS_4, (c) kesterite $Cu_2ZnSnSe_4$, and (d) stannite $Cu_2ZnSnSe_4$. (From [107].)

In Figure 2.8, we can see the sc*GW* band structures of the four compounds of the CZTS family in the energy region close to the fundamental gap. For kesterite Cu_2ZnSnS_4 and stannite $Cu_2ZnSnSe_4$, the experimental geometries were used [116–120]. For the remaining compounds, the experimental anion displacements were not reported [116–120], and it was therefore necessary to resort to the theoretical structures.

The choice of the theoretical framework for the geometry optimization is essential. In fact, the strong dependence of the gap on anion displacements is also present in CZTS compounds [107]. Also in this case slopes are substantially larger for sc*GW* calculations than for DFT–LDA, and even than for HSE06. Once again, they are controlled by the screening, which is essential to include in a self-consistent way, as it is done in a sc*GW* scheme. Botti *et al.* [107] verified that LDA/GGA-relaxed structures lead to unacceptably large errors (up to 40%) in the sc*GW* band gap, essentially related to the error in u: Cu-(S,Se) and Zn-(S,Se) bond lengths are slightly too small in DFT–LDA, while Sn-(S,Se) bond lengths are too large. Also, this time the solution to this issue relies on the use of the HSE06 functional, which yields extremely accurate values for u and the cell parameters, thanks to a better description of the localized states participating in bonding.

The major difference among the four band structures of Figure 2.8 is the width of the band gaps, which are in excellent agreement with experimental results. As in the CIGS family, Se compounds have a smaller gap with respect to S compounds. Furthermore, stannites have consistently smaller gaps than kesterites.

In conclusion, remarkable similarities were found in the electronic properties of stannites and kesterite CZTS, and, to a large extent, between CZTS and CIGS compounds. This fact suggests once more that CZTS materials are excellent candidates to replace more costly CIGS in the absorbing layer of thin-film PV devices.

We want to conclude this section with an example of optical spectra calculated from sc*GW* band structures, which is reported in Aguilera *et al.* [98].

Starting either from the KS or sc*GW* band structures of $CuGaS_2$, Aguilera *et al.* [98] calculated optical absorption spectra within linear response, both applying TDDFT and solving the BSE. Figures 2.9 and 2.10 [98] show calculations of the imaginary part of the macroscopic dielectric function for light polarized along the *c*-axis. The theoretical curves are also compared with the experimental data published in Levcenko *et al.* [86] and Rife *et al.* [128].

The experimental dielectric functions of Alonso *et al.* [99] were obtained with ellipsometry at room temperature, whereas the spectra of Levcenko *et al.* [86] were

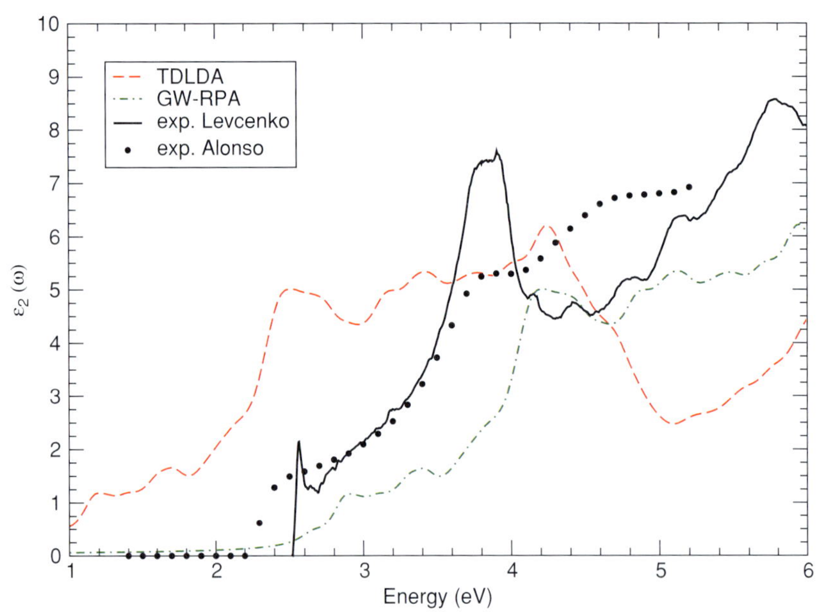

Figure 2.9 *Imaginary part of the dielectric function for light polarized along the c-axis obtained with TDDFT using the ALDA kernel (dashed line), RPA using GW eigenvalues (dot-dashed line) compared to experimental data of Alonso et al. [99] (dots) and Levcenko et al. [86] (solid line; in arbitrary units, see text). (From [98].)*

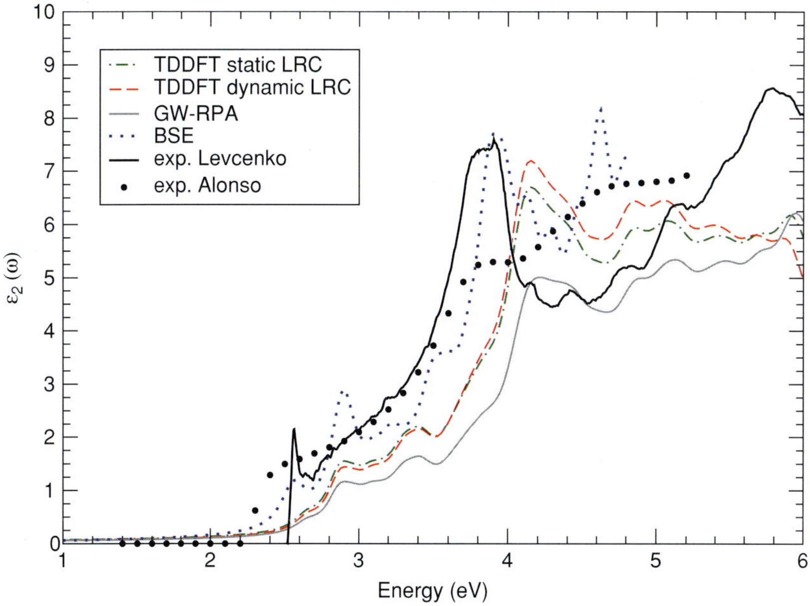

Figure 2.10 *Imaginary part of the dielectric function for light polarized along the c-axis obtained with TDDFT using the static (dot-dashed line) and the dynamic (dashed line) LRC kernel, GW-RPA (gray solid line) and BSE (dotted line), compared to experimental measurements of Alonso et al. [99] (larger dots) and Levcenko et al. [86] (solid line; in arbitrary units). (From [98].)*

extracted from reflection spectra measured at 77 K, using the Kramers–Kronig relations. Both results are for single-crystal samples. We can observe that the two curves are rather different. In particular, the absorption spectrum of Levcenko *et al.* [86] has a much smaller intensity than the experimental curve of Alonso *et al.* [99] and previous results by Rife *et al.* [128], probably due to the quality of the reflecting surface of the samples. In view of the above, the intensity of the absorption spectrum of Levcenko *et al.* [86] was rescaled. Another discrepancy among the available experimental data concerns the behavior of the spectra at around 5 eV. While the measurements of Alonso *et al.* [99] found a plateau in the absorption after the peak at 4 eV, in Levcenko *et al.* [86] and Rife *et al.* [128] a deep valley appears after 4 eV. Note that there are substantial discrepancies between the two experiments and that sufficiently accurate *ab initio* calculations are an essential tool to discriminate between the two sets of data.

In Figure 2.9, we compare results obtained within the RPA using scCOHSEX + G_0W_0 eigenvalues, and ALDA using KS eigenvalues. The RPA curve obtained using KS eigenvalues overlaps almost perfectly with the ALDA curve. This proves that the inclusion of the ALDA kernel does not have any relevant effect on the absorption, as the important contributions due to electron–electron and electron–hole interactions are not accounted for by the ALDA kernel

[39, 49, 129]. Both the RPA and ALDA spectra using KS-GGA eigenvalues are dramatically red-shifted. As a first conclusion, we can observe that the accuracy of RPA or ALDA *ab initio* calculations alone would not sufficient to explain the large discrepancies among experimental optical measurements.

If the scCOHSEX + G_0W_0 eigenvalues are used for an RPA calculation (dot-dashed line), the resulting spectra are blue-shifted, while the line shape does not change. As a result, the absorption peaks are now at too high energies. In fact, in the energy range of interest for absorption the GW correction is basically equivalent to applying a scissor operator.

What is still missing is a proper account of excitonic effects. For many intermediate-gap semiconductors, it was shown [49, 51, 84] that the LRC to the exchange–correlation kernel was enough to yield a good agreement with experimental absorption spectra. The LRC kernel works particularly well in the case of a continuum exciton, giving in general very good optical properties with the efficiency of a TDDFT calculation. In Figure 2.10, we show the spectra obtained using the static [49, 51] and the dynamic [84] versions of the LRC kernel. scCOHSEX + G_0W_0 eigenvalues were used in these calculations.

The TDDFT–LRC spectra shown in Figure 2.10 provide a clear improvement with respect to the curves displayed in Figure 2.9. The intensity of the peaks as well as the onsets of absorption is now closer to experiment. However, the peak at about 4 eV is still located at a higher energy than its experimental position. Note that the excitonic peak at the fundamental edge in $CuGaS_2$ has the nature of a bound exciton, even if the binding energy is rather small. This explains why the LRC models do not reproduce perfectly the BSE result.

If one wants a precise description of the bound excitonic peak of $CuGaS_2$ [85–88], it is necessary to resort to the solution of the BSE. In Figure 2.10, we see that—although a shifted mesh of 1000 k points is not sufficiently dense to smooth out the curve due to the strong dispersion of the lowest conduction state—the experimental onset at about 2.55 eV and the peak at about 3.8 eV are well reproduced in the BSE calculation [98]. At the absorption edge, the estimate of the excitonic binding energy of about 0.1 eV is also consistent with experimental data [130].

However, one cannot forget that TDDFT calculations using model exchange–correlation kernels are much less involved than the solution of the BSE. In many cases, for example when dealing with supercells with defects or doping, the only feasible approach that guarantees qualitatively correct results is to perform TDDFT calculations using model kernels derived from BSE.

2.3.2 Delafossite Transparent Conductive Oxides

Transparent conductive oxides (TCOs) are wide band-gap semiconductors characterized by large free carrier densities. These carriers are created by either intrinsic or extrinsic doping, giving to TCOs both low resistivity and transparency in the visible energy window. The most common examples of TCOs are n-doped semiconductors such as SnO_2, In_2O_3, and ZnO. Carrier concentrations up to 10^{21} cm^{-3} are commonly achieved with large mobilities of 30 cm^2/Vs, giving rise to conductivities of up to 3000 S/cm [131]. Up to now p-type conductivity has been much harder to achieve. The reason is related to well-known differences between holes and electrons in TCOs. First, holes have a much heavier mass than electrons and consequently much lower mobility. Second, doping rules point to the difficulty of p-type doping in oxides [34] because of the high electronegativity and the presence of hole killers, such as V_0, with relatively low formation energy.

Nevertheless, p-type conductivity was found in delafossite $CuAlO_2$ thin films [20]. A few years later, bipolar (both n- and p-types) conductivity was discovered in $CuInO_2$ [22, 132], another element of the same delafossite family. However, in p-type TCOs carrier concentrations remain smaller than 10^{18}cm^{-3} and mobilities limited to $\mu < 1$cm^2/V s, resulting in conductivities of 100 S/cm [131]. Other p-type oxides belonging to different families were found afterwards, namely $ZnRh_2O_4$ [133] and SnO [134]. Even though they have relatively low absorption onsets (at 2.1 eV and 2.9 eV, respectively) and, consequently, they can hardly be considered transparent, they display significantly higher p-type conductivity than delafossites. These spectacular findings opened the way for the fabrication of TCO p–n junctions [135], and to the development of a new technology entirely based on "invisible circuits" [136–138]. Many innovative applications require transparent electronics such as stacked solar cells, transparent screens, or functional windows that generate solar electricity.

$CuAlO_2$ is by far the most studied system of the family of delafossite TCOs, both theoretically and experimentally. For a long time, there has been no agreement either on the origin of the p-type conductivity or on the electronic band gap of the pure crystal. Measurements of the direct *optical* band gap (E_g^{dir}) of $CuAlO_2$ fall in the range from 2.9 eV to 3.9 eV [19, 20, 136, 139–149], with most values in the interval 3.4–3.7 eV. These experiments also yield a large dispersion of indirect gaps (E_g^{ind}), from 1.65 eV to 2.1 eV, with one experiment measuring 2.99 eV [146]. Unfortunately, there is only one photoemission experiment [148] that gives a value of 3.5 eV for the *quasi-particle* band gap. Note that the *optical* and *quasi-particle* gaps differ by the exciton binding energy. Concerning $CuInO_2$, optical experiments measured E_g^{dir} between 3.9 eV and 4.45 eV [22, 150, 151], with only one estimation of E_g^{ind} at 1.44 eV [150].

From the theoretical perspective, the situation is also quite complex, even if the full Cu 3*d* shell should exclude the strongly correlated electron regime. These materials are usually studied in literature within DFT, using the standard LDA or GGA [132, 144, 146, 148, 152–154]. We can expect that the KS band structures yield underestimated band gaps and the hybridization between the *d* electrons of the metal and *p* electrons of oxygen is misrepresented [14, 62].

Both LDA/GGA + U and hybrid functionals have been applied to study members of the delafossite family with some success [65, 153, 155–157]. Note that these calculations yield quasi-particle gaps, and not optical gaps. To evaluate these latter quantities, one mostly resorts to the solution of the BSE. For delafossite structures there is one such calculation starting from a GGA + U band structure [156]. It yields for $CuAlO_2$ a very large exciton binding energy of about 0.5 eV for the first direct transition. The choice of the parameter U was found to have significant consequences on the width of the band gap, but it did not affect significantly the exciton binding energy. We can thus assume that 0.5 eV is a reasonable estimate of this latter quantity.

In the following, we discuss calculations [65, 157] of the band structures of $CuAlO_2$ and $CuInO_2$, comparing the most accurate theoretical tools available in the community. These include standard LDA, hybrid functionals (namely B3LYP [158] and two flavors of Heyd–Scuseria–Ernzerhof, HSE03 and HSE06 [75, 76]), LDA+U, G_0W_0, and scGW. We expect scGW to be the most accurate *ab initio* approach.

In Figure 2.11, we show a summary of calculated direct and indirect photoemission gaps of $CuAlO_2$ [65]. The corresponding band structures are shown in Figure 2.12. The minimum E_g^{dir} of $CuAlO_2$ is always found at L, where the dipole transition between the band edge states is allowed [132]. All calculations, except scGW, give a fundamental E_g^{ind} between the conduction band minimum at Γ and the valence band maximum along the $\Gamma - F$ line. The experimental data for optical gaps are also presented with an error bar that reflects the dispersion of the most likely values found in literature. LDA exhibits, as expected,

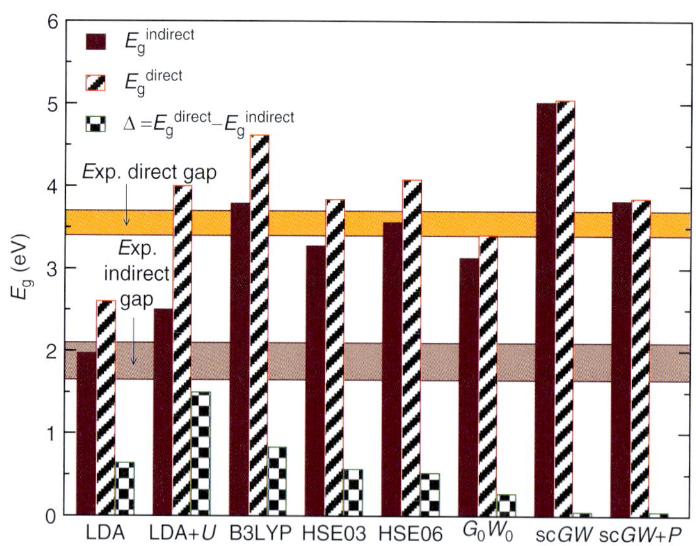

Figure 2.11 *Band gaps of $CuAlO_2$ using LDA, LDA + U, hybrid, G_0W_0, scGW, and scGW including model polaronic corrections. The horizontal zones contain data extracted from various optical experiments. (From [65].)*

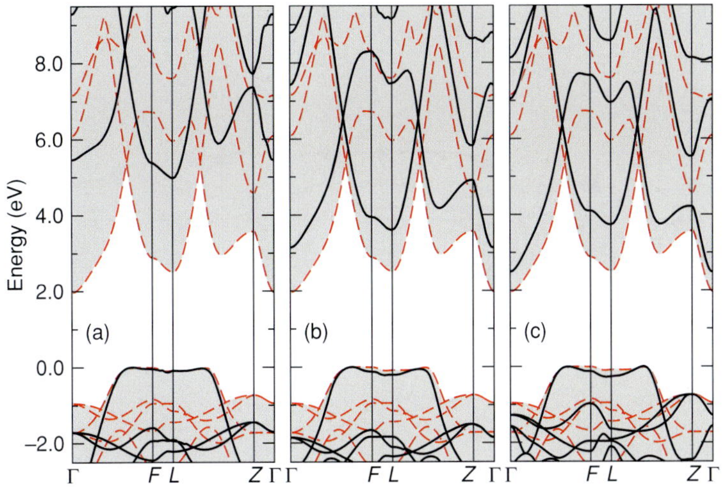

Figure 2.12 *Band structures for $CuAlO_2$: comparison of LDA (red dashed lines) with scGW (a), HSE03 (b), and LDA + U (c). (From [65].)*

the smallest gaps. Basically every approach beyond it opens up the gap by different amounts and modifies the band dispersions. The direct and indirect gaps have similar behaviors in the different theories, and both increase when going from LDA $<$ G_0W_0 $<$ HSE03 $<$ HSE06 $<$ B3LYP $<$ scGW.

Looking at the direct gap, we point out that most of the methods give results that are within the experimental range when an exciton binding energy of around 0.5 eV [156] is considered. This is true for LDA+U, G_0W_0, the hybrids HSE03, and HSE06. However, for scGW and even for B3LYP, the theoretical gap is larger by about 1–1.5 eV than the experimental findings.

We observe that the difference $E_g^{dir} - E_g^{ind}$ decreases with the sophistication of the method, reaching nearly zero for the scGW calculation. This is a consequence of the drastic change of the conduction band dispersion, which displaces the conduction minimum from Γ to L when scGW is applied (see Figure 2.12). The lowest conduction bands cross along the line between the high-symmetry points L and F, which results in a significant change in the character of the conduction band minimum. At Γ and Z the band has an E_g of Cu 3d character with a significant amount of s and p states from Al, O, and Cu. On the other hand, at L and F, the bottom conduction band is composed of Cu s and O p states. Therefore, it is not surprising that GW correction is so strongly k-point dependent. Figure 2.13 displays the difference between the expectation value of the Hartree potential and the exchange part of the self-energy, constructed with DFT wavefunctions or with scCOHSEX wavefunctions. The exchange self-energy changes up to 4 eV, while the Hartree term is rather constant. We often note that in semiconductor physics one assumes that the quasi-particle corrections can be modeled by a rigid shift (the so-called

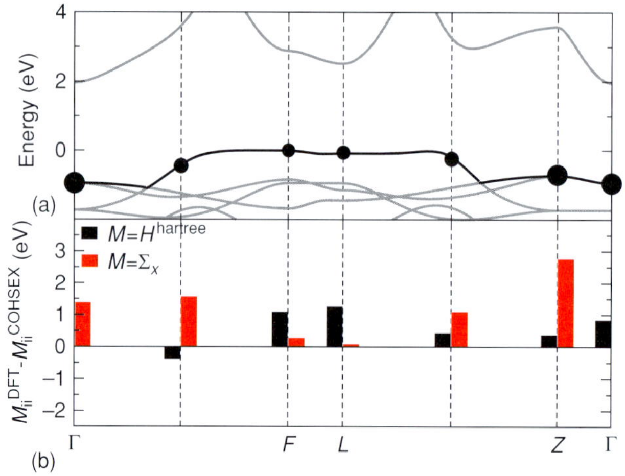

Figure 2.13 (a) Kohn–Sham conduction band structure of $CuAlO_2$. (b) Difference between the expectation value of the operator M ($M = H^{\text{Hartree}}$, Σ_x) constructed with DFT wavefunctions or with scCOHSEX wavefunctions for the bottom conduction band. (From [157].)

scissor operator). From these results it follows that one should refrain from using this simple approximation for delafossite TCOs.

Hybrid methods such as HSE03, HSE06, and B3LYP open the band gaps and reduce the difference $E_g^{\text{dir}} - E_g^{\text{ind}}$, following the trend observed for scGW calculations. The conduction band minimum (CBM) from HSE03 is still located at Γ, but the difference between the CBMs at L and Γ gets significantly smaller. The main effect of LDA+U is to open the LDA gap by an amount that can be controlled by the parameter U. The difference $E_g^{\text{dir}} + E_g^{\text{ind}}$ is increased within the LDA + U approximation, due to the change of character for the lowest conduction band along the symmetry line between Γ and F. We can conclude that hybrid calculations give a better description (i.e., closer to scGW) of band dispersions than LDA+U, even if the two approaches yield similar band gaps.

For $CuInO_2$ the smallest E_g^{dir} is located at Γ, where optical transitions are forbidden [132]. A meaningful comparison with optical experiments must then consider the gap at L. Both trends and quantitative results are analogous to those for $CuAlO_2$. In particular, scGW again yields E_g^{dir} larger by 1–1.5 eV than the experimental range [65, 157].

We stress again that, to date, scGW is arguably the best method available to estimate band gaps of wide-gap semiconductors, and that it gives excellent results for compounds like Cu_2O and CuIn(S, Se)$_2$ [14, 62]. It is unlikely that the presence of defects can explain the discrepancy between experiments and scGW gaps. However, there is another effect that has been neglected up to now—the change of screening due to the polarization of the lattice. The polaron constant defined as $\alpha_P = (1/\epsilon_\infty - 1/\epsilon_0)(\hbar/2ma_B^2\omega_{LO})^{1/2}$—where a_B is the Bohr radius, ω_{LO} the longitudinal optical frequency of the highest E_u phonon mode, ϵ_∞ and ϵ_0 the low frequency electronic and the static dielectric constants, respectively—measures the strength of the polaronic effect. According to the experimental data [159], unfortunately available only for $CuAlO_2$, the polaron constant for this system is large ($\alpha_P \sim 1$), indicating a nonnegligible contribution of the lattice polarization to the electronic screening. It is known that in other ionic compounds with similar polaron constants this can lead to shrinking of the band gap by about 1 eV [160]. A full scGW calculation including, in an *ab initio* framework, the effects of the lattice polarization is still beyond reach. However, a reliable estimate was obtained using the model proposed by Bechstedt *et al.* [160], which gives a static representation of the contribution of lattice polarization to the screening based on difference of experimental static dielectric constants. By performing a perturbative GW step, including a model contribution to the screening due to lattice polarization, on top of the scCOHSEX, Vidal *et al.* [65] found a uniform (k-independent) decrease of the band gap by 1.2 eV. As we can see in Figure 2.11, this correction brings the scGW results for E_g^{dir} well within the experimental range (once the excitonic correction of about 0.5 eV is also considered). As it is observed in Bechstedt *et al.* [160], the model employed to account for the lattice polarization can only overestimate the correction. All these results point to the conclusion that the agreement of the other methods with experiment was fortuitous and due to a cancellation of errors.

It is interesting to note that to design p-type TCOs the use of layered structures is recommended in order to increase the value of the band gap. Indeed, LaCuSO [161] and BaCu(S,Se,Te)F [162] display band structures very similar to the one of $CuAlO_2$. One can foresee that also in these materials the renormalization of the band gap due to the lattice contribution to the screening could then play an important role and significantly affect the interpretation of experiments.

To conclude, we focus now on the indirect gaps of Figure 2.11. Analogous conclusions could be obtained for $CuInO_2$ or $CuGaO_2$ [65, 157], but we rather focus on $CuAlO_2$ as more experimental data are available in literature. All hybrids and GW calculations yield indirect gaps much larger than the experimental range 1.65–2.1 eV, even taking into account any possible excitonic and polaronic effects. Moreover, scGW, the best method used in this work, yields the highest E_g^{ind} at around 5 eV, while the difference $E_g^{dir} - E_g^{ind}$ is in general much smaller than the experimental value ($\approx$2 eV), and even vanishing for scGW calculations.

These are very strong arguments in favor of Robertson *et al.* [153] who suggested that the experimental "indirect gap" absorption was due to defects, and should not be present in the defect-free compound. Also, Pellicer-Porres *et al.* [146] questioned the interpretation of the low energy peaks as indirect transitions, as the absorption coefficient is more than two orders of magnitude larger than in typical indirect absorption edges. It has been first asserted that the defects that could be responsible are oxygen interstitials O_i, as DFT calculations within the LDA predict low formation energies and the introduction of states in the gap at 0.7 eV and 1.4 eV [163]. More recently, Scanlon and Watson [164] proposed, based on screened hybrid calculations, that antisite Cu-on-Al Cu_{Al} is responsible for the features in the absorption spectra at 1.65–2.1 eV.

In summary, it is clear that the compounds of the delafossite family exhibit complex and unusual band-gap physics that cannot be captured by standard theoretical approximations. In particular, the direct band gap is well reproduced by the state-of-the-art many-body approaches if the contribution of lattice polarization to the screening is taken into account. We can expect that this situation, of a large gap that is reduced substantially by polaronic effects, is quite general and is present in many polar materials. In fact, the apparent good agreement between calculated gaps (with hybrid functionals or G_0W_0) and experimental gaps for materials as simple and widely studied as LiF can be accidental. The underestimation of the gap by these methods (the scGW gap is indeed 2 eV larger than the experimental and G_0W_0 gap) is compensated by the neglect of lattice polarization effects. Furthermore, the modifications with respect to the LDA KS bands are strongly k-dependent, which makes questionable the common practice of using a scissor operator. The band dispersion obtained by hybrid functional calculations is in between the LDA and scGW dispersions, while the LDA + U calculations open up the gap but do not give a significant improvement of the band dispersion. This is in agreement with what we have discussed in Section 2.3.1 for CIGS and CZTS compounds. Finally, scGW calculations rule out the interpretation of the low energy features in the absorption spectra as arising from a putative indirect band gap. These structures should rather come from intrinsic defects, as proposed in [146, 153]. However, a complete understanding of the electronic and excitation properties of these systems will only be achieved, in our opinion, by a high-level theoretical scheme (such as scGW) including defects and effects from the lattice polarization in an *ab initio* framework.

2.4 Conclusions

The response upon excitation of complex materials, such as materials for thin-film PVs, can nowadays be calculated from first principles with high accuracy. This was made possible by recent theoretical developments together with the optimization of computer codes.

When dealing with electronic excitations, the knowledge of the ground-state density of the system is no longer sufficient, which means that one has to find appropriate ways to go beyond DFT. Different paths have been explored for the past years, based on Green's functions, the time-dependent density, hybrid functionals in the generalized KS picture, and empirically modified Hamiltonians.

The state-of-the-art approach to calculate band structures and optical spectra, quantitatively comparable to experimental data, relies on the MBPT via the GW approximation and the solution of the BSE. However, one has to deal with one-particle and two-particle Green's functions, which require heavy calculations to be determined.

The *GW* approximation for the calculation of quasi-particle states, either in the perturbative or in the self-consistent schemes, cannot be applied at the moment to materials including *d* electrons if the unit cell is larger than 25–50 atoms.

As long as large supercells are out of reach for *GW* calculations, hybrid functionals promise to be a computationally efficient alternative with a satisfactory degree of accuracy. This fact motivates the search for improved approximations for the mixing parameter of hybrid functionals, which include screening as a physical ingredient.

In a similar way, whenever the solution of the BSE is out of reach to determine absorption spectra, one can obtain qualitatively reliable results by performing time-dependent DFT calculations, by applying model kernels derived from BSE.

The way is open now for many possible applications. Indeed, in the rapidly expanding field of PVs, the understanding of many problems of great complexity and technological interest is still lacking, especially concerning defects and interfaces. Even if great physical insight can already be obtained employing existing methods, the quest for improved approximations, as well as the search for novel theoretical approaches, cannot be abandoned. Hopefully new developments will appear in the next years.

References

1. Green, M. (2000) *Power to the People: Sunlight to Electricity Using Solar Cells*, New South Wales University Press.
2. Chapin, D., Fuller, C. and Pearson, G. (2004) A new silicon p-n junction photocell for converting solar radiation into electrical power, *J. Appl. Phys.* **25**, 676.
3. Sze, S. and Ng, K. (2007) *Physics of Semiconductor Devices*, Wiley-Blackwell.
4. Pearson, G. (1957) Conversion of solar to electrical energy, *Am. J. Phys.* **25**, 591.
5. Zhao, J., Wang, A., Green, M. and Ferrazza, F. (1998) 19.8% efficient honeycomb textured multicrystallineand 24.4% monocrystalline silicon solar cells, *Appl. Phys. Lett.* **73**, 1991.
6. Wadia, C., Alivisatos, A. and Kammen, D. (2009) Materials availability expands the opportunity for large-scale photovoltaics deployment, *Environ. Sci. Technol.* **43**, 2072.
7. Shah, A., Torres, P., Tscharner, R. *et al.* (1999) Photovoltaic technology: the case for thin-film solar cells, *Science* **285**, 692.
8. Wu, X., Keane, J., Dhere, R. *et al.* (2001) *16.5%-Efficient CdS/CdTe Polycrystalline Thin-Film Solar Cell.* Proceedings of the 17th European Photovoltaic Solar Energy Conference, vol. 995, James & James Ltd, London.
9. Jackson, P., Hariskos, D., Lotter, E. *et al.* (2011) New world record efficiency for Cu(In,Ga)Se_2 thin-film solar cells beyond 20%, *Prog. Photovolt. Res. Appl.* **19**(7), 894–897.

10. Repins, I., Contreras, M.A., Egaas, B. *et al.* (2008) 19.9%-efficient ZnO/CdS/ $CuInGaSe_2$ solar cell with 81.2% fill factor, *Prog. Photovolt.* **16**, 235.
11. Rau, U. and Schock, H. (1999) Electronic properties of Cu(In,Ga)Se_2 heterojunction solar cells-recent achievements, current understanding, and future challenges: thin film solar cells, *Appl. Phys. A* **69**, 131–147.
12. Partain, L., McLeod, P., Duisman, J. *et al.* (1983) Degradation of a cuxs/eds solar cell in hot, moist air and recovery in hydrogen and air, *J. Appl. Phys.* **54**, 6708–6720.
13. Guillemoles, J.F., Kronik, L., Cahen, D. *et al.* (2000) Stability issues of Cu(In,Ga)Se_2-based solar cells, *J. Phys. Chem. B* **104**, 4849–4862.
14. Vidal, J., Botti, S., Olsson, P. *et al.* (2010a) Strong interplay between structure and electronic properties in CuIn(S, Se)$_2$: a first-principles study, *Phys. Rev. Lett.* **104**, 056401.
15. Rediriger, A., Berg, DM., Dale, PJ. and Siebentritt, S. (2011) The consequences of kesterite equilibria for efficient solar cells, *J. Am. Chem. Soc.* **133**, 3320.
16. Todorov, K., Reuter, K.B. and Mitzi, D.B. (2010) High-efficiency solar cell with earth-abundant liquid-processed absorber, *Adv. Mater.* **22**, E156.
17. Nakamura, S., Senoh, M. and Iwasa, N. (1992) Thermal annealing effects on p-type Mg-doped GaN films, *Jpn. J. Appl. Phys.* **31**, L139.
18. Neugebauer, J. and Van de Walle, C.G. (1995) Hydrogen in GaN: Novel aspects of a common impurity, *Phys. Rev. Lett.* **75**, 4452–4455.
19. Benko, F. and Koffyberg, F. (1984) Opto-electronic properties of $CuAIO_2$. *J. Phys. Chem. Solids.* **45**, 57–59.
20. Kawazoe, H., Yasukawa, M., Hyodo, H. *et al.* (1997) P-type electrical conduction in transparent thin films of $CuAlO_2$, *Nature* **389**, 939.
21. Ueda, K., Hase, T., Yanagi, H. *et al.* (2001) Epitaxial growth of transparent p-type conducting $CuGaO_2$ thin films on sapphire (001) substrates by pulsed laser deposition, *J. Appl. Phys.* **89**, 1790–1793.
22. Yanagi, H., Hase, T., Ibuki, S. *et al.* (2001) Bipolarity in electrical conduction of transparent oxide semiconductor $CuInO_2$ with delafossite structure, *Appl. Phys. Lett.* **78**, 1583–1585.
23. Hohenberg, P. and Kohn, W. (1964) Inhomogeneous electron gas, *Phys. Rev.* **136**, B864–B871.
24. Kohn, W. and Sham, L.J. (1965) Self-consistent equations including exchange and correlation effects, *Phys. Rev.* **140**, A1133–A1138.
25. Godby, R.W., Schlüter, M. and Sham, L.J. (1987) Trends in self-energy operators and their corresponding exchange-correlation potentials, *Phys. Rev. B* **36**, 6497–6500.
26. Perdew, J.P., Burke, K. and Ernzerhof, M. (1996) Generalized gradient approximation made simple, *Phys. Rev. Lett.* **77**, 3865–3868. Erratum: ibid. **78**, 1396(E) (1997).
27. Fetter, AL. and Walecka, J.D. (1971) *Quantum Theory of Many-Particle Systems*, McGraw-Hill, New York.
28. Hedin, L. (1965) New method for calculating the one-particle Green's function with application to the electron-gas problem, *Phys. Rev.* **139**, A796.
29. Bruneval, F., Vast, N. and Reining, L. (2006a) Effect of self-consistency on quasiparticles in solids, *Phys. Rev. B* **74**, 045102.
30. Faleev, S.V., van Schilfgaarde, M. and Kotani, T. (2004) All-electron self-consistent GW approximation: application to Si, MnO, and NiO, *Phys. Rev. Lett.* **93**, 126406.

31. van Schilfgaarde, M., Kotani, T. and Faleev, S. (2006) Quasiparticle self-consistent GW theory, *Phys. Rev. Lett.* **96**, 226402.
32. Anisimov, V.I., Zaanen, J. and Andersen, O.K. (1991). Band theory and Mott insulators: Hubbard U instead of Stoner I, *Phys. Rev. B.* **44**, 943.
33. Lany, S. and Zunger, A. (2008) Assessment of correction methods for the band-gap problem and for finite-size effects in supercell defect calculations: case studies for ZnO and GaAs, *Phys. Rev. B* **78**, 235104.
34. Zunger, A. (2003) Practical doping sprinciples, *Appl. Phys. Lett.* **83**, 57.
35. Brothers, E.N., Izmaylov, A.F., Normand, J.O. *et al.* (2008) Accurate solid-state band gaps via screened hybrid electronic structure calculations, *J. Chem. Phys.* **129**, 011102.
36. Ernzerhof, M. and Scuseria, G.E. (1999) Assessment of the Perdew-Burke-Ernzerhof exchange-correlation functional, *J. Chem. Phys.* **110**, 5029.
37. Seidl, A., Görling, A., Vogl, P. *et al.* (1996) Generalized Kohn-Sham schemes and the band-gap problem, *Phys. Rev. B* **53**, 3764–3774.
38. Hanke, W. (1978) Dielectric theory of elementary excitations in crystals, *Adv. Phys.* **27**, 287–341.
39. Onida, G., Reining, L. and Rubio, A. (2002) Electronic excitations: density-functional versus many-body green's-function approaches, *Rev. Mod. Phys.* **74**, 601–659.
40. Sham, L.J. and Rice, T.M. (1966) Many-particle derivation of the effective-mass equation for the wannier exciton, *Phys. Rev.* **144**, 708–714.
41. Gross, E.K.U. and Kohn, W. (1985) Local density-functional theory of frequency-dependent linear response, *Phys. Rev. Lett.* **55**, 2850–2852.
42. Gross, E.K.U. and Kohn, W. (1986) Local density-functional theory of frequency-dependent linear response, *Phys. Rev. Lett.* **57**, 923–923(E).
43. Runge, E. and Gross, E.K.U. (1984) Density-functional theory for time-dependent systems, *Phys. Rev. Lett.* **52**, 997–1000.
44. van Leeuwen, R. (2001) Key concepts in time-dependent density-functional theory, *Int. J. Mod. Phys. B* **15**, 1969–2023.
45. Zangwill, A. and Soven, P. (1980) Resonant photoemission in barium and cerium, *Phys. Rev. Lett.* **45**, 204–207.
46. Petersilka, M., Gossmann, U.J. and Gross, E.K.U. (1996) Excitation energies from time-dependent density-functional theory, *Phys. Rev. Lett.* **76**, 1212–1215.
47. Botti, S., Schindlmayr, A., Del Sole, R. and Reining, L. (2007) Time-dependent density functional theory for extended systems, *Rep. Prog. Phys.* **70**, 357.
48. Adragna, G., Del Sole, R. and Marini, A. (2003) *Ab initio* calculation of the exchange-correlation kernel in extended systems, *Phys. Rev. B* **68**, 165108.
49. Botti, S., Sottile, F., Vast, N. *et al.* (2004) Long-range contribution to the exchange-correlation kernel of time-dependent density functional theory, *Phys. Rev. B* **69**, 155112.
50. Marini, A., Del Sole, R. and Rubio, A. (2003) Bound excitons in time-dependent density-functional theory: optical and energy-loss spectra, *Phys. Rev. Lett.* **91**, 256402.
51. Reining, L., Olevano, V., Rubio, A. and Onida, G. (2002) Excitonic effects in solids described by time-dependent density-functional theory, *Phys. Rev. Lett.* **88**, 066404.

52. Sharma, S., Dewhurst, J.K., Sanna, A. and Gross, E.K.U. (2011) Bootstrap approximation for the exchange-correlation kernel of time-dependent density functional theory, *Phys. Rev. Lett.* **107**, 186401.
53. Sottile, F., Olevano, V. and Reining, L. (2003) Parameter-free calculation of response functions in time-dependent density-functional theory, *Phys. Rev. Lett.* **91**, 056402.
54. Stubner, R., Tokatly, I.V. and Pankratov, O. (2004) Excitonic effects in time-dependent density-functional theory: an analytically solvable model, *Phys. Rev. B*. **70**, 245119.
55. von Barth, U., Dahlen, N.E., van Leeuwen, R. and Stefanucci, G. (2005) Conserving approximations in time-dependent density functional theory, *Phys. Rev. B* **72**, 235109.
56. Strinati, G., Mattausch, H.J. and Hanke, W. (1982) Dynamical aspects of correlation corrections in a covalent crystal, *Phys. Rev. B* **25**, 2867–2888.
57. Hybertsen, M.S. and Louie, S.G. (1985) First-principles theory of quasiparticles: calculation of band gaps in semiconductors and insulators, *Phys. Rev. Lett.* **55**, 1418–1421.
58. Hybertsen, M.S. and Louie, S.G. (1986) Electron correlation in semiconductors and insulators: Band gaps and quasiparticle energies, *Phys. Rev. B* **34**, 5390–5413.
59. Godby, R.W., Schlüter, M. and Sham, L.J. (1988) Self-energy operators and exchange-correlation potentials in semiconductors, *Phys. Rev. B* **37**, 10159–10175.
60. Del Sole, R., Reining, L. and Godby, R.W. (1994) GwΓ approximation for electron self-energies in semiconductors and insulators, *Phys. Rev. B* **49**, 8024–8028.
61. Aulbur, W.G., Jönsson, L. and Wilkins, J.W. (1999). Quasiparticle calculations in solids, *Solid State Phys.* **54**, 1.
62. Bruneval, F., Vast, N., Reining, L. *et al.* (2006b) Exchange and correlation effects in electronic excitations of Cu_2O, *Phys. Rev. Lett.* **97**, 267601.
63. Gatti, M., Bruneval, F., Olevano, V. and Reining, L. (2007) Understanding correlations in vanadium dioxide from first principles, *Phys. Rev. Lett.* **99**, 266402.
64. Kammerlander, D., Marques, M. and Botti, S. (2011) Band structures of Cu_2ZnSnS_4 and $Cu_2ZnSnSe_4$ from many-body methods, *Appl. Phys. Lett.* **98**, 241915.
65. Vidal, J., Trani, F., Bruneval, F. *et al.* (2010b) Effects of electronic and lattice polarization on the band structure of delafossite transparent conductive oxides, *Phys. Rev. Lett.* **104**, 136401.
66. Li, J.L., Rignanese, G.M., Chang, EK. *et al.* (2002) GW study of the metal-insulator transition of bcc hydrogen, *Phys. Rev. B* **66**, 035102.
67. Rinke, P., Qteish, A., Neugebauer, J. *et al.* (2005) Focus on photoemission and electronic structure combining gw calculations with exact-exchange density-functional theory: an analysis of valence-band photoemission for compound semiconductors, *New J. Phys.* **7**, 126.
68. Jiang, H., Gomez-Abal, R.I., Rinke, P. and Scheffler, M. (2009) Localized and itinerant states in lanthanide oxides united by GWLDA + U, *Phys. Rev. Lett.* **102**, 126403.
69. Kioupakis, E., Zhang, P., Cohen, M.L. and Louie, S.G. (2008) GW quasiparticle corrections to the LDA + U/GGA + U electronic structure of bec hydrogen, *Phys. Rev. B* **77**, 155114.
70. Fuchs, F. and Bechstedt, F. (2008) Indium-oxide polymorphs from first principles: Quasiparticle electronic states, *Phys. Rev. B* **77**, 155107.

71. Kresse, G. and Furthmüller, J. (1996a) Efficiency of ab-initio total energy calculations for metals and semiconductors using a plane-wave basis set, *Comput. Mater. Sci.* **6**, 15.
72. Kresse, G. and Furthmüller, J. (1996b) Efficient iterative schemes for *ab initio* total-energy calculations using a plane-wave basis set, *Phys. Rev. B* **54**, 11169–11186.
73. Paier, J., Marsman, M. and Kresse, G. (2007). Why does the B3LYP HF/DFT hybrid functional fail for metals?, *J. Chem. Phys.* **127**, 024103.
74. Paier, J., Marsman, M. and Kresse, G. (2008) Dielectric properties and excitons for extended systems from hybrid functionals, *Phys. Rev. B* **78**, 121201.
75. Heyd, J., Scuseria, G.E. and Ernzerhof, M. (2003) Hybrid functionals based on a screened coulomb potential, *J. Chem. Phys.* **118**, 8207.
76. Heyd, J., Scuseria, G.E. and Ernzerhof, M. (2006) Erratum: hybrid functionals based on a screened coulomb potential, *J. Chem. Phys.* **124**, 219906(E).
77. Gygi, F. and Baldereschi, A. (1989) Quasiparticle energies in semiconductors: Self-energy correction to the local-density approximation, *Phys. Rev. Lett.* **62**, 2160–2163.
78. Bylander, D.M. and Kleinman, L. (1990) Good semiconductor band gaps with a modified local-density approximation, *Phys. Rev. B* **41**, 7868–7871.
79. Clark, S. and Robertson, J. (2010) Screened exchange density functional applied to solids, *Phys. Rev. B* **82**, 085208.
80. Alkauskas, A., Broqvist, P. and Pasquarello, A. (2010) Defect levels through hybrid density functionals: Insights and applications, *Phys. Status Solidi B* **248**, 775.
81. Marques, M.A.L., Vidal, J., Oliveira, M.J.T. *et al.* (2011) Density-based mixing parameter for hybrid functionals, *Phys. Rev. B* **83**, 035119.
82. Gross, E.K.U., Dobson, F.J. and Petersilka, M. (1996) *Topics in Current Chemistry*, vol. 181 (ed. R.F. Nalewajski), Springer-Verlag, Heidelberg, p. 81.
83. Gonze, X., Ghosez, P. and Godby, R.W. (1995) Density-polarization functional theory of the response of a periodic insulating solid to an electric field, *Phys. Rev. Lett.* **74**, 4035–4038.
84. Botti, S., Fourreau, A., Nguyen, F. *et al.* (2005) Energy dependence of the exchange-correlation kernel of time-dependent density functional theory: a simple model for solids, *Phys. Rev. B* **72**, 125203.
85. Bellabarba, C., González, J. and Rincón, C. (1996) Optical-absorption spectrum near the exciton band edge in $CuGaS_2$ at 5 K, *Phys. Rev. B* **53**, 7792–7796.
86. Levcenko, S., Syrbu, N.N., Tezlevan, V.E. *et al.* (2007) Optical spectra and energy band structure of single crystalline $CuGaS_2$ and $CuInS_2$, *J. Phys. Condens. Matter.* **19**, 456222.
87. Ringeissen, J., Regolini, J.L. and Lewonczuk, S. (1973) Optical properties of $CuGaS_2$, *Surf. Sci.* **37**, 777–785.
88. Tell, B. and Kasper, H.M. (1973) Excitons and the spin-orbit splitting in $CuGaS_2$, *Phys. Rev. B* **7**, 740–742.
89. Gonze, X., Rignanese, G.M., Verstraete, M. *et al.* (2005) A brief introduction to the ABINIT software package, *Z. Kristallogr.* **220**, 558–562.
90. Troullier, N. and Martins, J.L. (1991) Efficient pseudopotentials for plane-wave calculations, *Phys. Rev. B* **43**, 1993–2006.
91. Blöchl, P.E. (1994) Projector augmented-wave method, *Phys. Rev. B* **50**, 17953–17979.

92. Kresse, G. and Joubert, D. (1999) From ultrasoft pseudopotentials to the projector augmented-wave method, *Phys. Rev. B* **59**, 1758–1775.
93. Marini, A., Hogan, C., Griming, M. and Varsano, D. (2009) Yambo: An ab initio tool for excited state calculations, *Comput. Phys. Commun.* **180**, 1392–1403.
94. Jaffe, J.E. and Zunger, A. (1984) Theory of the band-gap anomaly in ABC2 chalcopyrite semiconductors, *Phys. Rev. B* **29**, 1882.
95. Merino, J.M., Martín de Vidales, J.L., Mahanty, S. *et al.* (1996) Composition effects on the crystal structure of CuInSe, *J. Appl. Phys.* **80**, 5610.
96. Jaffe, J.E. and Zunger, A. (1983) Anion displacement and band gap anomaly in ternary ABC2 chalcopyrite semiconductors, *Phys. Rev. B* **27**, 5176.
97. Jiang, F.D. and Feng, J.Y. (2008) Electronic modification of Cu-based chalcopyrite semiconductors induced by lattice deformation and composition alchemy, *Semicond. Sci. Technol.* **23**, 025001.
98. Aguilera, I., Vidal, J., Wahnón, P. *et al.* (2011) First-principles study of the band structure and optical absorption of $CuGaS_2$, *Phys. Rev. B* **84**, 085145; D. Kammerlander, S. Botti, M.A.L Marques, A. Marini, C. Attaccalite, "Speeding up the solution of the Bethe-Salpeter equation by a double-grid method and Wannier interpolation", *Phys. Rev. B* **86**, 125203 (2012).
99. Alonso, M.I., Wakita, K., Pascual, J. *et al.* (2001) Optical functions and electronic structure of $CuInSe_2$, $CuGaSe_2$, $CuInS_2$, and $CuGaS_2$, *Phys. Rev. B* **63**, 075203.
100. Domain, C., Laribia, S., Tauniera, S. and Guillemoles, J.F. (2003) Ab initio calculation of intrinsic point defects in $CuInSe_2$, *J. Phys. Chem. Solids* **64**, 1657–1663.
101. Łdazewski, J., Jochym, P.T. and Parlinski, K. (2002) Band structure, Born effective charges, and lattice dynamics of CuInS from ab initio calculations, *J. Chem. Phys.* **117**, 2726.
102. Wei, S.H. and Zunger, A. (1988) Role of metal d states in II-VI semiconductors, *Phys. Rev. B* **37**, 8958–8981.
103. Abrahams, S.C. and Bernstein, J.L. (1973) Piezoelectric nonlinear optic $CuGaS_2$ and $CuInS_2$ crystal structure: Sublattice distortion in A1B3 and A2B4 type chalcopyrites, *J. Chem. Phys.* **59**, 5415–5422.
104. Zhang, S.B., Wei, S., Zunger, A. and Katayama-Yoshida, H. (1998) Defect physics of the $CuInSe_2$ chalcopyrite semiconductor, *Phys. Rev. B* **57**, 9642.
105. Xiao, H.Z., Yang, L.C. and Rockett, A. (1994) Structural, optical, and electrical properties of epitaxial chalcopyrite $CuIn_3Se_5$ films, *J. Appl. Phys.* **76**, 1503–1510.
106. Persson, C. and Zunger, A. (2005) Compositionally induced valence-band offset at the grain boundary of polycrystalline chalcopyrites creates a hole barrier, *Appl. Phys. Lett.* **87**, 211904.
107. Botti, S., Kammerlander, D. and Marques, M.A.L. (2011) Band structures of Cu_2ZnSnS_4 and $Cu_2ZnSnSe_4$ from many-body methods, *Appl. Phys. Lett.* **98**, 241915.
108. Chen, S., Gong, X.G., Walsh, A. and Wei, S.H. (2009a) Crystal and electronic band structure of Cu_2ZnSnX_4 ($X = S$ and Se) photovoltaic absorbers: first-principles insights, *Appl. Phys. Lett.* **94**, 041903.
109. Chen, S., Walsh, A., Luo, Y., Yang, J.H., Gong, X.G and Wei, S.H. (2010b) Wurtzite-derived polytypes of kesterite and stannite quaternary chalcogenide semiconductors, *Phys. Rev. B* **82**, 195203.

110. Persson, C. (2010) Electronic and optical properties of Cu_2ZnSnS_4 and $Cu_2ZnSnSe_4$, *J. Appl. Phys.* **107**, 053710.
111. Paier, J., Asahi, R., Nagoya, A. and Kresse, G. (2009) Cu_2ZnSnS_4 as a potential photovoltaic material: a hybrid Hartree-Fock density functional theory study, *Phys. Rev. B* **79**, 115126.
112. Chen, S., Gong, X.G., Walsh, A. and Wei, S.H. (2010a) Defect physics of the kesterite thin-film solar cell absorber Cu_2ZnSnS_4, *Appl. Phys. Lett.* **96**, 021902.
113. Nagoya, A., Asahi, R., Wahl, R. and Kresse, G. (2010) Defect formation and phase stability of Cu_2ZnSnS_4 photovoltaic material, *Phys. Rev. B* **81**, 113202.
114. Chen, S., Gong, X.G., Walsh, A. and Wei, S.H. (2009b) Electronic structure and stability of quaternary chalcogenide semiconductors derived from cation cross-substitution of II-VI and I-III-VI_2 compounds, *Phys. Rev. B* **79**, 165211.
115. Schorr, S. (2007) Structural aspects of adamantine like multinary chalcogenides, *Thin Solid Films* **515**, 5985.
116. Babu, G., Kumar, Y., Bhaskar, P. and Raja, V. (2008) Effect of post-deposition annealing on the growth of $Cu_2ZnSnSe_4$ thin films for a solar cell absorber layer, *Semicond. Sci. Technol.* **23**, 085023.
117. Hahn, H. and Schulze, H. (1965). Über quaternäre Chalkogenide des Germaniums und Zinns, *Naturwiss* **52**, 426.
118. Hall, S., Szymanski, J. and Stewart, J. (1978) Kesterite $Cu_2(Zn,Fe)SnS_4$ and stannite $Cu_2(Fe,Zn)SnS_4$ structurally similar but distinct minerals, *Can. Mineral.* **16**, 131.
119. Olekseyuk, D., Gulay, L., Dydchak, I. *et al.* (2002) Single crystal preparation and crystal structure of the $Cu_2Zn/Cd,Hg/SnSe_4$ compounds, *J. Alloys Compd.* **340**, 141.
120. Schorr, S., Hoebler, H.J. and Tovar, M. (2007) A neutron diffraction study of the stannite-kesterite solid solution series, *Eur. J. Mineral.* **19**, 65.
121. Ahn, S., Jung, S., Gwak, J. *et al.* (2010) Determination of band gap energy of $Cu_2ZnSnSe_4$ thin films: On the discrepancies of reported band gap values, *Appl. Phys. Lett.* **97**, 021905.
122. Kamoun, N., Bouzouita, H. and Rezig, B. (2007) Fabrication and characterization of Cu_2ZnSnS_4 thin films deposited by spray pyrolysis technique, *Thin Solid Films.* **515**, 5949.
123. Katagiri, H., Saitoh, K., Washio, T. *et al.* (2001) Development of thin film solar cell based on Cu_2ZnSnS_4 thin films, *Sol. Energy Mater. Sol. Cells* **65**, 141.
124. Nakayama, N. and Ito, K. (1996) Sprayed films of stannite Cu_2ZnSnS_4, *Appl. Surf. Sci.* **92**, 171.
125. Seol, J.S., Lee, S.Y., Lee, J.C. *et al.* (2003) Electrical and optical properties of Cu_2ZnSnS_4 thin films prepared by rf magnetron sputtering process, *Sol. Energy Mater. Sol. Cells* **75**, 155.
126. Tanaka, T., Nagatomo, T., Kawasaki, D. *et al.* (2005). Preparation of Cu_2ZnSnS_4 thin films by hybrid sputtering, *J. Phys. Chem. Solids* **66**, 1978.
127. Zhang, J., Shao, L., Fu, Y. and Xie, E. (2006) Cu_2ZnSnS_4 thin films prepared by sulfurization of ion beam sputtered precursor and their electrical and optical properties, *Rare Met.* **25**, 315.
128. Rife, J.C., Dexter, R.N., Bridenbaugh, P.M. and Veal, B.W. (1977) Optical properties of the chalcopyrite semiconductors $ZnGeP_2$, $ZnGeAs_2$, $CuGaS_2$, $CuAlS_2$, $CuInSe_2$, and $AgInSe_2$, *Phys. Rev. B* **16**, 4491–4500.

129. Gavrilenko, V. and Bechstedt, F. (1997) Optical functions of semiconductors beyond density-functional theory and random-phase approximation, *Phys. Rev. B* **55**, 4343–4352.
130. Syrbu, N.N., Tiginyanu, I.M., Nemerenco, L.L. *et al.* (2005) Exciton spectra, valence band splitting, and energy band structure of $CuGa_XIn_{1X}S_2$ and $CuGa_XIn_{1X}Se_2$ crystals, *J. Phys. Chem. Solids* **66**, 1974–1977.
131. Ginley, D. (2010) *Handbook of Transparent Conductors*, Springer Verlag.
132. Nie, X., Wei, S. and Zhang, S. (2002) Bipolar doping and band-gap anomalies in delafossite transparent conductive oxides, *Phys. Rev. Lett.* **88**, 066405.
133. Mizoguchi, H., Hirano, M., Fujitsu, S. *et al.* (2002) $ZnRh_2O_4$: A p-type semiconducting oxide with a valence band composed of a low spin state of Rh^{3+} in a $4d^6$ configuration, *Appl. Phys. Lett.* **80**, 1207.
134. Ogo, Y., Hiramatsu, H., Nomura, K. *et al.* (2008) p-channel thin-film transistor using p-type oxide semiconductor, SnO, *Appl. Phys. Lett.* **93**, 032113–032113.
135. Yanagi, H., Ueda, K., Ohta, H. *et al.* (2002) Fabrication of all oxide transparent p-n homjunction using bipolar $CuInO_2$ semiconducting oxide with delafossite structure, *Solid State Comm.* **121**, 15–17.
136. Banerjee, A., Ghosh, C., Das, S. and Chattopadhyay, K. (2005) Electro-optical characteristics and field-emission properties of reactive dc-sputtered p-$CuAlO_{2+x}$ thin films, *Physica B.* **370**, 264–276.
137. Ohta, H., Nomura, K., Hiramatsu, H. *et al.* (2003) Frontier of transparent oxide semiconductors, *Solid State Electronics* **47**, 2261–2267.
138. Thomas, G. (1997) Materials science: Invisible circuits, *Nature (London)* **389**, 907.
139. Alkoy, E. and Kelly, P. (2005) The structure and properties of copper oxide and copper aluminium oxide coatings prepared by pulsed magnetron sputtering of powder targets, *Vacuum* **79**, 221–230.
140. Banerjee, A. and Chattopadhyay, K. (2005) Size-dependent optical properties of sputter-deposited nanocrystalline p-type transparent $CuAlO_2$ thin films, *J. Appl. Phys.* **97**, 084308.
141. Banerjee, A., Kundoo, S. and Chattopadhyay, K. (2003) Synthesis and characterization of p-type transparent conducting $CuAlO_2$ thin film by dc sputtering, *Thin Solid Films* **440**, 5–10.
142. Banerjee, A., Maity, R. and Chattopadhyay, K. (2004) Preparation of p-type transparent conducting $CuAlO_2$ thin films by reactive dc sputtering, *Mat. Lett.* **58**, 10–13.
143. Dittrich, T., Dloczik, L., Guminskaya, T. *et al.* (2004) Photovoltage characterization of $CuAlO_2$ crystallites, *Appl. Phys. Lett.* **85**, 742–744.
144. Gilliland, S., Pellicer-Porres, J., Segura, A. *et al.* (2007) Electronic structure of $CuAlO_2$ and $CuScU_2$ delafossites under pressure, *Phys. Stat. Sol. (b)* **244**, 309–314.
145. Ong, C. and Gong, H. (2003) Effects of aluminum on the properties of p-type Cu-Al-O transparent oxide semiconductor prepared by reactive co-sputtering, *Thin Solid Films* **445**, 299–303.
146. Pellicer-Porres, J., Segura, A., Gilliland, A. *et al.* (2006) On the band gap of $CuAlO_2$ delafossite, *Appl. Phys. Lett.* **88**, 181904.
147. Tate, J., Ju, H.L., Moon, J.C. *et al.* (2009) Origin of *p*-type conduction in single-crystal $CuAlO_2$, *Phys. Rev. B* **80**, 165206.

148. Yanagi, H., Inoue, S., Ueda, K. *et al.* (2000) Electronic structure and optoelectronic properties of transparent p-type conducting $CuAlO_2$, *J. Appl. Phys.* **88**, 4159–4163.
149. Yu, R.S., Liang, S.C., Lu, C.J. *et al.* (2007) Characterization and optoelectronic properties of p-type n-doped $CuAlO_2$ films, *Appl. Phys. Lett.* **90**, 191117.
150. Sasaki, M. and Shimode, M. (2003) Fabrication of bipolar $CuInO_2$ with delafossite structure, *J. Phys. Chem. Solids* **64**, 1675–1679.
151. Teplin, C.W., Kaydanova, T., Young, D.L. *et al.* (2004) A simple method for the preparation of transparent p-type Ca-doped $CuInO_2$ films: Pulsed-laser deposition from air-sintered Ca-doped $Cu_2In_2O_5$ targets, *Appl. Phys. Lett.* **85**, 3789–3791.
152. Ingram, B.J., Mason, T.O., Asahi, R. *et al.* (2001) Electronic structure and small polaron hole transport of copper alumínate, *Phys. Rev. B* **64**, 155114.
153. Robertson, J., Peacock, P., Towler, M. and Needs, R. (2002) Electronic structure of p-type conducting transparent oxides, *Thin Solid Films* **411**, 96–100.
154. Shi, L., Fang, Z. and Li, J. (2008) First-principles study of p-type transparent conductive oxides $CuXO_2$ (X = Y, Sc, and Al), *J. Appl. Phys.* **104**, 073527.
155. Huda, M.N., Yan, Y., Walsh, A. *et al.* (2009) Group-IIIA versus IIIB delafossites: Electronic structure study, *Phys. Rev. B* **80**, 035205.
156. Laskowski, R., Christensen, N.E., Blaha, P. and Palanivel, B. (2009) Strong excitonic effects in $CuAlO_2$ delafossite transparent conductive oxides, *Phys. Rev. B* **79**, 165209.
157. Trani, F., Vidal, J., Botti, S. and Marques, M.A.L. (2010) Band structures of delafossite transparent conductive oxides from a self-consistent GW approach, *Phys. Rev. B* **82**, 085115.
158. Stephens, P., Devlin, F.J., Chabalowski, C.F. and Frisch, M.J. (1994) Ab initio calculation of vibrational absorption and circular dichroism spectra using density functional force fields, *J. Phys. Chem.* **98**, 11623.
159. Pellicer-Porres, J., Segura, A. and Kim, D. (2009) Refractive index of the $CuAlO_2$ delafossite, *Semicond. Sci. Technol.* **24**, 015002.
160. Bechstedt, F., Seino, K., Hahn, P.H. and Schmidt, W.G. (2005) Quasiparticle bands and optical spectra of highly ionic crystals: AlN and NaCl, *Phys. Rev. B* **72**, 245114.
161. Ueda, K., Inoue, S., Hirose, S. *et al.* (2000) Transparent p-type semiconductor: LaCuOS layered oxysulfide, *Appl. Phys. Lett.* **77**, 2701.
162. Zakutayev, A., Kykyneshi, R., Schneider, G. *et al.* (2010) Electronic structure and excitonic absorption in BaCuChF (Ch=S, Se, and Te), *Phys. Rev. B* **81**, 155103.
163. Hamada, I. and Katayama-Yoshida, H. (2006) Energetics of native defects in $CuAl0_2$, *Physica B* **376–377**, 808–811.
164. Scanlon, D. and Watson, G. (2010) Conductivity limits in $CuAlO_2$ from screened-hybrid density functional theory, *J. Phys. Chem. Lett.* **1**, 3195–3199.

3

Energy Generation: Nuclear Energy

Dorothy Duffy
Department of Physics and Astronomy, University College London, London, UK

3.1 Introduction

The challenge of generating sufficient electricity to satisfy the ever-growing demands for energy, without contributing to excessive greenhouse gas production, has led to a resurgence in interest in nuclear power generation. The current generation of nuclear power plants is making a significant contribution to clean power generation, and modern (generation III) reactors have gained from earlier experience to produce more efficient and even safer designs. Looking ahead, the next-generation (generation IV) reactors could produce hundreds of times more energy from the same mass of fuel, produce minimal radioactive waste and, possibly, even use existing waste to produce electricity. Further in the future the dream of many is that fusion power will make a meaningful contribution to power generation. The fuel for fusion power is deuterium, which is found in seawater, and tritium, which can readily be made from lithium; therefore, there is a plentiful supply for many millennia. Fusion power is also intrinsically safe, as any disruption to the containment vessel would result in rapid cooling of the plasma, and it would produce minimal long-lived radioactive waste.

As with all large-scale electricity generation, nuclear power is not without its challenges. There is the overriding concern about safety, particularly in the aftermath of Fukushima, although lessons have been learnt from past experience and modern designs have much improved safety systems. There is also the issue of the disposal of the spent fuel, which remains radioactive for many thousands of years,

Computational Approaches to Energy Materials, First Edition. Edited by Aron Walsh, Alexey A. Sokol and C. Richard A. Catlow.

in such a way that the fuel does not contaminate the environment. This chapter focuses on the effect that radioactivity has on the properties of the materials that are used in nuclear technology. Under certain conditions radiation creates defects in the material and, over time, these defects accumulate and alter the microstructure, and hence the mechanical properties, of the materials. Metals can, for example, become brittle and more susceptible to fracture. The local alloy composition may change due to radiation-enhanced diffusion and segregation to grain boundaries, which can also degrade properties. The microstructure of nuclear fuel changes dramatically during burn-up, with the formation of gas bubbles that cause swelling and weaken the structure.

The effects of radiation on materials are varied and complex. Experimental measurements of properties following radiation give information about the macroscopic effects, but without a thorough understanding of the fundamental mechanisms that are responsible for the changes, it is impossible to extrapolate to untested scenarios. Modeling has a very important role to play in the task of achieving this understanding, particularly when used in conjunction with experimental techniques that can study changes in microstructure at the nanoscale. In this chapter, we start by discussing the effects that radiation has on materials used in fission technology, future fusion technology and radioactive waste disposal. In Section 3.3, we describe the various modeling techniques that are currently used to gain an understanding about the mechanisms and effects of radiation damage and give examples of the application of these techniques to nuclear materials. In Section 3.4, we discuss the future challenges and obstacles to progress.

3.2 Radiation Effects in Nuclear Materials

3.2.1 Fission

The design of nuclear reactors has evolved substantially since the early (generation I) reactors in the 1950s. Most current operating power plants use generation II reactors, which were built up until the early 1990s. Generation III reactors, which started to be commissioned from the mid-1990s, have improved safety and economics. The next-generation reactors (generation IV) are still very much in the research stage with several potential technologies, but these would aim to make substantial improvements in safety, efficiency, economics, and above all, a substantial reduction in the volume of high-level radioactive waste. Nuclear reactors are broadly classified by the cooling technology (light water, gas, boiling water) and the energy of the neutrons that are used for fission. Thermal reactors use neutrons with average energies around 0.025 eV, and fast reactors use neutrons with average energies in the range 0.2–0.5 MeV. In general, the materials in fast reactors experience much more hostile conditions than those in thermal reactors because of higher neutron energies, neutron fluxes and operating temperatures. In addition, the amount of energy obtained per unit mass of fuel (the burn-up) is significantly higher in fast reactors. In all types of reactors, the materials that cause most concern are the structural materials (both in the core and the pressure vessel), the fuel and the cladding (the material that contains the fuel).

3.2.1.1 Structural Materials

Although the details of irradiation effects in structural materials in nuclear power plants vary between the different designs and types of material, the primary issues of concern are common to all scenarios. The materials must maintain dimensional stability (they must resist swelling), they must maintain acceptable mechanical properties (tensile strength and toughness) and they must resist corrosion. Materials in a high-temperature environment must also be resistant to creep to ensure they maintain dimensional stability under the influence of stress.

The primary radiation effects in structural materials are caused by neutron irradiation. Neutrons interact with atoms of the material and transfer some of their kinetic energy. If the energy is greater than a threshold value, known as the displacement energy (E_d), the atom is displaced from its equilibrium site and moves through the material. The atom may then collide with other atoms, knocking them out of position, and a collision cascade is initiated, which will terminate when all atoms have energies lower than the displacement energy. The net result is that a number of Frenkel (vacancy-interstitial) pairs are created by the radiation event and this number increases with the energy of the initial impact. The number of defects created by a radiation event can be estimated using various models, the simplest of which is the Kinchin–Pease model [1]. In this model the damage cascade is considered to develop as a series of collisions. The number of atoms involved in the cascade doubles after each collision and the energy is shared between the participating atoms. If a uniform probability is assumed for the energy transfer for each collision, then the number of defect pairs (N_{KP}) created by a radiation event with energy E_p is given by

$$N_{KP} = E_p/2E_d. \tag{3.1}$$

The Kinchin–Pease model overestimates the number of stable defects because it neglects two important factors. When an atom or ion moves through a material, it loses some energy by interactions with the electrons; therefore, this energy is not available for the creation of defects. In addition, a proportion of the Frenkel pairs created by the radiation event recombine within a few picoseconds of the event. An improved model, suggested by Norgett *et al.* [2], employs only the proportion of the energy not lost to electronic excitations (E') and includes a numerical factor k ($\sim$0.8) to account for defect recombination to give the number of Frenkel pairs (N_{NRT}):

$$N_{NRT} = kE'/2E_d. \tag{3.2}$$

The Norgett, Robinson and Torrens (NRT) model, as equation (3.2) is generally referred to, is widely used to give an estimate of the number of defects created by a radiation event. It is used to calculate the dpa (displacements per atom) value for a material that has been subjected to a radiation dose over an extended time period. The dpa gives an estimate of the average number of times each atom has been displaced from its lattice site after a radiation dose. Note that this is not the same as the number of defects because many atoms will return to a different lattice site after a radiation event and these do not count as defects.

The Kinchin–Pease and NRT models are very useful starting points; however, as more sophisticated modeling techniques have been developed, it has become apparent that the linear relationship between the number of defects and the knock-on energy is not generally valid. In many situations the collision events are not well separated; the atoms interact with many atoms simultaneously, resulting in the correlated motion of many atoms. In these cases a thermal spike view is more appropriate. The thermal spike is a local region of disordered energetic atoms, which resembles a molten, approximately spherical, volume of material. The energy is transported away from the hot (thermal spike) region by thermal conductivity, and recrystallization of the disordered region follows. However, recrystallization is rarely perfect, and point defects or small defect clusters are created in the crystal. The number of residual defects is generally lower than the number predicted by the NRT model, and there is a sublinear increase with knock-on energy (E_{p}).

The creation of vacancies, interstitials and clusters following a radiation event is generally complete a few tens of picoseconds after a radiation event, but it is the evolution of these defects over time periods of days, weeks or even years that eventually degrades the material properties. In close-packed crystal structures, interstitials have low migration barriers; therefore, they diffuse rapidly through the crystal and get adsorbed by dislocations or grain boundaries. They may also interact with other interstitials to form clusters, which collapse to dislocation loops. Vacancies also diffuse, albeit at a much slower rate, and get absorbed at dislocations or grain boundaries or they create clusters. The increase in volume associated with the interstitials is greater than the decrease associated with the vacancies; therefore, the volume of the material increases due to radiation damage. Such swelling can induce stress under constrained conditions.

The main effect on the mechanical properties of the defects and defect clusters is embrittlement. Irradiated metals have higher yield strength and lower ductility than nonirradiated metals. The motion of dislocations through the material is inhibited by interactions with interstitial loops and vacancy clusters, and this makes the material more brittle and susceptible to fracture. These effects are particularly significant at low temperatures.

Another significant effect that radiation has on metal alloys is that it can induce changes in the local composition. Metallic alloys are generally not in equilibrium because the solute atoms that are added to improve the material properties are often in supersaturated solution. Radiation events increase the local lattice temperature for a few picoseconds, and the high temperature results in increased diffusion in the volume close to the event. Radiation increases the number of vacancies, which also enhance solute diffusion. Radiation-enhanced diffusion may drive the composition toward equilibrium and result in phase separation or the precipitation of particles, which have a different phase from the bulk metal. One problematic example is the α' precipitates in Fe–Cr steel. Cr is added to steel to protect against corrosion, and alloys with greater than 10% wt Cr decompose into a mixture of α (an Fe-rich solid solution) and α' (a Cr-rich solid solution) at high temperatures. The decomposition contributes to embrittlement and, as it is accelerated by radiation-enhanced diffusion, it occurs at ambient temperatures in a radiation environment.

Radiation-enhanced diffusion can also be responsible for enhanced segregation or depletion of solute atoms at grain boundaries and dislocations. If the solute primarily diffuses via a vacancy mechanism, it will tend to diffuse away from the sink (grain boundary), whereas interstitial solutes will diffuse toward the sink and segregation will be enhanced. The depleted region will have degraded properties as the composition is modified with respect to the design value. A particular example again occurs in austenitic steel, where the Ni concentration is enhanced near grain boundaries and the Cr concentration is reduced. The Cr-depleted regions near grain boundaries are susceptible to corrosion [3]. Radiation-induced segregation is a problem only in a limited temperature range for each material. At low temperatures diffusion is minimal, and at high temperatures normal diffusion processes dominate.

Next-generation (generation IV) nuclear reactors will be designed to be safer, more sustainable and more economical than current designs. The designs aim to produce several hundred times more power from the same amount of fuel as current designs, produce less waste and waste that only needs to be stored for hundreds (not thousands) of years. Some designs would have the added advantage of using waste from existing power plants as fuel. There are several different proposed technologies for generation IV, and they each have widely different demands on the structural materials (Figure 3.1). In general, however, they operate at higher temperatures and the materials must endure higher levels of radiation damage (higher dpa values) than current designs.

Austenitic stainless steels are common structural materials in current nuclear power plant designs because of their good resistance to corrosion, ductility and high-temperature strength. These steels have a face-centered cubic (fcc) crystal structure, and they are more susceptible to swelling under radiation than body-centered cubic (bcc) structured metals. They have large thermal expansion and low thermal conductivity, which makes them susceptible to thermal stresses. They are also susceptible to stress corrosion cracking. Ferritic–martensitic steels have a bcc crystal structure, and their improved resistance to swelling makes them the favored choice in many applications. They have a serious disadvantage for high-temperature applications as they rapidly lose strength at around 550 °C, as the material approaches the α- to γ-phase transition boundary. The drive to increase the operating temperatures in next-generation designs has highlighted the requirement

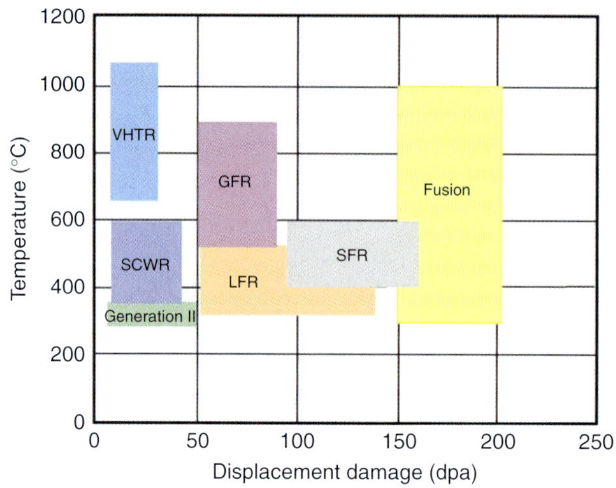

Figure 3.1 *Operating temperatures and radiation dose (dpa) for current (generation II) and future (generation IV) fission and fusion reactors. Generation IV reactors are the supercritical water reactor (SCWR), the very high-temperature reactor (VTHR), the gas fast reactor (GFR), the lead fast reactor (LFR) and the sodium fast reactor (SFR). (After [4].)*

for a ferritic (bcc) steel with good high-temperature strength. With this objective a significant research effort is currently focused on the development of oxide dispersion-strengthened (ODS) steels. ODS steels have nanometer-sized precipitates of Y and Ti oxide particles that are highly insoluble in the Fe matrix and stable at high temperatures. The particles inhibit dislocation motion and thereby counteract the softening that occurs at high temperatures. It is anticipated that the interface between the oxide particles and the Fe matrix will act as a sink for radiation defects, which would enhance defect recombination and improve radiation resistance.

3.2.1.2 Fuel

The microstructure and the composition of nuclear fuel undergo dramatic changes during the useful lifecycle due to the plethora of fission events that occur in the material. The most common elements used for fuel in fission power plants are ^{235}U and ^{239}Pu, typically in the oxide form. One example of the many possible neutron-induced fission events is

$$^{235}\mathrm{U} + {}^{1}\mathrm{n} \longrightarrow {}^{134}\mathrm{Xe} + {}^{100}\mathrm{Sr} + 2\,{}^{1}\mathrm{n}.$$

Around 200 MeV is released from a typical fission event, and most of this energy appears as kinetic energy of the daughter nuclei (Xe and Sr in this case) and neutrons. A small proportion of the energy is released as high-energy photons (gamma rays).

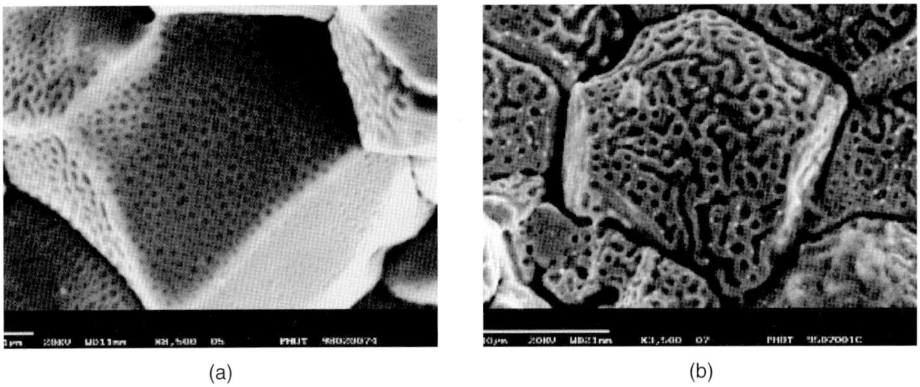

Figure 3.2 Fracture surface of fuel after (a) 13 GT day/t U, showing small isolated bubbles, and (b) 21 GT day/t U and held at 1800 °C for 30 min, showing the growth and coalescence of bubbles. (From [5].)

The change in the microstructure induced by fission events in nuclear fuel has two primary sources: the radiation effects caused by the released energy and the solubility of the fission products. The broad spectrum of fission events creates a variety of fission products. Some of these products (lanthanum for example) will be dissolved in the lattice and they will form substitutional impurities. Some will be insoluble and form precipitates. Gaseous products, such as Xe and Kr, are highly insoluble in the fuel, and they have, therefore, the most dramatic effect on the microstructure. The interaction between gaseous products and vacancy clusters, created by radiation damage, results in the formation of bubbles, which may precipitate preferentially at grain boundaries (Figure 3.2). A high density of bubbles on a grain boundary can result in the formation of microcracks, which weaken the material and are a potential source of radioactive gas release in an accident scenario.

The daughter ions produced by the fission event typically have energies of the order of 80 MeV, and they will, therefore, produce substantial knock-on damage and electronic excitations. The defects created by the high-energy products diffuse through the material and accumulate at sinks, such as grain boundaries and dislocations, or interact with other defects to form clusters. Some of the deposited energy goes into phonon creation, resulting in a substantial temperature increase. High lattice temperatures result in enhanced diffusion and, under certain conditions, grain growth and recrystallization. Fuels that have experienced high burn-up have been observed to have a distinctive microstructure (known as high burn-up structure or HBS). The original micron-sized grains subdivide into submicron (0.1–0.5 μm), and the fission gas is expelled from the fuel matrix and retained in closed pores. The restructuring starts at the rim of the fuel and extends inwards as burn-up progresses. It is believed that the HBS will not compromise fuel performance under normal operation, but the release of fission gas from the pores could be a concern under accident conditions.

The primary requirements for nuclear fuel during burn-up are that it should retain the fission products and fuel nuclides, it should maintain its dimensions within acceptable limits and it should be possible to cool it. The first requirement is to avoid radioactivity entering the environment should containment be ruptured. The second is to avoid unacceptable levels of stress on the containment (cladding) material, and the third is necessary to ensure that the fuel does not reach unacceptably high temperatures and that heat can be efficiently transported away from the fuel to be used for electricity production. The total burn-up of the fuel is a measure of how much fissionable material has been used up, and it is measured either as a percentage of the fissionable atoms or as energy per unit mass (GWd/tHM or GW days per tonne of heavy metal). One of the main aims for future fuel design is to increase burn-up significantly. High burn-up would reduce downtime for refueling and also reduce the volume of radioactive waste produced.

One very important property of fuel is the thermal conductivity. The thermal energy that is produced continuously in the fuel needs to be transported to the edge of the material and removed by the coolant. The temperature profile of the fuel pellet is influenced by the thermal conductivity, with lower thermal conductivities resulting in strong temperature gradients. The differential thermal expansion resulting from such temperature gradients can cause cracking and, consequently, the release of fission gas. The thermal conductivity of the fresh fuel is relatively high, so heat transport is efficient. However, as the fuel burns the microstructure changes strongly affect the thermal conductivity and the complex nature of the changes makes this difficult to estimate accurately. Some microstructure changes, such as the formation of defect clusters and gas bubbles, decrease the thermal conductivity substantially, whereas others, such as the formation of metallic precipitates and grain growth, increase the thermal conductivity. The net effect generally is that the thermal conductivity decreases with increasing burn-up, which can be a limiting factor for the in-service lifetime of the fuel.

The most common form of nuclear fuel is UO_2, sometimes mixed with PuO_2. The oxide is used, in preference to the metal, because of its superior dimensional stability, resistance to corrosion and higher melting temperature. UO_2 powder is sintered into cylindrical pellets, which are packed into metal rods. More complex designs, such as the TRISO (tristructural-isotropic) fuel, allow for higher temperature operation and burn-up. The TRISO particle is a (~60 mm) fuel sphere of UO_2 or UCO coated by layers of pyrolytic graphite, SiC and a further layer of dense pyrolytic graphite, all a few millimeters thick (Figure 3.3) [6]. The inner graphite layer shrinks under radiation, which accommodates for the swelling of the fuel. The SiC layer retains the fission products at high temperatures and provides the structural strength, and the outer graphite layer protects against damage during the handling and filling process. Significant research is being carried out worldwide on TRISO-coated fuel, with the aim of extending burn-up to 10–20%, compared to a

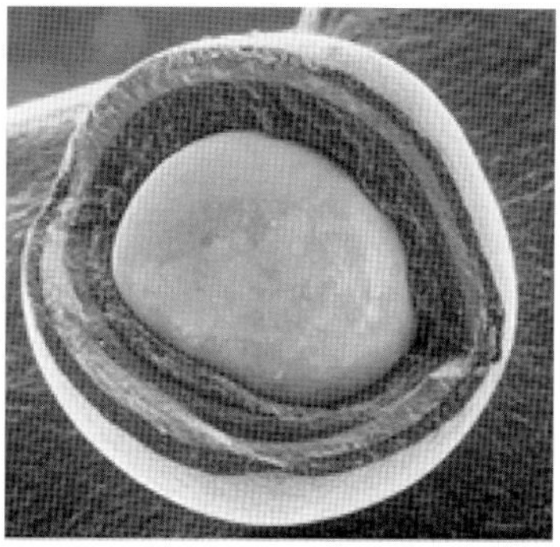

Figure 3.3 *TRISO fuel particle showing the three coating layers.*

small percentage for conventional fuels, and it is one of the fuels being considered for some generation IV designs. In addition to oxides, there is research into mixed carbide and nitride ceramic fuels, which have higher intrinsic thermal conductivities than oxides. This would allow for higher concentrations of fissile elements to be incorporated.

3.2.1.3 Cladding

The cladding is the material that separates the fuel from the coolant in order to retain the fissile material and radioactive fission products in the core of the reactor vessel. In several fuel element designs, the cladding takes the form of a hollow metal rod into which cylindrical fuel pellets are packed. A gap is left between the fuel and the cladding in order to allow for thermal expansion and swelling. The gap is generally filled with helium to maximize thermal transport from the fuel to the cladding and enhance the cooling. As burn-up proceeds, the fission gasses are transported down the thermal gradient to the exterior of the pellet, and this tends to increase the pressure in the rod. The microstructure change, due to fission products and the formation of bubbles of trapped fission gasses in voids created by radiation damage, results in increased swelling as burn-up progresses. This swelling, combined with thermal expansion, can close the gap and bring the fuel into direct contact with the cladding, allowing interdiffusion of the two materials. The wide range of fission products present in the outer rim of the fuel results in complex fuel–clad interactions. A complex oxide layer may be formed between the fuel and the cladding, which can result in thermal stresses, corrosion and cracking.

The cladding has to endure increasing stress during lifetime, due to the fuel swelling and fission gas pressure, and the ability of the irradiated material to withstand these stresses limits the useful life of the fuel rod and may prevent the fuel reaching its maximum possible burn-up. Like the structural materials, discussed in Section 3.2.1.2, cladding materials suffer embrittlement from neutron damage, which renders it susceptible to brittle failure under pressure. The cladding is also susceptible to swelling, which will contribute to the closing of the gap between the fuel and the rod. Corrosion, due to interactions between the cladding material and the coolant or the fission products, can degrade the mechanical properties. At the fuel–clad interface, fission products, such as iodine, may make the cladding susceptible to stress corrosion cracking.

Zirconium alloy (zircalloy) is a common cladding material in thermal reactors, mainly because it is transparent to thermal neutrons. It has good corrosion resistance in aqueous coolants, but it does react with water at high temperatures and the released hydrogen reacts with the metal to form hydrides. The hydride precipitates inhibit dislocation motion and contribute to the embrittlement. Stainless steels, both austenitic and ferritic–martensitic, are also used as cladding materials, particularly for fast reactors where liquid sodium is the preferred coolant. One issue for these materials is radiation-induced segregation because the depletion of the chromium near grain boundaries reduces the corrosion protection of the material.

With the improvements in uranium-enrichment technology and increasing burn-up, degradation of the cladding material, due to radiation damage and the fuel–clad interactions, may be the main factors that limit the useful life of a fuel element. The fuel rod must be removed before there is any significant risk of failure, and this is generally well before most of the fissionable material has been used up. Improving the lifetime of the cladding material would, therefore, make a significant contribution to the efficiency of a nuclear power plant as it would increase fuel burn-up and also increase the time interval between refueling.

3.2.2 Fusion

Fusion technology has, in many respects, the potential for the development of an ideal power source because it is clean, safe and the raw materials are plentiful. The fuel for deuterium–tritium (DT) fusion is plentiful in seawater (deuterium) or it can readily be obtained from lithium (tritium). It does not produce CO_2 and, with careful selection of materials, it will have only short-lived radioactive waste. There are, however, many challenges associated with the path from the current plasma experiments, which produce minimal fusion reactions, to a full-scale power plant that will contribute electricity to the grid. One is these challenges is the selection and design of materials that will withstand the harsh radiation environment that would be present in a fusion power plant [7].

Current plasma experiments use hydrogen or deuterium and, consequently, the fusion cross-section is extremely low. In order to get significant fusion rates, it is necessary to maintain a mixed DT plasma at a temperature of around 1.5×10^8 °C. DT fuel has only been used at two experimental facilities (IFTR and JET), and JET currently holds the record for power produced. However, this power is lower than the power consumed to heat the plasma to the required temperature; therefore such reactions are not self-sustaining. Further technological advances are required to achieve a burning (self-sustaining) plasma, and the ITER experiment, due to commissioned in Cadarache in 2018, will aim to cross this fundamental threshold and produce ten times as much power as it consumes.

The exceptionally high temperature (ten times hotter than the core of the sun) required for fusion means that the plasma must be kept away from the vessel wall, and this is achieved by confining the charged plasma particles by a magnetic field. The products of DT fusion reaction are an α particle (3.5 MeV) and a neutron (14.1 MeV). The charged α particle will be confined in the plasma by the magnetic bottle and, in a burning plasma, the collisions between the energetic α particles and the DT fuel will heat the plasma and sustain the fusion reaction. The neutral neutrons will escape the confines of the magnetic field and lose energy to the surrounding material. It is this energy that must be extracted to produce electricity. The materials that experience the harshest radiation environment are the structural materials of the first wall and the materials that are used to line the wall, sometimes referred to as the plasma-facing materials.

3.2.2.1 Structural Materials

The first wall of a fusion power plant will enclose a blanket module that will contain a coolant, to extract thermal energy, and a lithium-based breeding material that will be used to produce tritium for the fuel. The high-energy neutrons produced by the fusion reaction will penetrate several tens of centimeters into the wall and interact with the atoms of the structure. The issues causing concern for the fusion structural materials are similar to those of the fission structural materials, discussed in Section 3.2.1.1. However, there are two very important differences that make material selection and design for fusion much more challenging than that for fission. The first is the neutron spectrum that, due to the 14 MeV neutron, reaches much higher energies than those of even fast reactors. Parts of the first wall can be expected to experience as much as 150 dpa during the lifetime of a reactor. The second problem that is a particular issue for fusion reactors is the transmutation events that are induced by neutron interactions, many of which result in the formation of insoluble gasses, such as H or He. An example of such a reaction is

$$^{58}\mathrm{Ni} + \mathrm{n} = {}^{59}\mathrm{Ni} + \gamma; {}^{59}\mathrm{Ni} + \mathrm{n} = {}^{56}\mathrm{Fe} + \alpha.$$

As both H and He are highly insoluble, the gas atoms rapidly diffuse through the material to various sinks, which can be dislocations, grain boundaries or even vacancy clusters that have been created by radiation damage events. The formation of gas bubbles in the lattice and at grain boundaries is an important source of embrittlement in fusion structural materials, particularly at high temperatures. The stabilization of vacancy clusters by gas atoms results in much greater level of swelling than would be obtained by vacancy clusters alone.

The most likely material that will be chosen for the first wall of demonstration fusion power plants is a special ferritic–martensitic steel. In spite of its superior high-temperature strength and corrosion resistance, austenitic steel can be ruled out because its high Ni content would result in copious amounts of He production due to transmutations. Other elements are problematic due to activation (i.e., the production of long-lived radioactive isotopes); therefore, reduced activation ferritic–martensitic (RAFM) steels have been developed that replace the elements of concern (mainly Nb and Mo), with alternative elements (Ta and W), and these steels have been demonstrated to have mechanical properties that are at least as good as conventional steels.

The main concern about ferritic–martensitic steel is its high-temperature strength. As the bcc-structured ferritic steel approaches the temperature for the transition to the fcc structure (910 °C in pure Fe), there is a rapid loss of strength. This restricted operating temperature precludes the use of this material in the more efficient high-temperature designs. One possible solution would be the successful development of reduced activation ODS steels, as discussed in Section 3.2.1.2. Other suggested materials for improved high-temperature strength are vanadium and SiC/SiC fibre composites. The joining and shaping technologies for these materials are much less well established than those of steel; therefore, steel, in its various forms, will remain the favored material for the foreseeable future.

3.2.2.2 *Plasma-Facing Materials*

The materials that line the first wall will experience the harshest conditions of all in a fusion power plant. In addition to the high heat and neutron fluxes experienced by the first wall, the liner materials experience high fluxes of various other particles [8]. In a plasma experiment or fusion reactor vessel, the plasma is confined by the magnetic field, although some plasma particles inevitably diffuse across the field lines and escape confinement. In the ITER design, the magnetic field is configured in such a way that the outer field lines are open and cross in a circular region near the base of the vessel (Figure 3.4). The purpose of this configuration is to direct (or divert) any plasma particles or fusion products that escape the confinement of the closed magnetic field to the base of the vessel where they are extracted. The region where the magnetic field lines intersect with the vessel walls is known as the divertor, and the combination of exceptionally high heat load and particle (α particles, H atoms and neutrons) flux means that the divertor materials experience exceptionally harsh conditions and the design of materials that will withstand these conditions, for a commercially viable period of time, is extremely challenging.

An additional restriction on the choice of plasma-facing materials is their effect on the plasma. Atoms knocked off the surface of the material by stray particles may enter the plasma, and the energy required to ionize these atoms will remove energy from (or cool) the plasma. Low atomic number (low Z) atoms require less energy to ionize; therefore, low-Z materials are the preferred choice to face the plasma. Low-Z materials usually have low cohesive energies and they are, therefore, susceptible to surface erosion by particle bombardment, so the requirements for plasma-facing materials are somewhat contradictory.

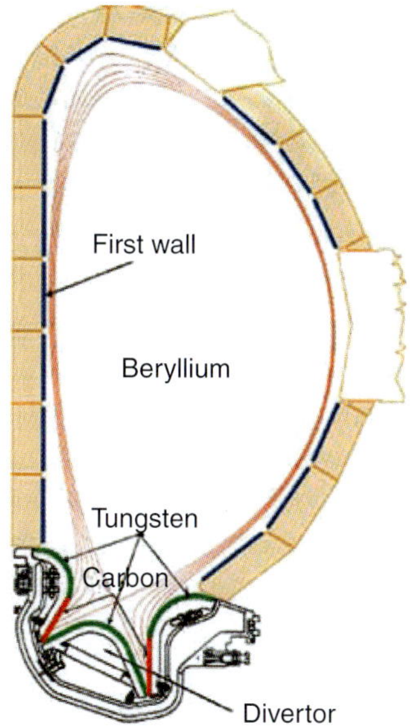

Figure 3.4 *Schematic cross-section of ITER, showing the outer magnetic field lines (red), which cross near the base of the vessel. The favored materials to line the vessel (Be, C, W) are marked [9].*

Graphitic carbon has been the material of choice for plasma experiments in the past because it is a low-*Z* material with strong bonds. It has, however, a serious disadvantage that would severely restrict its use in a burning plasma, that is its reactivity with hydrogen isotopes. Low-energy hydrogen atoms react with the surface atoms of carbon to form volatile hydrocarbons, which are easily desorbed from the surface and redeposited elsewhere in the vessel. In a DT plasma this would be a serious problem both for safety and resource reasons. Tritium is highly radioactive and its short half-life means that it is a scarce resource; therefore, most of the tritium used in an experiment needs to be recoverable. Chemical sputtering of carbon would render the tritium difficult to recover. For this reason the ITER design uses beryllium to line the walls of the main vessel and tungsten for most of the divertor (Figure 3.4). It is possible that carbon, in the form of fibre-reinforced composites, will need to be employed for the strike points, where the flux lines intersect the walls, because even tungsten could melt under the exceptionally high heat loads experienced at these points.

3.2.3 Waste Disposal

The safe, long-term disposal of radioactive waste is one of the main reasons cited by environmental groups for their opposition to nuclear power, and it is clear that the radioactive material produced by decommissioned nuclear power plants presents a very real challenge. Some current technologies use a "once-through" fuel cycle, which means that the spent fuel is treated as waste and no attempt is made to recover any remaining fissile material. A closed fuel cycle, on the other hand, uses reprocessing to recover the majority of the fissile nuclides from the spent fuel. The amount of radioactive waste produced is, therefore, strongly dependent on the amount of reprocessing undertaken before storage.

Radioactive waste management depends on immobilizing radioactive constituents and isolating them from the environment for a sufficient time period for the radioactivity to fall below acceptable levels. Depending on the half-lives of the constituents, this time can be as long as a million years. The desirable characteristics of an encapsulating material for immobilizing nuclear waste are the ability to accommodate a range of different radioactive nuclides and a high density of these nuclides. The durability of the encapsulating material is also important, and this includes factors such as the dissolution rate and the susceptibility of the material to radiation damage caused by the energetic products of decay events. The leach rate is the most important measure of the durability of a waste form. It is necessary to gain a fundamental understanding of the processes that cause an increase in the leach rate with increasing radiation damage in order to make predictions about the long-term durability of the encapsulating material.

The principal sources of radiation in high-level waste are β decay of fission products (^{137}Cs and ^{90}Sr for example) and α decay of actinides (U, Np, Pu, Am and Cm). β decay produces electrons with energies of the order of 0.5 MeV, and the relatively short half-lives of these fission products mean that this is the primary source of radiation in the first few hundred years. In general, the actinides have much longer half-lives than the fission products. The α particles from actinide decay have energies in the range 4.5–5.5 MeV and the recoil nuclei have energies in the range 70–100 keV [10]. The energetic heavy recoil ions produce substantial radiation damage in the encapsulating material, and there is some concern that, over a time scale of several thousand years, this radiation damage will enhance the leach rate of encapsulating materials.

Glasses, particularly borosilicate and phosphate glasses, are currently the material of choice for encapsulating radioactive waste and vitrification is an established technology [11]. Vitrified waste is stored in metal containers, or canisters, which act as a barrier between the waste form and groundwater. The canisters would be expected to be an effective barrier for several hundred years, but corrosion will eventually make this barrier ineffective [12]. Glass is susceptible to corrosion via diffusion-controlled ion exchange and glass network hydrolysis. Ion exchange involves alkali ions leaving the glass and protons entering it, resulting in a hydrated alkali-deficient layer. Hydrolysis involves the interaction of hydroxyl ions with the SiO_2 network, resulting in dissolution and the possible precipitation of a silica gel layer. Such processes depend on both the temperature and the pH of the aqueous environment. As it is already amorphous, the structure of the glass is not expected to change significantly due to radiation. The most important radiation effect in glass is radiation-enhanced diffusion, which can result in the formation of gas bubbles and the redistribution of cations.

Although vitrification is currently the preferred technology for nuclear waste disposal, there is considerable interest in the development of alternative materials that will encapsulate a high density of actinides, particularly Pu, but also Np, Am and Cm [13]. The idea is that radioactive nuclides would substitute for cations at specific sites in a ceramic crystal; therefore, materials with complex crystal structures, which have a range of cation coordination, atomic volumes and possible charge states, are favored. A broad range of ceramic materials have been suggested as promising candidates for encapsulation of radioactive actinides, with pyrochlores ($A_2B_2O_7$; A and B are metal cations) possibly receiving the most attention. The pyrochlore structure is related to the fluorite structure (AO_2), but there are two different cation sites and one-eighth of the anions are absent. Pyrochlores can easily incorporate cations with valences up to 6 by O vacancy compensation. Zirconolite ($CaZrTi_2O_7$) is also a favored candidate ceramic as it can incorporate nearly all the radioactive nuclides in a single phase. SYNROC, a synthetic rock developed in 1978, is a multiphase ceramic-composed of naturally occurring titanates (including zirconolite) that have immobilized uranium and thorium for billions of years. Different types of SYNROC have been developed to encapsulate different types of radioactive waste.

One important issue for radioactive waste storage materials is the effects that the decay of the stored elements have on the microstructure evolution and phase stability of the encapsulating material. The light, high-energy α particles have low cross-sections for interactions with the atomic nuclei; therefore, they travel long distances in the materials, losing energy mainly to electronic excitations. The heavy recoil nuclei, on the other hand, lose their energy mainly to the atomic nuclei, and this creates a molten volume of material, which may recrystallize as the thermal energy is transported away via thermal conduction. The damage (defects or amorphized material) created by each decay event gradually degrades the material properties and, possibly, the ability to encapsulate the waste atoms.

3.3 Modeling Radiation Effects

The effects of radiation on the wide range of materials used in nuclear technology are varied and complex, and they have the potential to have very serious consequences if they are not properly accounted for. It was recognized in the early days of the nuclear industry that a fundamental understanding of radiation effects was necessary in order to make predictions about the long-term behavior of nuclear materials. Some of the earliest atomistic modeling programs were developed in nuclear research laboratories in order to calculate the properties of radiation-produced defects in materials.

Radiation damage is an inherently multiscale problem, both in terms of time and length scales. The energy associated with a radiation event is deposited either as electronic excitations, as kinetic energy to the atoms, or both electronic excitations and atomic kinetic energy. These atoms may receive sufficient energy to displace them from their equilibrium positions, but the deposited energy will dissipate by thermal conductivity and within a few tens of picoseconds the atoms come to rest, although possibly at different positions from they started from. This time scale (a few tens of picoseconds) is typical of the time taken to create the primary residual defect structure of the radiation event, which usually consists of point defect (vacancies, self-interstitials, antisite defects and gas atoms) and small defect clusters, located within a few tens of nanometers of the original event. Over much longer time periods (hours, days or even years), these defects and clusters migrate through the lattice and form larger clusters, dislocation loops, or get absorbed at sinks such as grain boundaries or dislocations. It is the long-term changes in the microstructure caused by numerous radiation events that result in changes to the macroscopic materials properties, such as dimensions, strength and ductility.

In order to model the effect of radiation damage, it is necessary to understand and model effects that occur at femtosecond and picosecond time scales and use the information gained to predict the resultant property changes. Such a task is challenging and not yet fully achievable, but the fundamental understanding gained from modeling using a broad range of methodologies has made, and continues to make, a significant contribution to the development of materials that will withstand radiation.

3.3.1 BCA Modeling

In Section 3.2.1.1, we introduced the Kinchin–Pease (KP) and NRT models as convenient methods for estimating the number of stable defects created by a radiation event. Both models assume that cascades are created by a series of independent collision events between two atoms. Such binary collision approximation (BCA) models can be improved substantially by using numerical models with realistic scattering potentials. The method can be easily applied to amorphous targets by using Monte Carlo (MC) techniques to randomly assign the distance between collisions and the impact parameters of the collisions. Accounting for the crystalline structure is computationally more challenging and expensive, but it captures effects that would be absent in amorphous models. One such effect is channelling, which is the name given to the situation where a knock-on atom travels between lattice planes where it experiences little energy loss and, consequently, moves long distances before coming to rest. In BCA models random two-body collisions are considered, and after each collision the energies of the target atom (E_t) and the projectile (E_p) are calculated. Depending on the energies of these atoms, in comparison with the displacement energy E_d (the minimum energy required to create a stable defect) and the cut-off energy E_{cut} (the energy at which the target or projectile atoms cease to be part of the cascade), there are four possible outcomes:

1. $E_t > E_d$ *and* $E_p > E_{cut}$: A displacement event where the target atom will join the cascade.

2. $E_t > E_d$ and $E_p < E_{cut}$: A replacement event where the target atom joins the cascade, but the projectile atom replaces the target atom and ceases to be part of the cascade.
3. $E_t < E_d$ *and* $E_p > E_{cut}$: The target atom remains on the lattice site and the projectile atom continues in the cascade with modified energy.
4. $E_t < E_d$ *and* $E_p < E_{cut}$: The target atom remains on the lattice site and the projectile atom comes to rest on an interstitial site, and therefore a defect pair is formed.

The value of E_{cut} is not necessarily the same as E_d, but it is chosen to give agreement with experiment or other modeling techniques.

BCA codes such as SRIM [14], for amorphous materials, and MARLOWE [15], for crystalline materials, are widely used to estimate the range of ions in materials and, to a lesser extent, damage production. They should be used with caution, however, because the BCA model, which assumes separated two-body collisions, is only appropriate for energetic light atoms. The displacement energy (E_d) is an essential parameter of BCA models, as well as the KP and NRT models, and can be calculated using molecular dynamics (MD) simulations; however, it is sensitive to the interatomic potentials used in the simulations.

3.3.2 Molecular Dynamics

The primary assumption of the BCA models is that the collisions are isolated independent events that do not interfere with each other. This is a reasonable approximation in situations where the interaction cross-section is low, as it is for light energetic ions, but for many important radiation events the collision cross-sections are high and a moving atom interacts with many atoms simultaneously. These low- to intermediate-energy (<50 keV) events are well suited to modeling using MD because the energy is dissipated on picosecond time scales and the defects are created within a volume with a diameter of a few tens of nanometers.

3.3.2.1 Cascade Simulations

Radiation events are modeled using MD by employing a technique generally referred to as a cascade simulation. The idea is that one atom, known as the primary knock-on atom (PKA), in a simulation cell is initiated with a finite velocity in a particular direction to simulate the kinetic energy imparted to an atom by a radiation event, such as a neutron collision (Figure 3.5). When this atom moves through the simulation cell, it interacts with neighboring atoms and some of these interactions result in atoms leaving their lattice sites and moving through the cell—hence the name cascade. The cascade has two main phases. In the first, the ballistic phase, the kinetic energy exceeds the potential energy as the PKA energy is redistributed by multiple collisions between atoms. In the second phase, known as the thermal spike, the potential and kinetic energy approach equilibrium and a local thermal distribution is established. A disordered region of liquid-like atoms forms in the crystal. As time progresses, the energy in the thermal

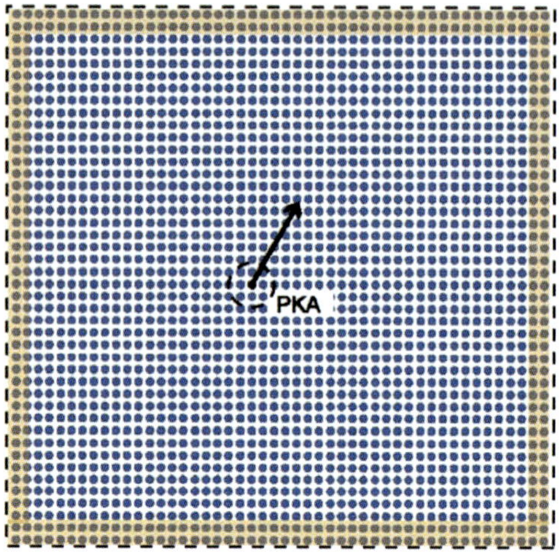

Figure 3.5 *Schematic illustration of a cascade simulation. The PKA is initiated with a high velocity. The dashed line represents periodic boundary conditions and the orange region represents the boundary (pseudo) thermostat.*

spike spreads out by thermal conductivity and the temperature of the spike falls. The disordered material generally, but not always, recrystallizes; however, some defects remain [16]. The variation in the number of residual defects with the PKA energy is an important output from these simulations as these can be used as input for the longer time scale simulations, described in Sections 3.3.2 and 3.3.3. The non-equilibrium nature of cascade simulations raises issues that are of particular concern to this class of simulation; therefore, these are discussed in the following sections.

Interatomic Potentials for Cascade Simulations The accuracy of all types of MD simulations is limited by the accuracy of the interatomic potentials that are used to represent the interactions between the atoms in the simulation, as discussed in Chapter 1. Interatomic potentials vary enormously in complexity, from simple two-body Lennard Jones and Buckingham potentials to very complex potentials that aim to reproduce the effects of bond forming and breaking, such as the reactive bond order (REBO) [17] potential for carbon and the REAXFF [18] series of potentials. It is important to bear in mind that all potentials are an approximation to the real interactions and as such they will reproduce some properties of materials better than others. Interatomic potentials are derived by fitting to experimental data or to *ab initio* calculations. For radiation damage simulations, it is important that the defect energies are reproduced accurately by the potentials; therefore, in some cases defect conformations and energies, calculated using density functional theory (DFT), are used as fitting parameters for potentials. It is also important that the thermal expansion coefficient, the melting temperature and the latent heat are reproduced with reasonable accuracy as these parameters will determine the response of the material to the thermal spike and the spatial extent of the molten region created in the material.

Interatomic potentials are commonly derived by fitting to equilibrium, or near equilibrium, material properties (see Chapter 1 for further details), whereas radiation damage is a highly nonequilibrium process. In particular, during cascade simulations, atoms often experience close encounters with other atoms, and thus, they explore regions of the potential energy surface where the interatomic potential bears little relationship to the true interactions. In order to address this problem, the standard short-range interactions are replaced by a screened Coulomb potential, with the screening function fit to reasonably accurate quantum mechanical calculations. The most commonly used form is the Ziegler, Biersack, Littmark (ZBL) potential [19], which has the following functional form:

$$V(r) = \frac{Z_1 Z_2 e^2}{4\pi \varepsilon_0 r}\varphi(r/a). \tag{3.3}$$

where $\phi(x) = 0.1818e^{-3.2x} + 0.5099e^{-0.9423x} + 0.2802e^{-0.4024x} + 0.02817\, e^{-0.2016x}$ and $a = \frac{0.8854a_0}{(Z_1^{0.23}+Z_2^{0.23})}$. Here, a_0 is the Bohr radius (0.5292 Å) and Z_1 and Z_2 are the atomic numbers of the two atoms involved in the interaction.

The ZBL potential is joined to the standard potential with a spline function at short interatomic separations (r), typically around 1–2 Å, which ensures that both the potential and its first derivative are continuous at all interatomic separations.

Size of the Simulation Cell and Boundary Conditions The energy of knock-on events that can be modeled in cascade simulations is restricted by the size of the simulation cell because the number of atoms in the simulation cell increases linearly with the PKA energy. Typically around 25 atoms/eV of the PKA energy are required in order to limit the temperature increase and contain the damage within the cell. As a general rule, the length of the simulation cell should be at least four times greater than the average range of the PKA.

In order to avoid edge effects, periodic boundary conditions are generally employed for simulation cells of cascade simulations. The energy deposited by the PKA thermalizes within a few picoseconds and creates in a nanometer-sized volume of material near the deposition site in which the temperature is very high. The thermal expansion associated with this region creates a pressure wave, which spreads to the boundary of the simulation cell, where it interacts with the periodic image of the wave on the opposite face. Ideally, the box should be large enough so that the thermal phase of the simulation is complete before the heat and pressure waves return in order to avoid anomalous effects. The heat wave can be partially absorbed by introducing a thermostat in the outer layer of the simulation cell. Such a thermostat is known as a pseudo-thermostat because it acts only on the boundary atoms. The thermostat reduces the temperature of the boundary layer either by simple rescaling of the velocities or by a Langevin thermostat. Ideally, the boundary thermostat should mimic thermal conductivity out of the simulation cell to the rest of the material, but this is challenging to achieve in practice.

Simulation Time Step An additional issue associated with highly nonequilibrium nature of cascade simulations is the simulation time step. Typically, MD time steps are of the order of 1 fs as this provides a reasonable balance between energy conservation, which is degraded by long time steps due to the numerical integration of the equations of motion, and attaining a reasonable simulation time with the available computational resources. In the early stages of a cascade simulation, the velocities of the PKA and surrounding atoms are very high; therefore, the atomic configuration changes substantially over a femtosecond time scale. Energy conservation can only be ensured by using much shorter time steps, sometimes orders of magnitude shorter than normal. As the simulation progresses and the energy is dissipated, the maximum atomic velocities fall to more reasonable values. Cascade simulations generally employ a variable time step methodology to permit very small time steps to be employed in the early stage of the simulation, when forces are large, with gradually increasing values as the simulation approaches equilibrium. A maximum value is chosen for the displacement during one time step (of the order of a few hundredth of an angstrom), and the relevant time step is calculated using the current values of the atomic velocities and accelerations. An upper limit of around 1–3 fs is imposed to ensure reasonable energy conservation for the whole simulation. Most MD programs that are commonly used for radiation damage simulations have a facility for employing variable time steps, including the widely used DL_POLY [20, 21] code.

Analysis of Results The primary information that can be obtained from cascade simulations is the damage function, which is the number of Frenkel pairs created by a PKA with a given energy. This number is sensitive to the crystallographic direction of the PKA; therefore, it is necessary to perform several simulations with different knock-on directions to determine the mean number of defects. Defects are identified by two main methods. The simplest approach is to label an atom as an interstitial if it is located greater than a given distance from a lattice site in a crystal and a vacant lattice site as one where there is no atom within a fixed distance (e.g., 0.3 of the lattice parameter for a bcc structure) from the site. With this method a dumbbell, or split interstitial, will be identified by most analysis programs as two interstitial atoms and one vacancy. An alternative, more rigorous, method involves defining a Wigner Seitz cell around each lattice site. A cell containing two atoms signals an interstitial and an empty cell signals a vacancy. If a perfect crystal is used as a starting configuration then the number of vacancies should equal the number of interstitials. Antisite defects in materials with more than one type of atom refer to the wrong type of atom sitting on a lattice site.

MD simulations are clearly an approximate method for modeling radiation damage; therefore, it is necessary, if possible, to validate the model by comparison with experiment. This is clearly challenging, given that the incompatible time and length scales of modeling and experiment, nevertheless it is possible to calculate some quantities that can also be measured experimentally [22]. The ionic mixing parameter (ζ_{IM}) is defined as

$$\zeta_{\mathrm{IM}} \equiv \frac{\langle R^2 \rangle}{nE},$$

where $\langle R^2 \rangle$ is the mean square displacement of the atoms in the simulation cell, n is the number density of radiation events and E is the deposited energy per radiation event. The mixing parameter can be easily calculated in an MD simulation and it can also be measured, at least approximately, by experiment. ζ_{IM} is typically in the range ten to several hundred Å^2/eV.

The displacement threshold energy, the minimum energy required to create a stable defect pair, can be calculated by MD and measured experimentally although, since it is particularly sensitive to the crystallographic direction, many simulations are necessary to give a reliable average value. The damage function, or the number of defects produced by a single cascade of PKA energy E, is a particularly important parameter as it represents the number of defects created by a particular damage event. It is often expressed as a ratio to the number predicted by the NRT model (Equation 3.2). This ratio decreases rapidly as the cascade energy increases, from a value close to 1 at low energy to 0.3–0.4 at high energy. The damage function can be measured experimentally by techniques such as resistivity measurements, but it is difficult to make a direct connection between modeling and experiment.

Electronic Effects Classical MD considers interactions between atoms, and any effect of the electrons is considered only at the level of the interatomic potentials. However, some types of radiation excite electrons and the effects of these excitations are challenging to model using current simulation techniques [23]. Irradiation by lasers and irradiation by electrons are two such examples, and ion irradiation also excites electrons to some extent. An ion moving in a solid loses energy by elastic collisions with atoms and inelastic collisions involving electronic excitations. The proportion of energy lost to inelastic collisions increases as the energy of the moving ion increases. For low-energy events of the type normally modeled by cascade simulations, most energy is lost to collisions with other atoms and it is safe to neglect the inelastic component. However, as computing power increases, ever-higher PKA energies are modeled and the electronic losses become more significant. This is usually dealt with by including a friction term in the equations of motion for atoms above a particular energy (of the order of 10 eV) to represent the energy lost to electronic excitations [24]. A more sophisticated method tracks the electronic energy by calculating the electronic temperature increase and evolving the electronic temperature via the heat diffusion equation [25]. The electronic energy is returned to the lattice via a Langevin thermostat, using the local electronic temperature as the thermostating temperature, to represent electron–phonon coupling. This methodology is included in both the LAMMPS and DL_POLY MD programs [26].

Results The detailed results from cascade simulations are sensitive to the material being modeled, in particular the type of bonding and the crystal structure, nevertheless some features are common to many types of simulations. In crystalline materials events such as channeling and replacement collision sequences can occur. Channeling occurs when an ion is deflected in such a way that it travels between the crystallographic planes and avoids close collisions with atoms. Energy loss is primarily by inelastic collisions with electrons, and the range of a channeling ion is significantly higher than an ion with equivalent energy that collides with atomic nuclei. Another type of event that is commonly observed in cascade simulations is the replacement collision sequence, which occurs when the knock-on direction corresponds to a close-packed crystallographic direction. In this case the moving atom knocks an atom out of its position and occupies the vacant site. The knocked-out atom replaces the next atom in the chain, and this sequence of events continues until the energy falls below the displacement energy threshold, at which time an interstitial atom is created. Such events are efficient mechanisms for the creation of separated vacancy–interstitial pairs.

A notable result derived from MD simulations is that defect production is much less efficient than the NRT model predicts, with the number of defects 20–40% N_{NRT} for PKA energies greater than 1 keV. In the NRT model, the number of defects produced increases linearly with the PKA energy; however, MD simulations invariably show a sublinear increase with the following form [27]:

$$N_{\mathrm{FP}} = A(E_{\mathrm{p}})^{m}. \tag{3.4}$$

Here, A and m are material-dependent parameters, but A is approximately 5 (when the PKA energy, E_{p}, is expressed in keV) and m is in the range 0.7–0.8 for several metals. The reduced defect creation efficiency is attributed to the enhanced recombination of vacancies and interstitials due to thermal energy. In general, vacancies are located near the center of the thermal spike region and interstitials are found near the periphery, due mainly to the pressure wave that is formed by the rapid thermal expansion and, to some extent, the rapid diffusion of interstitials and replacement collision sequences [28, 29]. The defects from a typical 10 keV collision cascade in Fe are shown in Figure 3.6.

The cross-section for nuclear interactions decreases with increasing PKA energy, and therefore the distance between close collisions increases. In practice, this means that a high-energy PKA creates a number of subcascades that are largely independent of each other. This effect has been used to argue that very high PKA energies can effectively be modeled by considering only the lower energy events. The threshold energy for subcascade formation is strongly dependent on the material (∼20 keV for Cu and ∼200 keV for Au) [30]. Subcascades form when

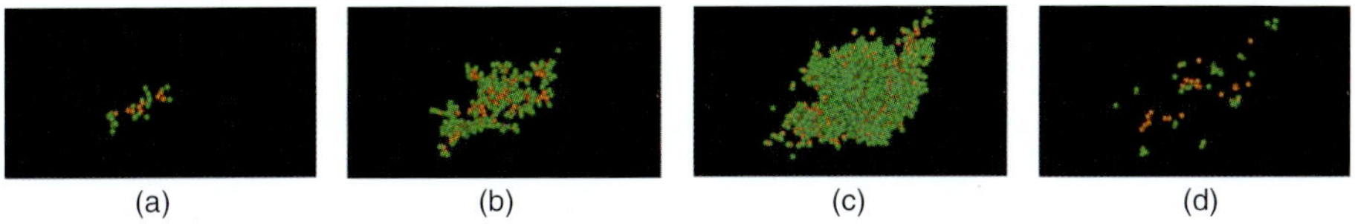

Figure 3.6 Snapshots of defects (green, interstitials; gold, vacancies) during a 10 keV cascade in Fe: (a) The collision regime, (b and c) the development of the thermal spike and (d) the final (residual) defect structure.

the mean free path of the PKA is greater than the diameter of the damaged zone created by a typical collision event.

Metallic materials invariably recrystallize following a radiation event, albeit with some residual damage, but this is not the case for ceramic materials, particularly those with complex structures or strongly covalent bonds [31]. In these materials a radiation event can result in a volume of the material being rendered disordered or amorphous. The resistance to amorphization is an important material property for ceramic materials that could potentially be used to encapsulate radioactive waste, as amorphous material would offer lower resistance to leaching.

3.3.2.2 Sputtering Simulations

The process by which atoms or ions are ejected from a material by ion or particle bombardment of surfaces is known as sputtering, and a comprehensive theory for this process was developed by Sigmund [32]. Such processes are particularly relevant to plasma-facing materials in fusion technology (Section 3.2.2.2) because the particles and ions that escape the plasma impinge on the plasma-facing materials directly, leading to surface erosion and plasma contamination.

Sputtering simulations have strong similarities to cascade simulations, but in the case of sputtering simulations the atom that is given the initial velocity is located outside the material and the velocity is directed toward the material surface (Figure 3.7). As with standard cascade simulations, the resulting damage is strongly

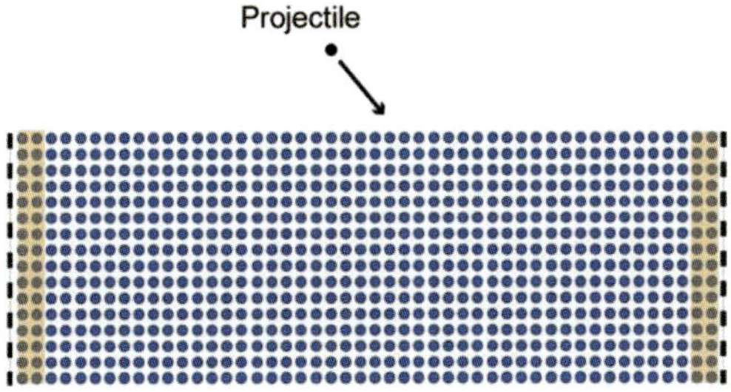

Figure 3.7 Schematic representation of a sputtering simulation. The dashed lines show periodic boundary conditions and the orange area represents the volume where the pseudo-thermostat is applied. The atoms on the top and bottom surfaces are free to move.

dependent on the angle of the projectile atom and also the point of initial contact with respect to the surface atoms. For example, a projectile atom colliding directly with a surface atom will have a different effect from the one that penetrates into the material before its first collision.

The measured sputtering yield (the average number of emitted atoms per incident particle) is typically in the range between 1 and 10. The bombarding ion transfers energy to target atoms and some of the recoil atoms gain sufficient energy to escape the surface. MD simulations of sputtering have the advantage over analytic and binary collision approaches in that, not only can the sputtering yield be calculated, but the detailed surface damage can be modeled, which can include craters and adatoms. In cases where a thermal spike forms in the subsurface region with a molten volume extending to the surface, the molten material expands and flows over the surface, which forms a crater after resolidification [33]. A deeper thermal spike creates much less surface damage. As with cascade simulations, channeling and replacement collision events result in defects at larger depths than random events.

A particular issue for plasma-facing materials is chemical sputtering. Chemical sputtering occurs when there is a chemical interaction between the bombarding ions and the substrate, with the result that atoms are ejected from the surface at much lower energies than would occur with physical sputtering alone. One important example is the chemical sputtering of a carbon surface with hydrogen. The hydrogen interacts with surface carbon atoms and reduces the number of surface bonds. The remaining bonds are easily broken by further impinging atoms and volatile hydrocarbons are ejected from the surface. This is an issue for fusion because resource and safety issues mean that the tritium fuel needs to be recoverable. Tritium–carbon molecules ejected from the surface will redeposit elsewhere on the vessel wall and will be difficult to recover.

Chemical sputtering is challenging to study using MD techniques because of the difficulty in representing chemical interactions using classical interatomic interaction potentials. In the case of the carbon–hydrogen (tritium) system, chemical interactions are described well by the REBO [17] and the associated AIREBO (adaptive intermolecular reactive bond order) potential [34]. A potential has also been developed for the beryllium–hydrogen [35] interaction, which has enabled chemical sputtering to be modeled for beryllium, another important plasma-facing material for fusion reactors.

3.3.3 Monte Carlo Simulations

With carefully developed interatomic potentials, MD simulations can make accurate predictions about the number of primary defects created by radiation events that interact with atomic nuclei. The primary defects are those which survive for a few tens of picoseconds, by which time the excess energy will have dissipated and the lattice temperature will have returned to the ambient level. The microstructure will continue to evolve, albeit at a much slower rate, as the defects diffuse through the crystal. Diffusion, the hopping of atoms from one low energy site to another, in the solid state is an "infrequent event" because there are long time periods in between events when not much happens. In general, particularly for metals, the diffusion rates for interstitials are much higher than those for vacancies; therefore, interstitials will cluster and migrate to sinks more readily than vacancies. Defect clusters have a much stronger effect on the mechanical properties than isolated defects; therefore, the time evolution of defect clusters, over periods of months and years, is an important aspect of radiation damage.

The disconnect between the picosecond time scale of MD and the lifetimes of nuclear power plants is challenging to fill, but MC techniques can extend MD time scales to some extent. MC techniques, as discussed in Chapter 1, are event-based modeling methods that calculate the temperature-dependent probability of an event occurring. The acceptance or rejection of an event at a particular temperature is determined by a random number generator. Metropolis MC is not related to a particular timeframe, as whether an event is accepted depends only on the energy difference between the initial and final states; however, kinetic MC (KMC) is concerned with transition rates, therefore can be associated with a real time period. KMC is, therefore, useful for modeling the microstructure evolution of radiation damage [36, for example].

3.3.3.1 Kinetic Monte Carlo

If we consider a collection of interacting atoms then the potential energy surface of the system will have a number of low energy basins and most time will be spent close to the bottom of these basins. Occasionally the thermal fluctuations will be such that the system traverses an energy barrier and descends into a neighboring basin. There may be a number of possible paths out of each basin and each will be associated with a particular rate constant. The rate constant (k_{ij}) is the probability per unit time that the system will migrate from state (basin) i to state j. KMC is implemented by associating a line with length k_{ij} with each possible transition from i to j and laying these lines end to end to give a line with total length $k_i^{\text{tot}} = \sum_j k_{ij}$. At each KMC step a random number (r) between 0 and 1 is generated. The event corresponding to rk_i^{tot} on the line is selected for acceptance at that KMC step (Figure 3.8). Thus, the probability of acceptance of a particular transition at each step is equal to k_{ij}/k_i^{tot}, which gives the correct ratio of acceptance events for a large number of steps. The time is progressed by a value, t_s, which is determined by choosing a random time from the exponential distribution of the total rate constant (k_i^{tot}), using the expression $t_s = \ln(r)/k_i^{\text{tot}}$, where r is again a random number (different from before) lying between 0 and 1. Note that the time progressed by each step is not directly related to the selected transition event, but it is a separate random process [37].

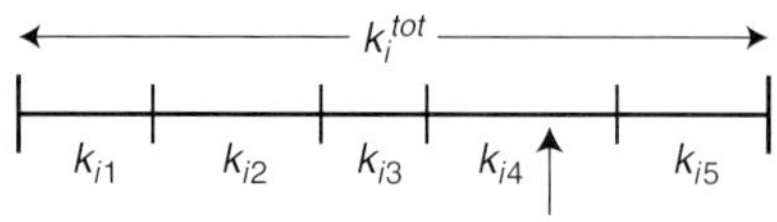

Figure 3.8 Illustration of the selection of an accepted event in a KMC simulation. There are five possible events (j), each associated with a transition rate k_{ij}. The accepted event is chosen by selecting a random number (r) and choosing the event that corresponds to the position rk_i^{tot}.

The time scale that can be modeled using KMC is strongly dependent on the transition rate of the most probable event. A typical advance time is the inverse of the fastest possible transition rate, which means that for a migration barrier of 0.1 eV and a typical attempt frequency of $10^{13}\,s^{-1}$ the time achieved for 1 KMC step is around 5 ps at 300 K and 1 ps at 600 K. If the lowest barrier is 0.5 eV then these times increase dramatically (25 μs and 1 ns, respectively). The low barrier problem is an issue for many methodologies that attempt to extend simulation time scales. An additional problem for KMC is that, in order to describe the dynamics successfully, a complete catalogue of possible transitions must be identified. This is a challenging task because many transitions are counterintuitive and it is difficult to predict their existence, let alone calculate the transition rate. One of the earliest examples comes from the diffusion of adatoms on metal surfaces, where it was assumed that the dominant mechanism was hopping from one surface site to a neighboring site. In fact, it was found that the exchange mechanism, where an adatom replaced a surface atom and the replaced atom became the new adatom, was the dominant mechanism. There have been many other cases where complex diffusion mechanisms, for which it is often difficult to calculate rate constants, have been identified. In fact, it is not possible to guarantee that, for a particular system, all possible transitions have been identified, and this may introduce errors to the predicted system dynamics.

3.3.3.2 *Object Kinetic Monte Carlo*

A variation of KMC that has proved to be very effective for extending the time scales for radiation damage modeling is object KMC (OKMC). In this method, rather than considering individual atoms, the diffusive motion of defects and clusters of defects on a lattice are considered to be the events and the corresponding rate constants for these events are calculated. Only the motion of these objects is tracked and the background lattice of atoms is ignored. Additional events, such as defect emission from and adsorption at clusters, are defined to allow for transitions between types of clusters (Figure 3.9). Object KMC can reach much longer length and time scales than atom-based KMC; however, it must be used with caution. The dynamics of defect clusters is significantly more complex than those of isolated defects. A good example has been identified in MgO, where an interstitial cluster was found to have a particularly low migration barrier [38]. Each cluster may have several different

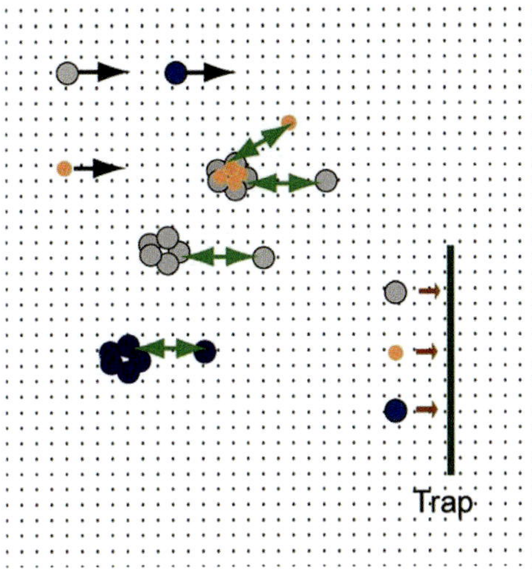

Figure 3.9 *Objects (vacancies, grey; interstitials, blue; He atom; orange) and events (defect migration, black arrows; defect absorption or emission from clusters, green arrows; defect trapping, red arrows) that are included in a typical OKMC calculation.*

transition pathways, and the true dynamics can only be reproduced if all of these are accounted for.

A variation of OKMC, known as event-based MC, ignores the lattice completely and assigns continuous spatial coordinates to the objects. The events are the encounters between two objects and the emission of mobile objects. Several events occur within one simulation time step at a rate that is influenced by the spatial separation between the objects [39].

3.3.3.3 *Transition Rates*

The main parameters affecting the accuracy of both KMC and OKMC techniques are the transition rates for the catalogue of events that have been identified [37]. Transition rates can be determined using harmonic transition state theory (otherwise known as Vineyard theory) [40] in which the transition point is characterized by the saddle point on the potential energy surface. In the harmonic theory, the transition rate is given by the following expression:

$$k = \nu_o \exp(-E_b/k_b T). \quad (3.5)$$

The energy barrier (E_b) is the difference between the energy of the minimum energy configuration and the energy at the saddle point. The prefactor (ν_o) is related to the 3N normal mode frequencies (where N is the number of atoms in the system) at in the basin ($\nu_i^{\min}$) and the $3N - 1$ nonimaginary normal mode frequencies at the saddle point (ν_i^{sad}) by the following expression:

$$\nu_0 = \frac{\prod_i^{3N} \nu_i^{\min}}{\prod_i^{3N-1} \nu_i^{\text{sad}}}. \tag{3.6}$$

The prefactor (ν_0) generally lies in the range 10^{12}–10^{13} s^{-1}, which corresponds to a typical lattice vibration frequency. Indeed, a value of 10^{13} s^{-1} is often used without performing the full calculation; however, it should be noted that this value has been found to be strongly erroneous in some scenarios, so it should be used with caution.

3.3.3.4 *Examples*

OKMC simulations have been used to compare the damage accumulation as a function of dose in Cu and Fe [41]. The aim of the study was to elucidate the difference between the defect density observed in fcc metals, such as Cu, and bcc metals, such as Fe. Transition electron microscope (TEM) experiments revealed defects in Cu at very low doses, but much higher doses were required before any defects were visible in Fe. The list of events includes

- jumps to neighboring lattice site;
- cluster growth and decay by adsorption/emission of defects of the same/opposite type;
- cluster decay by emission of a defect with the same type;
- trapping or annihilation at impurities or sinks.

The one-dimensional diffusion of the dislocation loops in Cu was included in the model and found to influence the results. Good agreement was found with the number of visible defect clusters observed in TEM.

Domain *et al.* [42] used OKMC to model the evolution of vacancies and interstitials in pure Fe and Fe–Cu alloys. The objects were point defects (vacancies and interstitials) and point defect clusters.

Including the solute (Cu) atoms and clusters of solute atoms almost doubles the number of events that need to be considered. Dissociation and jump probabilities are considered as thermally activated processes. Trapping and annihilation take place spontaneously when the defects come within the capture radius. External events, corresponding to radiation events, are modeled by inserting defects with a probability corresponding to the desired dose rate. Both neutron irradiation and electron irradiation are considered, the difference being the spatial distribution of the defect production.

3.3.4 Cluster Dynamics

An alternative method for extending the dynamics beyond the time scales that can be achieved by MD methods is known as cluster dynamics. Cluster dynamics is based on rate theory modeling and, as it requires less computational time, it can be used to extend simulation times beyond that achievable using KMC. Like OKMC, cluster dynamics is concerned with the time evolution of an ensemble of defect clusters, and there is characteristic transition rate associated with each change in cluster size. The variables in the cluster dynamics methodology, as applied to radiation damage, are the concentrations of defect clusters of various types. These defects are created, by radiation damage, at a predetermined rate, and annihilated, by diffusion to sinks (grain boundaries or dislocations). They can also transform from one type to another by, for example, emitting or absorbing a defect cluster. The defects considered are generally vacancies and interstitials, but it is possible to include impurities, such as helium. The time evolution of the cluster distribution can be calculated by solving a set of coupled rate equations:

$$\frac{\partial C_j}{\partial t} = K_j + \sum_k w(k, j)C_k - \sum_k w(j, k)C_j - L_j. \tag{3.7}$$

In this equation, C_j represents the concentration of j-type clusters that could, for example, be a cluster with n interstitials or vacancies; $w\ (k,j)$ is the transition rate from a k-type cluster to a j-type cluster; and K_j and L_j are the creation and annihilation rates of j-type clusters, respectively. A relatively simple case would occur when only vacancy and interstitial defect clusters are present and transitions can only change the size of the clusters by ± 1. Thus, an interstitial cluster with n interstitials can increase/decrease in size by absorbing/emitting a single interstitial, and it can decrease in size by absorbing a single vacancy. Equation (3.7) has, therefore, three terms, in addition to the source/sink terms, and there is one equation for each type of defect cluster considered. However, even in this case modeling large voids and interstitial clusters will involve many thousands of coupled equations. Further efficiency can be gained by considering the cluster size, n, to be a continuous variable in which case Equation (3.7) can be replaced by a Fokker–Planck equation with a drift term and a diffusion term. This extension is beyond the scope of this chapter, but details and further references are given by Mansur [43,44].

This methodology has had notable success in modeling void swelling and dislocation density evolution in response to irradiation [45]. Cluster dynamics, particularly in its continuum form, has the potential to extend the time scales of microstructure evolution modeling to reactor lifetimes.

3.3.4.1 Examples

Hardouin Duparc *et al.* [46] employed cluster dynamics to model the microstructure evolution of ferritic steel under electron irradiation. Their model assumed that only isolated vacancies and interstitials were mobile and that clusters grew by the capture of defects of the same type and shrank by the capture of defects of the opposite type. The equation for the rate of change of the concentration of clusters with n interstitials (C_{ni}) is, therefore,

$$\frac{\partial C_{ni}}{\partial t} = \left(\beta^{i}_{(n-1)i} C_{1i}\right) C_{(n-1)i} + \left(\beta^{v}_{(n+1)i} C_{1v} + \alpha^{i}_{(n+1)i}\right) C_{(n+1)i} - \left(\alpha^{i}_{ni} + \beta^{v}_{ni} C_{1v} + \alpha^{v}_{nv}\right) C_{ni}. \quad (3.8)$$

Here, $\beta^{i}{}_{nd} C_{1d}$ is the rate of capture of a d-type defect by a cluster of n d-type defects and $\alpha^{\mathrm{d}}{}_{\mathrm{nd}}$ is the emission rate of d-type defects from clusters with n d-type defects, and d represents vacancies or interstitials. There is a set of equivalent equations for vacancies. The creation and annihilation terms are included for the equations with $n = 1$. The time evolution of the cluster size is calculated by the numerical solution of $N_{\mathrm{v}} + N_{\mathrm{i}}$ coupled equations, where N_{v} and N_{i} are the maximum size of the vacancy and interstitial clusters considered in the system. The rate coefficients for clustering, recombination and capture by sinks—all depend on the defect diffusion coefficients and a number of other parameters, and the derivation of a good set of parameters is, in itself, a challenging problem. Nevertheless, the method appears to give good agreement between the temperature variations of the number of defects clusters calculated by the model and experimental observations.

Ortis *et al.* [47] included He atoms, in addition to vacancy and interstitial clusters, in a cluster dynamics model of He diffusion in irradiated Fe. As discussed in Section 3.2.2.1, the formation of He bubbles in the first walls of fusion reactors, due to the interaction between transmutation He and vacancy clusters, makes a significant contribution to embrittlement and swelling. It is, therefore, important to be able to estimate the time evolution of the bubble concentration. In [47], 41 different reactions were considered. Of these, three involved vacancy cluster growth by adsorption of single vacancies, three involved interstitial cluster growth and the remainder involved interactions between He atoms and vacancies or interstitials. Single vacancies, single interstitials, di-interstitials and He interstitials were all considered to be mobile. All other defects were considered to be immobile. The results were compared with experimental He desorption data. Good agreement with experiment could only be obtained by taking a higher value for the activation energy for vacancy diffusion than calculated values. This was attributed to trapping of vacancies by impurities, particularly carbon, in the metal; however, given the number of necessary approximations involved in this type of approach, it is difficult to rule out all other possible explanations.

3.3.4.2 Comparison with OKMC

The main difference between cluster dynamics and OKMC is that the rate theory approach employed by cluster dynamics is a mean field methodology, which is concerned only with average defect concentrations in the material and it has no knowledge of the local concentrations. The implication of this is that no spatial correlation between the defects can be included in the model, nor can it distinguish between different microstructures (e.g., grain shapes). OKMC can create defects and defect clusters at particular points in space and time, to represent particular radiation events, whereas defect creation in cluster dynamics is considered to occur uniformly in space and time. In OKMC, the microstructure (grain boundaries and dislocations) can be represented by a particular distribution of sinks, but in cluster dynamics, defect annihilation occurs uniformly in space and time. The cluster dynamics methodology has, however, a significant advantage in computational efficiency. OKMC can only cope with small volumes of material and the defect density in these volumes must be relatively high in order to achieve accurate statistics, which means that it can only be used to model high dose rates. Rate theory is a mean field theory; therefore, there is no length scale associated with cluster dynamics – the results produce the average concentrations of the relevant defects throughout the volume of the material. The attainable time scale depends strongly on the number of defect species and transition rates between the defect species but, by employing the continuum approach for the cluster size, time scales of tens of years can readily be attained.

Detailed comparisons between cluster dynamics and OKMC models have been carried out for a limited number of scenarios, and close agreement was found in regimes where the two methods could be applied reliably, that is for high radiation doses and high densities of sinks [48]. The effect of spatial correlation was examined by comparing damage evolution for 30 keV Fe irradiation, which produces a high concentration of defects and clusters in a local region of the material, with electron irradiation, which produces a uniform distribution of vacancies and interstitials. The difference in the initial defect distribution had an effect on the evolution of the cluster size for low temperature scenarios, but the results were indistinguishable at high temperatures. The implications from such tests were that, under certain conditions, the spatial correlation had minimal effect and cluster dynamics modeling would be a more efficient technique [49].

3.3.5 Density Functional Theory

The nonequilibrium nature of radiation damage means that the drive has been to extend length and time scales of the simulations; therefore, classical methods are generally employed. *Ab initio* methods do, however, have a significant role to play as they are used to calculate various parameters that are required as input to the longer time scale models. In particular, DFT has been widely used to calculate parameters for fitting interatomic potentials and for calculating the activation barriers for the various transitions involved in OKMC and cluster dynamics.

3.3.5.1 Interatomic Potentials

As discussed in Section 3.3.1, the accuracy of cascade and other MD simulations relies strongly on the quality of the interatomic potentials employed. One of the important parameters for such simulations is the relative energy of the defect conformations. Early interatomic potentials for Fe gave erroneous values for the relative order of the energies of split (dumbbell) interstitials with different orientations. In contrast to the early interatomic potentials, DFT calculations predicted the <110> oriented dumbbell interstitials to have lower energy than the <111> orientations. This has significant implications because different dumbbell orientations have different mobilities, which would affect the microstructure evolution. Modern interatomic potentials, derived specifically for radiation damage simulations, use calculated defect energies as fitting parameters [50, 51].

DFT calculations were also used to develop reliable interatomic potentials for He in Fe. Calculations [52] indicated that a He atom in the tetrahedral interstitial site had a lower energy than He in an octahedral site, whereas the early potentials [53] found the octahedral interstitial site to be more stable. New potentials were, therefore, derived that used the DFT values as fitting parameters to get the correct relative stability [54].

3.3.5.2 *Transition Rates*

Another area where DFT calculations have played a crucial role is in the calculation of energy barriers and binding energies for transition rates for both OKMC and cluster dynamics models. The energy barriers for the rate constants, E_b in Equation (3.5), of a transition can be calculated directly using DFT if the saddle point is known. For example, for vacancy migration in a rock salt-structured oxide, such as MgO, the saddle point is known to be in the center of the face of a cube of ions; therefore, the energy barrier can be obtained by calculating the energies of 1 and 2 and taking the difference. If periodic methods are used, care must be taken to employ sufficiently large unit cells to minimize interactions between the defect and its periodic images. Nevertheless, these methods are well established and have recently been used to calculate migration energies of defects with varying charge states with some surprising results [55]. If the saddle point is not known *a priori* then methods such as the nudged elastic band (NEB) technique can be used to identify the saddle point and calculate the barrier. The NEB method can be used to locate the minimum energy pathway between two known local minima. A detailed description of the NEB method is beyond the scope of this chapter but is discussed in [56] and included in several MD and DFT codes.

Standard DFT codes can also be used to calculate the normal mode frequencies, and these can be used to estimate the prefactor (ν_0) for the transition rates in KMC calculations. This is only possible for relatively simple transition events, such as single vacancy or interstitial hopping.

3.4 Summary and Outlook

Modeling has an important role to play in the science of radiation damage of materials for nuclear applications, not least because experiments are difficult and expensive to perform. Modeling can contribute in three main ways:

1. It contributes to the fundamental understanding of radiation damage processes and recovery.

2. It can help with the interpretation of experimental results.
3. It can be used to predict the long-term effects on material properties.

To date its largest contribution has been in the area of fundamental understanding of mechanisms involved in defect creation and microstructure evolution. There is a substantial and increasing effort into the cross-fertilization between modeling and experiment, particularly as the length scales considered by these two approaches converge. Making predictions about the long-term effects is a goal worth striving for, but there are currently large uncertainties associated with such methods.

There is still much to be gained from using the full range of modeling techniques to study of the fundamental radiation processes and material properties. Next-generation fission and fusion technologies will place severe demands on materials, and a thorough understanding will be essential in order to select or design appropriate materials for the various functions. Current modeling is mostly done on pure materials, with simple microstructures, but real engineering metals are alloys with several elements and unintentional impurities. The development of reliable interatomic potentials for the important alloying elements would aid progress toward modeling real materials. The effect of the microstructure has not been fully explored to date. The microstructure of an engineering material dominates its mechanical properties and, indeed, modification of the microstructure, by changing the density of grain boundaries, dislocations and precipitates, is the primary mechanism by which metals are strengthened. The microstructure also has a dominating effect on long-term damage evolution, and a thorough understanding of these effects could enable the design of microstructures that would minimize damage and enhance the lifetime of the materials in radiation environments.

In the past it has been difficult to make direct comparison between modeling and experiment because of the large discrepancy in the time and length scales. As computers become ever more powerful and the resolution of some experimental techniques approaches the nanometer and femtosecond scales, the gap is certainly narrowing, but direct comparison between measurements and calculated parameters is still only possible in a very limited number of cases. Maximum value is obtained from both experiment and modeling when the two communities work together to design experiments and, perhaps more significantly, interpret experimental results. Often an iterative process is required to arrive at an interpretation that fits both the modeling and the experimental observations.

One of the major challenges associated with nuclear design is the prediction of long-term effects of radiation on materials. Many experimental facilities exist that are used to irradiate materials with ions with a range of energies, and the task is then to extrapolate the results from such experiments to realistic radiation scenarios. Ion irradiation penetrates only a few millimeters into the surface, whereas neutrons can penetrate several centimeters. In order to achieve a realistic fluence from experimental measurements it is necessary to employ high ion flux. In an ideal situation, modeling and theory would be used to obtain an accurate interpretation of the effects observed by ion and electron irradiation experiments, and the knowledge and parameters obtained would then be used to extrapolate to the radiation type, flux and time scale relevant to a reactor. Several research programs are currently taking this type of approach, but the challenges involved remain formidable.

In summary, modeling methods have made, and continue to make, a vital contribution to the field of nuclear materials and much more remains to be done. With the development of novel computational techniques and increased computer power, the field will be able to tackle evermore challenging problems and progress toward the ultimate goal of the prediction of long-term behavior and the design of radiation-resistant materials.

References

1. Kinchin, G.H. and Pease, R.S. (1955) The displacement of atoms in solids by radiation, *Rep. Prog. Phys.* **18**, 1.
2. Norgett, M.J., Robinson, M.T. and Torrens, I.M. (1975) A proposed method of calculating displacement dose rates, *Nucl. Eng. Des.* **33**, 50.
3. Allen, T.R. and Busby, J.T. (2009) Radiation damage concerns for extended light water reactor service, *JOM* **61**, 29.
4. Zinkle, S. and Busby, J.T. (2009) Structural materials for fission and fusion energy, *Mater. Today* **12**, 11, 12.
5. White, R.J. (2004) The development of grain-face porosity in irradiated oxide fuel, *J. Nucl. Mater.* **325**, 61.
6. Petti, D., Crawford, D. and Chauvin, N. (2009) Fuels for advanced nuclear energy systems, *MRS Bull.* **34**, 40.
7. Zinkle, S. (2005) Fusion materials science: overview of challenges and recent progress, *Phys. Plasmas* **12**, 058101.
8. Federici, G., Skinner, C.H., Brooks, J.N. *et al.* (2001) Plasma-material interactions in current tokamaks and their implications for next-step fusion reactors, *Nucl. Fusion* **41**, 1967.
9. Roth, J., Tsitrone, E., Loarer, T. *et al.* (2008) Tritium inventory in ITER plasma-facing materials and tritium removal procedure, *Plasma Phys. Control. Fusion* **50**, 103001.
10. Weber, W.J., Navrotsky, A., Stevanovsky, S. *et al.* (2009) Materials science of high-level nuclear waste immobilization, *MRS Bull.* **34**, 46.
11. Weber, W.J., Ewing, R.C., Angell, C.A. *et al.* (1997) Radiation effects in glasses used for immobilization of high-level waste and plutonium disposition, *J. Mater. Res.* **12**, 1946.
12. Ojovan, M.I. and Lee, W.E. (2005) *An Introduction to Nuclear Waste Immobilization*, Elsevier Science Publishers B.V., Amsterdam.
13. Weber, W.J., Ewing, R.C., Catlow, C.R.A. *et al.* (1998) Radiation effects in crystalline ceramics, *J. Mater. Res.* **13**, 1434.
14. Ziegler, J.F. *SRIM—The stopping and Range of Ions in Matter*, http//www.srim.org (accessed 10 November 2012).
15. Robinson, M.T. and Torrens, T.M. (1974) Computer simulation of atomic-displacement cascades in solids in binary collision approximation, *Phys. Rev. B* **9**, 5008.

16. Bacon, D.J. and de la Rubia, T.D. (1994) Molecular dynamics computer simulations of displacement cascades in metals, *J. Nucl. Mater.* **216**, 275.
17. Brenner, D.W. (2000) The art and science of an analytic potential, *Phys. Stat. Sol. (b)*, **217**, 23.
18. van Duin, A.C.T., Dasgupta, S., Lorant, F. and Goddard, W.A. III (2001) ReaxFF: a reactive force field for hydrocarbons, *J. Phys. Chem. A* **105**, 9396–9409.
19. Ziegler, J.F., Biersack, J.P. and Littmark, U. (1985) *The Stopping and Range of Ions in Matter*, Pergamon, New York.
20. Smith, W. and Forester, T.R. (1996) The DL_POLY molecular simulation package, *J. Molec. Graphics* **14**, 136.
21. Todorov, I.T. and Smith, W. (2004) DL_POLY_3: the CCP5 national UK code for molecular-dynamics simulations, *Phil. Trans. R. Soc. Lond. A* **362**, 1835.
22. Averback, R.S. and de la Rubia, T.D (1998) Displacement damage in irradiated metals and semiconductors, *Solid State Phys.* **51**, 281–402.
23. Race, C.P., Mason, D.R., Finnis, M.W. *et al.* (2010) The treatment of electronic excitations in atomistic models of radiation damage in metals, *Rep. Prog. Phys.* **73**, 116501.
24. Nordlund, K., Ghaly, M., Averbach, R.S. *et al.* (1998) Defect production in collision cascades in elemental semiconductors and fcc metals, *Phys. Rev. B* **57**, 7556.
25. Duffy, D.M. and Rutherford, A.M. (2007) Including the effects of electronic stopping and electron-ion interactions in radiation damage simulations, *J. Phys. Condens. Matt.* **19**, 016207.
26. LAMMPS, *LAMMPS Molecular Dynamics Simulator*, http://lammps.sandia.gov/ (accessed 10 November 2012).
27. Bacon, D.J., Gao, F. and Osetsky, Y.N. Computer simulation of displacement cascades and the defects they generate in metals, (1999) *Nucl. Instr. Meth. B* **153**, 87.
28. Nordlund, K. and Averback, R.S. (2000) Collision cascades in metals and semiconductors: defect creation and interface behavior, *J. Nucl. Mater.* **276**, 194–201.
29. Stoller, R.E. and Calder, A.F. (2000) Statistical analysis of a library of molecular dynamics cascade simulations in iron at 100 K, *J. Nucl. Mater.* **283–287**, 746–752.
30. Ryazanov, A.I., Metelkin, E.V. and Semenov, E.V. (2009) Modeling of cascade and sub-cascade formation at high PKA energies in irradiated fusion structural materials, *J. Nucl. Mater.* **386**, 132.
31. Trachenko, K., Pruneda, J.M., Artacho, E. and Dove, M.T. (2005) How the nature of the chemical bond governs resistance to amorphization by radiation damage, *Phys. Rev.* **B71**, 184104.
32. Sigmund, P. (1981) *Sputtering by Particle Bombardment I: Physical Sputtering of Single Elemental Solids in Topics in Applied Physics*, vol. 47 (ed. R. Behrisch), Springer Verlag, Berlin, pp. 9–71.
33. Ghaly, M. and Averbach, R.S. (1994) Effect of viscous flow on ion damage near solid surfaces, *Phys. Rev Let.* **72**, 364.
34. Stuart, S.J., Tutein, A.B. and Harrison, J.A. (2000) A reactive potential for hydrocarbons with intermolecular interactions, *J. Chem. Phys.* **112**, 6472.
35. Björkas, C., Vörtler, K., Nordlund, K. *et al.* (2009) Chemical sputtering of Be due to D bombardment, *New J. Phys.* **11**, 123017.

36. Soneda, N. and de la Rubia, T.D. (1998) Defect production, annealing kinetics and damage evolution in alpha-Fe: an atomic-scale computer simulation, *Phil. Mag. A* **78**, 995.
37. Voter, A.F. (1986) Classically exact overlayer dynamics—diffusion of rhodium clusters on rh(100), *Phys. Rev. B* **34**, 6819–6829.
38. Uberuaga, B.P., Smith, R., Cleave, A.R. *et al.* (2005) Dynamical simulations of radiation damage and defect mobility in MgO, *Phys. Rev. B* **71**, 104102.
39. Dalla Torre, J., Bocquet, J.-L., Doan, N.V. *et al.* (2005) JERK, an event-based Kinetic Monte Carlo model to predict microstructure evolution of materials under irradiation, *Phil. Mag.* **85**, 549, 2005.
40. Vineyard, G.H. (1957) Frequency factors and isotope effects in solid state rate processes, *J. Phys. Chem. Solids* **3**, 121.
41. Caturla, M.J., Soneda, N., Alonso, E. *et al.* (2000) Comparative study of radiation damage accumulation in Cu and Fe, *J. Nucl. Mater.* **276**, 13.
42. Domain, C., Becquart, C.S. and Malerba, L. (2004) Simulation of radiation damage in Fe alloys: an object kinetic Monte Carlo approach, *J. Nucl. Mater.* **335**, 121.
43. Mansur, L.K. (1987) Mechanisms and kinetics of radiation effects in metals and alloys, in *Kinetics of Non-Homogeneous Processes* (ed. G.R. Freeman), Wiley-Interscience, New York, pp. 377–463.
44. Mansur, L.K. (1994) Simulation of radiation damage in Fe alloys: an object kinetic Monte Carlo approach, *J. Nucl. Mater.* **216**, 97.
45. Stoller, R.E. and Odette, G.R. (1987) A composite model of microstructural evolution in austenitic stainless steel under fast neutron irradiation. *Radiation-Induced Changes in Microstructure, 13th International Symposium, ASTM STP 955* (eds F.A. Garner, N.H. Packan and A.S. Kumar), American Society of Testing and Materials, Philadelphia, pp. 371–392.
46. Hardouin Duparc, C., Moingeon, N., Smetniansky-de-Grande and Barbu, A. (2002) Microstructure modelling of ferritic alloys under high flux 1 MeV electron irradiations, *J. Nucl. Mater.* **302**, 143.
47. Ortis, C.J., Caturla, M.J., Fu, C.C. and Williame, F. (2007) He diffusion in irradiated alpha-Fe: an ab-initio-based rate theory model, *Phys. Rev. B* **75**, 100102(R).
48. Stoller, R.E., Golubov, S.I., Domain, C. and Becquart, C.S. (2008) Mean field rate theory and object kinetic Monte Carlo: a comparison of kinetic models, *J. Nucl. Mater.* **382**, 77–90.
49. Dalle Torre, J., Fu, C.C., Williame, F. *et al.* (2006) Resistivity recovery simulations of electron-irradiated iron: kinetic Monte Carlo versus cluster dynamics, *J. Nucl. Mater.* **352**, 42.
50. Derlet, P.M., Nguyen-Manh, D. and Dudarev, S.L. (2007) Multiscale modeling of crowdion and vacancy defects in body-centered-cubic transition metals, *Phys. Rev. B* **76**, 054107.
51. Chiesa, S., Derlet, P.M., Dudarev, S.L. and Van Swygenhoven, H. (2011) Optimization of the magnetic potential for alpha-Fe, *J. Phys. Conden. Matter* **23**, 206001.
52. Seletskaia, T., Osetsky, Y., Stoller, R.E. and Stocks, G.M. (2005) Magnetic interactions influence the properties of helium defects in iron, *Phys. Rev. Lett.* **94**, 046403.
53. Wilson, W.D., Bisson, C.L. and Baskes, M.I. (1981) Self-trapping of He in metals, *Phys. Rev. B* **24**, 5616.

54. Juslin, N. and Nordlund, K. (2008) Pair potential for Fe-He, *J. Nucl. Mater.* **382**, 143.
55. Mulroue, J. and Duffy, D.M. (2011) An ab initio study of the effect of charge localisation on oxygen defect formation and migration energies in magnesium oxide, *Proc. R. Soc. A* **467**, 2054.
56. Henkelman, G., Uberuaga, B.P. and Jonsson, H. (2000) A climbing image nudged elastic band method for finding saddle points and minimum energy paths, *J. Chem. Phys.* **113**, 9901.

4

Energy Storage: Rechargeable Lithium Batteries

M. Saiful Islam[1] *and Craig A.J. Fisher*[2]
[1]*Department of Chemistry, University of Bath, Bath, UK*
[2]*Nanostructures Research Laboratory, Japan Fine Ceramics Center, Nagoya, Japan*

4.1 Introduction

A range of different energy storage devices are being developed, but in all cases their performance is crucially dependent on the properties of their component materials. Indeed, innovative materials science lies at the core of advances that have already been made in this field, an excellent example being the rechargeable lithium battery. Lithium-ion (Li–ion) batteries have since become instrumental in powering the revolution in portable electronics (including mobile phones, laptops and digital audio players), as their energy density is superior to all other secondary battery chemistries exploited so far.

This chapter provides a timely opportunity to highlight some of the exciting advances made in computational studies of lithium battery materials in recent years. Given the vast number and wide range of studies being carried out worldwide, it is beyond the scope of this chapter to provide an exhaustive summary. Nor is it intended to be a comprehensive review of lithium battery technology and materials, as excellent reviews can be found elsewhere [1–12]. Rather, we discuss several examples to illustrate the kinds of significant contributions atomistic simulation can make, focusing on recent progress in analyzing electronic structure,

Computational Approaches to Energy Materials, First Edition. Edited by Aron Walsh, Alexey A. Sokol and C. Richard A. Catlow.

lithium transport, structural disorder (including defect chemistry), as well as surface configurations, in two important classes of cathode materials, namely, layered oxides and olivine phosphates.

4.2 Overview of Computational Approaches

Our description of the computational techniques is brief since these are discussed in detail in Chapter 1 and more comprehensive reviews are available elsewhere [13–15], including in relation to Li–ion batteries [16, 17]. In general, three main classes of technique have been employed in the study of lithium battery materials: *energy minimization* (also known as geometry optimization), *molecular dynamics* and the *cluster expansion* method (together with Monte Carlo techniques). These techniques are applied to a system of atoms contained within a (usually three-dimensional) simulation box representing the material of interest. Interactions between atoms can be treated within one of two paradigms: either classically using effective potentials (typically in the form of analytical equations, and sometimes referred to as force fields) or quantum mechanically using electronic structure methods such as density functional theory (DFT). In this chapter, we are concerned primarily with energy minimization calculations, but brief mention is made to examples of use of other techniques as well.

As introduced in Chapter 1, geometry optimization is the most common technique used in atomistic simulations. It provides the lowest energy configuration of a given structure (i.e., its local energy minimum) through a search of the potential energy surface using efficient minimization routines for the particular conditions specified by the user. As only the potential energy surface is probed, this is essentially an athermal or a zero kelvin method, although vibrational entropy effects can be included, for example, using the quasi-harmonic approximation, if thermal properties such as phonon dispersion curves need to be calculated. For bulk systems, three-dimensional periodic boundary conditions (PBCs) are normally applied to the simulation box, in effect making the structure infinite in extent. This structure and its energy then serve as the reference state for comparison with any perturbations made to it or indeed entirely different structures or compounds while remaining within the same modeling paradigm. The accuracy of the calculations, of course, is contingent on the accurate description of the potential energy of the system as a function of the nuclear coordinates.

For polar solids, the Born model framework (see Chapter 1), which partitions the total energy into long-range Coulombic interactions and a short-range term to describe the repulsions and van der Waals forces between electron charge clouds, is commonly employed in classical simulations. The shell model is often added as a simple description of polarisability effects, and has proven to be effective in simulating dielectric and lattice dynamical properties. In most cases formal charges are assumed for the ions, as this aids transferability, although partial charge models are also sometimes used. The parameters used in the potential functions can be derived empirically by fitting to observed crystal structures and properties, or by fitting to energy surfaces calculated using *ab initio* methods. Either way, the general validity of this interatomic potential model is assessed primarily by its ability to reproduce observed properties of the crystal lattice, such as unit cell parameters and bond lengths, dielectric and elastic constants. The most widely used code for geometry optimization (and related defect calculations) using effective potentials is GULP [18].

The advantage of a classical model is that it is computationally inexpensive and thus allows large ensembles of atoms, from thousands to millions, to be handled efficiently within a reasonable time frame on today's computers. It also simplifies the treatment of one- and two-dimensional defects (such as point defects and surfaces) because the discrete nature of the atomic particles allows boundary conditions other than three-dimensional periodic to be applied. Individual defects can thus be treated in isolation (i.e., infinite dilution), for example, using the Mott–Littleton method, rather than as groups of two or more (i.e., with their mirror images) as is the case when using a three-dimensional supercell method.

While classical simulations can provide valuable information on crystal properties and atom-scale phenomena important for Li–ion battery materials, it is also crucial to know the electronic structure of the solid. In this case more rigorous (but computationally expensive) techniques based on quantum theory are employed. These techniques are predicated on the Hellmann–Feynman theorem, which states that once the spatial distribution of the electrons has been determined, all the forces in the system can be calculated using classical electrostatics. For inorganic solids, the most widely used theoretical framework today is DFT, which avoids solving the quantum mechanical many-body problem explicitly by describing the ground-state energy of the system as a function only of the electron density. The utility of this method is evidenced by the explosion in the number of studies carried out on all manner of solids, including Li–ion battery materials, over the past 15 or so years.

A widely used implementation of DFT combines a plane-wave basis set with the pseudopotential method, in which an effective interaction, or pseudopotential, replaces the potential term for core electrons. This approximation greatly reduces the computational burden, so that today it is possible to treat systems of a few to several hundreds of atoms in a reasonable time on a high-performance computing cluster. One limitation of the DFT approach, however, is that the exchange–correlation component of the potential energy cannot be calculated exactly, so approximations such as the local density approximation (LDA) or the generalized-gradient approximation (GGA) are used. DFT+U methods are often used for open-shell transition metal compounds, where U is the on-site Coulomb interaction parameter. There is also the increasing use of hybrid exchange functionals, which include a fraction of the exact Hartree–Fock exchange. There are many *ab initio* codes available today (see Table 1.1), with two of the most widely used general-purpose codes being the VASP [19] and CASTEP packages [20].

4.3 Li–Ion Batteries

To provide a general starting point for discussing computational advances, it is apposite to outline the main components of a conventional rechargeable Li–ion battery (illustrated in Figure 4.1). The battery cell comprises a positive electrode (cathode); a nonaqueous liquid electrolyte, typically $LiPF_6$ salt in an organic solvent; and a negative electrode (anode), normally of graphite. The good intercalation behavior of the conventional cathode, layered $LiCoO_2$, was first reported in the early 1980s [21, 22], and this material has been studied intensively since its successful commercialization in 1991.

In the ground state, $LiCoO_2$ adopts the α-$NaFeO_2$ structure (Figure 4.2) with consecutive alternating CoO_2 and Li layers; the Co^{3+} and Li^+ ions are octahedrally coordinated in a cubic close-packed O^{2-} lattice, resulting in a hexagonal structure with $R\bar{3}m$ symmetry. During charging, a voltage applied to the external circuit forces lithium ions to be extracted from the layer-structured $LiCoO_2$ crystal. These diffuse through the electrolyte, and are intercalated between the graphite sheets in the anode material. During discharge, this process is reversed, with electrons passing around the external circuit providing useful power.

The cathode is one of the limiting components in conventional Li–ion batteries because the amount of lithium that can be reversibly extracted and re-inserted into the crystal structure influences the overall capacity, while its Fermi energy affects the voltage of the battery cell. Current efforts to develop next-generation lithium batteries for large-scale applications are focusing on replacing $LiCoO_2$ with intercalation materials that are lower cost and environmentally benign. The

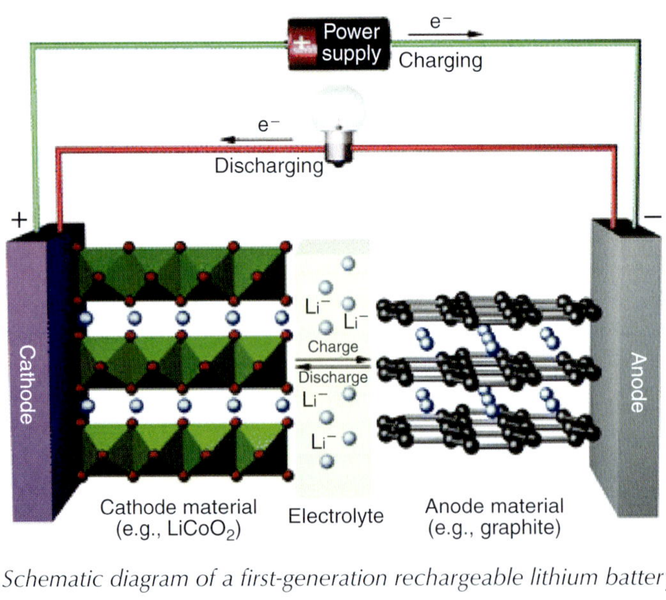

Figure 4.1 *Schematic diagram of a first-generation rechargeable lithium battery. (From [9].)*

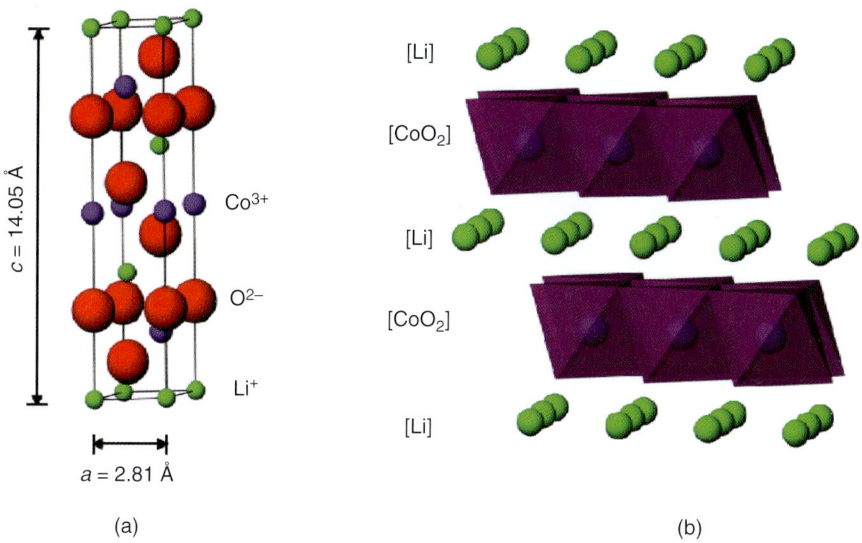

Figure 4.2 Crystal structure of α-$LiCoO_2$ (a) hexagonal unit cell (spacegroup $R\bar{3}m$) and (b) portion of structure showing Li ions intercalated between CoO_6 octahedral layers.

cost and toxicity of Co in particular makes it desirable to replace it with other multivalent elements such as Fe and Mn. For large-scale lithium batteries, thermal and chemical stability and high energy density (and high voltage) are two other important considerations. Two candidate materials systems that could achieve these aims include mixed-metal layered oxides (e.g., $LiNi_{0.5}Mn_{0.5}O_2$) and polyanionic materials (e.g., $LiFePO_4$, Li_2FeSiO_4, and $LiFeSO_4F$). In particular, the olivine-structured orthophosphate $LiFePO_4$ is a highly promising and much-studied cathode material because of its good electrochemical performance (particularly its high operating voltage and large theoretical gravimetric capacity), as well as its low cost and safety advantages [23–30]. The $LiFePO_4$ olivine structure (Figure 4.3) consists of PO_4 tetrahedra with Fe^{2+} ions on corner-sharing octahedral positions and Li^+ ions on edge-sharing octahedral positions, the latter running parallel to the crystallographic *b*-axis.

Although computational methods are being used to study all aspects of battery operation, in the following sections we focus on atomic-level studies of important properties of cathode materials such as cell voltages, phase stability, lithium diffusion, defect chemistry and surface structures to highlight recent trends in this field.

4.4 Cell Voltages and Structural Phase Stability

Early computational work [31,32] demonstrated how the lithium insertion voltage (relative to a lithium metal anode) can be derived for $LiCoO_2$ and other layered

Figure 4.3 *Unit cell of olivine-structured $LiFePO_4$ showing Li ions (green), PO_4 tetrahedra (yellow) and FeO_6 octahedra (orange).*

transition metal oxides from DFT calculations with good accuracy. Within this theoretical framework, lithium intercalation into a cathode host can be represented by the following equation:

$$\mathrm{Li}_{x_1}\ \text{[cathode host]} + (x_2 - x_1)\ \text{Li metal [anode]} \rightarrow \mathrm{Li}_{x_2}\text{[cathode host]}, \quad (4.1)$$

where x_1 and x_2 are the limits of the intercalation reaction. The average equilibrium voltage, $V(x)$, is related to the difference in the Gibbs free energy (ΔG) between the delithiated phase (charged state) and lithiated phase (discharged state) by

$$V(x) = \frac{-\Delta G}{(x_2 - x_1)F}, \quad (4.2)$$

where F is the Faraday constant. ΔG can be approximated by the internal (potential) energy change per intercalated Li^+ ion since the vibrational and configurational entropy contributions to the cell voltage at room temperature are expected to be small. The change in total energy of the system per intercalated lithium atom leads to a predicted cell voltage that is an average value for compositions between x_1 and x_2. This method is now widely used to estimate cell voltages of batteries for a range of different cathode materials.

This approach was first used to investigate the trends of lithium intercalation voltages for a number of $LiMO_2$ (M = Co, Ni, Ti, V, Cu, Al, Mn) layered compounds assuming the α-$NaFeO_2$ structure type for all compositions [31]. These studies showed that moving to the right of the periodic table there is an increase in the electronic charge that is transferred to the anionic band when Li ions are inserted in the MO_2 host. Calculations in which oxygen was replaced with other anion species showed that the anion (O, S or Se) has a very strong influence on the voltage, with oxygen clearly giving the highest voltages. The important role of the anion can be explained in terms of the significant degree of charge transfer to the anion that occurs upon intercalation of lithium. In the oxides, more charge is transferred to the oxide ions than to the metal ions. Early DFT calculations of doped $LiCoO_2$ [32] also identified new candidate materials in which nontransition metals are substituted for Co. For one such material, $Li(Co,Al)O_2$, it was predicted that aluminium substitution would raise the cell voltage while decreasing both the density of the material and its cost, and this was soon verified experimentally [32].

The above results for DFT calculations within the LDA or GGA illustrate the success of these techniques for computing cell voltages and phase stabilities. However, for transition metal-containing compounds they were often found to underestimate redox potentials. This is because an electron transferred from a transition metal ion to a lithium ion experiences significantly less self-interaction in the latter state, leading to underestimation of the energy required. A commonly used method for improving reproduction of these properties is the DFT+U method. As introduced in Chapter 1, on-site Coulomb interactions are modified by adding a Hubbard-type term, U. Using self-consistently derived U values [33], the experimental lithium intercalation voltages and band gaps of a number of transition metal compounds, including layered Li_xMO_2 (M = Co, Ni), olivine Li_xMPO_4 (M = Mn, Fe, Co, Ni) and spinel $Li_xM_2O_4$ (M = Mn, Co), can be reproduced with better accuracy (Figure 4.4). Such a computational approach has been applied to a wide range of

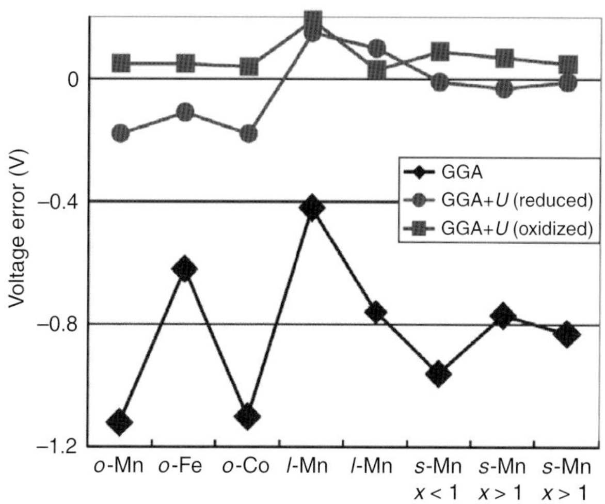

Figure 4.4 *Difference between calculated and experimental voltage for GGA and GGA+U, at the calculated U of the oxidized (delithiated) and reduced (lithiated) states, respectively (o, olivine $LiMPO_4$; l, layered $LiMO_2$; s, spinel LiM_2O_4). (From [33].)*

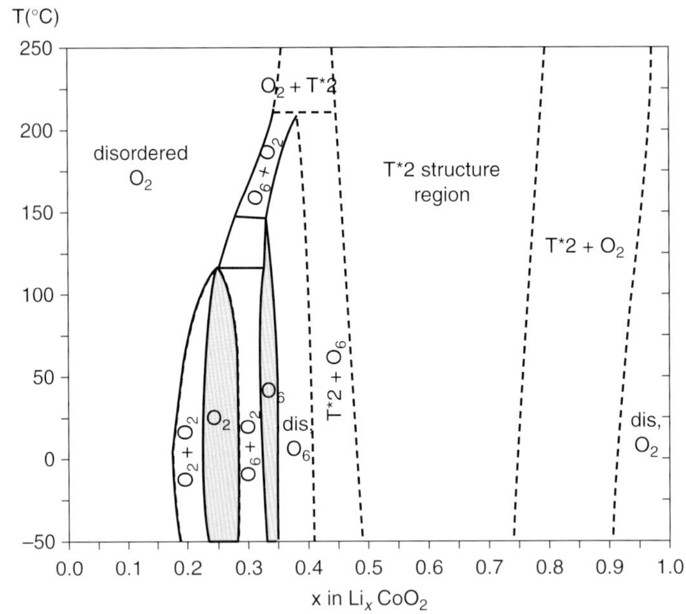

Figure 4.5 *Calculated Li_xCoO_2 (O_2 system) phase diagram. The dashed lines indicate the approximate limits of the $T^{\#}2$ domain, and the grey areas indicate ordered phases. (From [40].)*

other lithium battery electrode materials including Li_2MnO_3 [34], vanadium oxides [35, 36] and iron silicates [37, 38].

DFT has also been used to provide insights into structural changes and stabilities on lithium extraction. Quite early on [39], the phase diagram of Li_xCoO_2 was predicted based on DFT calculations, indicating a tendency for Li ordering at $x = 1/2$, in close agreement with experiment. At low Li concentration, it was found that a staged compound in which Li ions selectively segregate to every other Li plane, leaving the remaining Li planes vacant, was stable; however, no evidence was found for the two-phase region observed at high Li concentration. The same methods have been used to investigate the relative stabilities of different layer stackings in $LiCoO_2$, such as the T(#)2 and O6 polymorphs (Figure 4.5), using a cluster expansion method combined with Monte Carlo simulations [40]. These studies show that two tetrahedral sites for the Li ions need to be considered in the delithiated structure for qualitative agreement with experiment, indicating that this structural phase transformation is induced by enhanced configurational entropy and not by a metal–insulator transition as had been previously proposed.

4.5 Li–Ion Diffusion and Defect Properties

Li–ion diffusion pathways and activation energies that govern Li–ion transport within cathode materials are of considerable interest when considering rates at

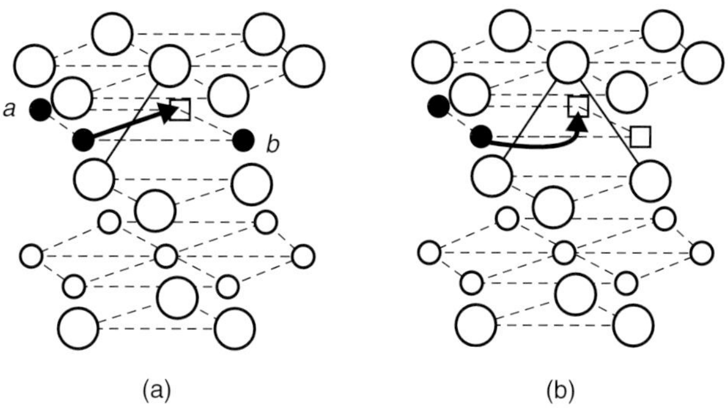

Figure 4.6 *Two lithium migration paths in layered Li_xCoO_2. The filled circles are lithium ions, the empty squares are lithium vacancies, the large empty circles are oxygen ions and the small empty circles are cobalt ions. (a) The oxygen dumbbell hop (ODH) occurs when the sites a and b adjacent to the end points of the hop are simultaneously occupied by lithium ions. (b) The tetrahedral site hop (TSH) occurs when one or both of the sites adjacent to the endpoints of the hop are vacant. (From [42].)*

which a battery can be charged and discharged. While interatomic potentials have a proven record in this area, the majority of contemporary studies are performed within the DFT framework where the nudged elastic band method can be used to derive the maximum energy barrier (transition state) for an ion diffusing between adjacent lithium sites. DFT studies [41, 42] of lithium diffusion in the layered Li_xCoO_2 system have shown that lithium transport is mediated by a divacancy mechanism between $x = 0$ and $x < 1$ but by single isolated vacancies at infinite vacancy dilution. Two different migration paths (Figure 4.6) are associated with each scenario, with the former having a significantly lower migration barrier energy than the latter. This reveals that the activation barrier is very sensitive to the lithium concentration due to the strongly varying c parameter of the host lattice and the change in effective valence of the cobalt ions, so that the Li–ion diffusion coefficient varies by several orders of magnitude with state of charge.

Turning to the alternative mixed-metal layered cathode based on $Li(Ni,Mn)O_2$, DFT calculations demonstrate that its increased stability relative to $LiCoO_2$ is due to the particular valence distribution on the transition metals [43, 44], with Mn in oxidation state +4 independent of the Li content. Examination of delithiation of $Li_xNi_{0.5}Mn_{0.5}O_2$ indicates that the electrochemical behavior is linked to the oxidation of Ni^{2+}, while the Mn valence state remains unchanged. The high insertion voltage for $LiNi_{0.5}Mn_{0.5}O_2$ can be partly attributed to the change in Mn–Ni interaction upon Li cycling [43, 44]. It is known, however, that as-prepared $Li(Ni_{0.5}Mn_{0.5})O_2$ contains 8–12% site exchange between Li and Ni [45]. DFT has been used to explore the effect of Li–Ni site exchange on lithium mobility [46], with Figure 4.7 showing the calculated activation energy as a function of the distance between the oxygen layers on each side of the Li plane; this clearly

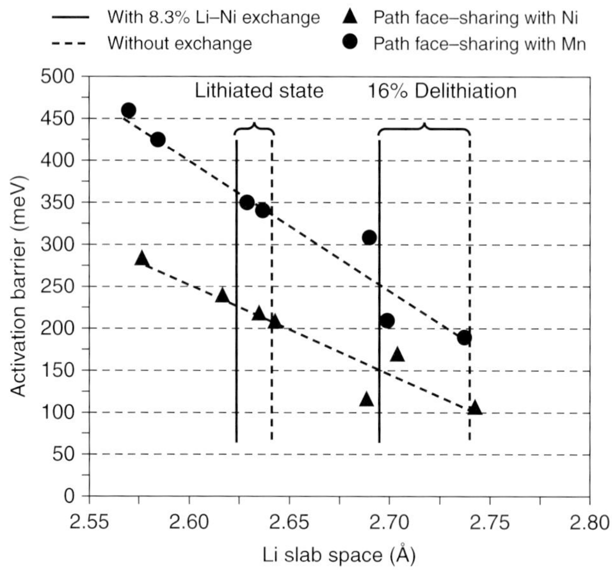

Figure 4.7 *Calculated activation barrier for Li migration in $Li(Ni_{0.5}Mn_{0.5})O_2$ as a function of the Li slab space. Triangles and circles represent the activated state that face-shares with Ni and Mn, respectively. The activation barriers have been calculated for a hypothetically perfect layered system, for a system with 8.3% excess Ni present in the Li layer without a change in the transition metal layer, and for a system with 8.3% Li–Ni exchange. (From [46].)*

indicates that more space between these oxygen layers substantially reduces the activation energy. The Li–Ni disorder has also been suggested as the reason why $Li(Ni_{0.5}Mn_{0.5})O_2$ does not exhibit high rate performance. The calculations indicate that Li–Ni disorder greatly limits the opening of the Li slab space upon delithiation, and hence a key conclusion is that Li diffusivity would be greatly improved by reducing the amount of Li–Ni exchange in $Li(Ni_{0.5}Mn_{0.5})O_2$; this was subsequently achieved by modifying its crystal structure, and resulted in high measured rate capability [46].

As noted, $LiFePO_4$ is a highly promising and much-studied alternative cathode material. Activation barriers to Li–ion motion in Li_xMPO_4 (M = Mn, Fe, Co, Ni) olivines have been calculated using DFT methods [47], indicating that Li diffuses rapidly down one-dimensional channels. In this case, an unusually low activation barrier (~0.1 eV) was first calculated for migration along the b-axis, perhaps because the lack of a U term underestimated the repulsion between charged species, and the intrinsic Li diffusivity was calculated to be much higher than found experimentally. Atomistic defect modeling studies [48, 49] of $LiFePO_4$, on the other hand, predicted an activation barrier in better agreement with experiment (~0.5 eV), and also revealed that lithium ions followed a nonlinear, curved trajectory down the [010] channel (Figure 4.8). High barriers for other pathways suggest that lithium ions cannot readily span the large jump (>4.5 Å) between

Figure 4.8 Schematic diagram of the calculated pathway for Li–ion migration down [010] channels of $LiFePO_4$ viewed perpendicular to the ab plane. (Adapted from [48].)

channels, in agreement with the DFT results; such a one-dimensional transport mechanism is consistent with the strongly anisotropic nature of the orthorhombic olivine structure. This example provides a good illustration of the utility of atomistic modeling, as neutron diffraction measurements of $LiFePO_4$ [50] later confirmed the one-dimensional nature of Li^+ diffusion in this material (Figure 4.9),

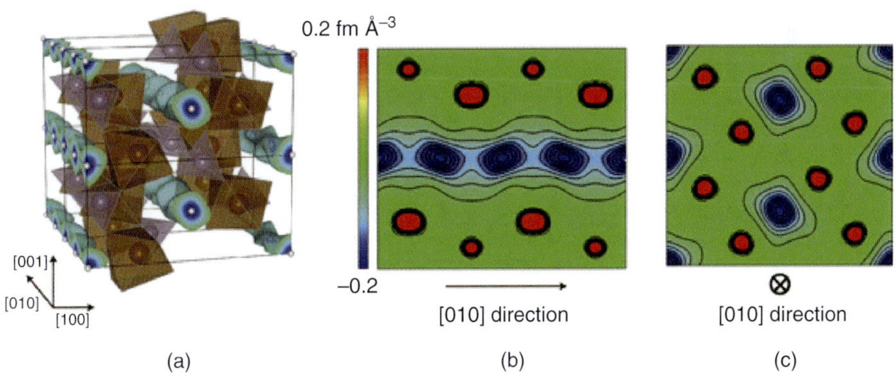

Figure 4.9 Experimental visualization of lithium diffusion in Li_xFePO_4 from combined powder neutron diffraction and the maximum entropy method: (a) Three-dimensional Li nuclear density data shown as blue contours. The brown octahedra represent FeO_6 and the purple tetrahedra represent PO_4 units. (b) Two-dimensional contour map sliced on the (001) plane at $z = 0.5$; lithium delocalizes along the curved one-dimensional chain along the [010] direction. (c) Two-dimensional contour map sliced on the (010) plane at $y = 0$; all atoms remain near their original positions. (From [50].)

with a curved migration pathway between adjacent lithium sites as predicted. More recently, the diffusion constant has been shown to depend on particle size, with diffusion in large $LiFePO_4$ crystals being much slower than in nanoparticles because of the presence of a larger number of channel-blocking defects in the former [51].

To understand the electrochemical behavior of olivine phosphates, knowledge of the underlying defect properties is needed, and this is most readily obtained by potential-based methods. Classical simulation studies [48, 49] suggest that the most favorable intrinsic defect in $LiMPO_4$ (M = Mn, Fe, Co, Ni) systems is the "antisite" or cation-exchange defect, in which a small population (~2%) of Li^+ and Fe^{2+} ions are expected to sit on the "wrong" cation sites; this defect is temperature dependent and hence sensitive to experimental synthesis conditions. Since reporting the simulation results, structural analysis of hydrothermally synthesized $LiFePO_4$ found 3 mol% Fe on lithium sites [29], while a scanning transmission electron microscopy study [52] provided evidence of antisite defects in $LiFePO_4$, quoting a concentration of around 1%.

Charged point defects such as occur in cathode materials can also associate to form localized clusters, which can have significant effects on transport behavior. These defect clusters can be probed using atomistic simulation methods, for example, by calculating binding energies between the component defects. Such binding energy calculations for $LiFePO_4$ suggest clustering of antisite defects as well as impurity-vacancy defects (rather than a random distribution); such phenomena may be important as precursors to local ordering or nanodomain formation [53]. Energetically favorable nanoclusters include neutral one-dimensional clusters of antisite defects ($Fe^{\bullet}_{Li}$) and Li vacancies orientated along the b-axis channel (Figure 4.10), which could result in trapping of migrating Li^+ vacancies. Computer modeling results thus lead us to expect that defect clustering will inhibit Li extraction.

In addition to intrinsic defects, potential-based methods can be used to investigate cation doping by generating quantitative estimates of the energies of different dopant substitution reactions. This can provide a useful systematic guide to the site selectivity of different dopant species and to trends in dopant solubility. Reports that low-level (<3%) doping of $LiFePO_4$ with cations (such as Al^{3+}, Zr^{4+} and Nb^{5+}) produces orders-of-magnitude improvements in electronic conductivity [25] have stimulated considerable debate about the precise substitution mechanisms, nature of the conductive species, and, crucially, whether the dopant goes into the crystal lattice, or if the enhanced conductivity is due to secondary effects such as carbon contamination and/or phosphide formation. A wide range of dopants in $LiFePO_4$ from monovalent to pentavalent have been simulated [48] constituting a wider survey than experimental reports. Low favorable energies were found only for Na^+ substitution on the Li^+ site, and divalent dopants (e.g., Zn, Cu, and Mg) on the transition metal site. In contrast, supervalent doping (especially Ti^{4+} and Nb^{5+}) appears unfavorable on both Li^+ and Fe^{2+} sites; moreover, the

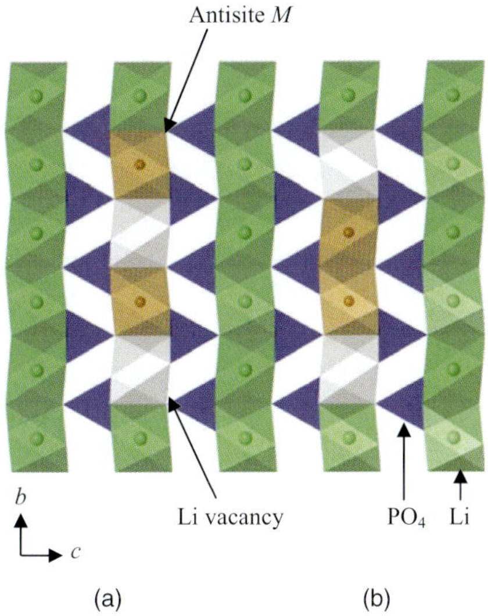

Figure 4.10 *Schematic diagram of a structural plane of $LiFePO_4$ containing two neutral defect cluster arrangements within b-axis lithium channels composed of two antisite defects (Fe on Li sites) and two Li vacancies: (a) antisite cation and Li vacancy on alternating sites and (b) antisite cations on adjacent sites. (From [53].)*

charge-compensation mechanism found for such doping does not alter the Fe^{2+} valence state and hence is unlikely to contribute to high electronic conductivity. These results are in accord with experimental reports of unsuccessful incorporation of significant levels ($>3\%$) of aliovalent dopants to enhance electronic conductivity.

4.6 Surfaces and Morphology

It is clear that morphological control of bulk crystalline or nanomaterials has an impact on their performance, as many extensive properties are highly shape and size dependent [2, 4]. For example, the formation of $LiFePO_4$ particles of submicron or nanometer size is thought to enhance electrochemical performance by reducing transport path lengths, as well as being important for the electrode–electrolyte interface [54–56]. Knowledge of these features on the atomic level thus provides valuable information for understanding electrochemical mechanisms. The variation in shape and sizes of particles as a result of different synthesis techniques and conditions, however, makes it difficult to consistently produce crystals with identifiable surface planes for which such fundamental details can be extracted by experiment alone.

As with bulk simulations, advanced atomistic methods can be used to examine surface structures and from their relative energies construct ideal crystal morphologies. These techniques have recently been applied to lithium battery materials such as $LiMn_2O_4$ [57] and $LiFePO_4$ [58,59]. Focusing on the $LiFePO_4$ system, both first-principles and interatomic potential methods have recently been used to identify the same stable (low energy) surface structures by examining systematically different terminations of low index planes [58, 59]. The advantage of interatomic potential methods is demonstrated by the large number of different surface planes and terminations that can be examined individually, quickly and efficiently (when two-dimensional PBCs are used). First-principles methods, in contrast, can more accurately model the subtle changes in electron densities in the surface regions, as well as allow for deviations in stoichiometry and charge state under different conditions (i.e., chemical potentials). Despite the difference in methodologies, there is often good overall agreement between classical and quantum mechanical simulations.

Simulation studies have shown that surface structures of $LiFePO_4$ exhibit a complex, uneven topology on account of the different sizes of the three constituent moieties, Li^+, Fe^{2+} and $PO_4{}^{3-}$. The majority of the surfaces undergo considerable relaxation, which confirms that the surface chemistry and electrochemical activity cannot be reliably predicted by assuming rigid, unrelaxed terminations of the bulk lattice. Low energy (010) and (201) surfaces identified by both potential-based and DFT calculations dominate the equilibrium morphology based on a Wulff construction (Figure 4.11). In addition to the surface energies, potential-based calculations readily provide the relative attachment energies (the energy released on attaching a new slice of crystal to the surface), from which so-called growth morphologies can be constructed. In the case of $LiFePO_4$, such a nonequilibrium crystal is capped by (010) faces (Figure 4.12); this morphology is consistent with some samples of pure $LiFePO_4$ [55], as evidenced by SEM analysis, although different synthesis routes have produced a variety of crystallite morphologies such as hexagonal platelets and block-type shapes [54, 56].

In olivine-structured $LiFePO_4$ crystals, the exposure of the (010) surface is significant since it is normal to the most facile pathway for Li–ion conduction (the [010] channel), shown in Figure 4.13, and hence important for the (de)intercalation rate

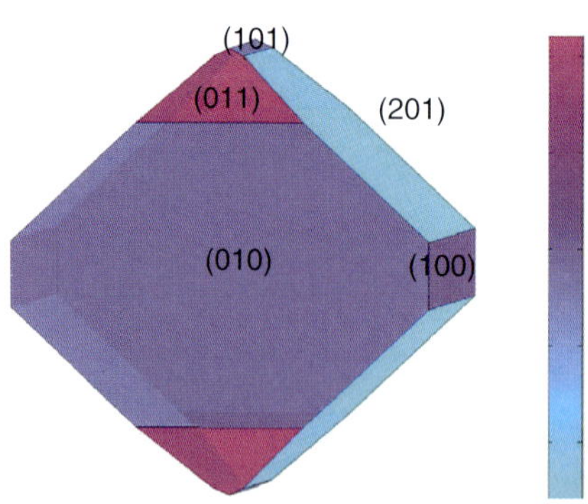

Figure 4.11 *Equilibrium (Wulff) shape of $LiFePO_4$ using surface energies calculated with DFT. (From [59].)*

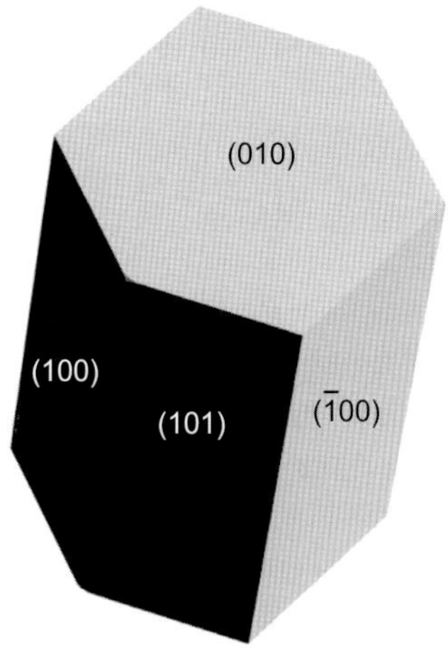

Figure 4.12 *Calculated growth morphology of $LiFePO_4$ using attachment energies. (From [58].)*

of lithium ions. GGA+U calculations [59] show that the surface redox potential for the extraction and insertion of Li from the (010) surface of $LiFePO_4$ is significantly lower (~0.6 V) than the bulk value. A reduction in diffusion path length of lithium ions by increasing the proportion of (010) surface expressed is expected to enhance the electrochemical performance of a cathode prepared from this material. Future

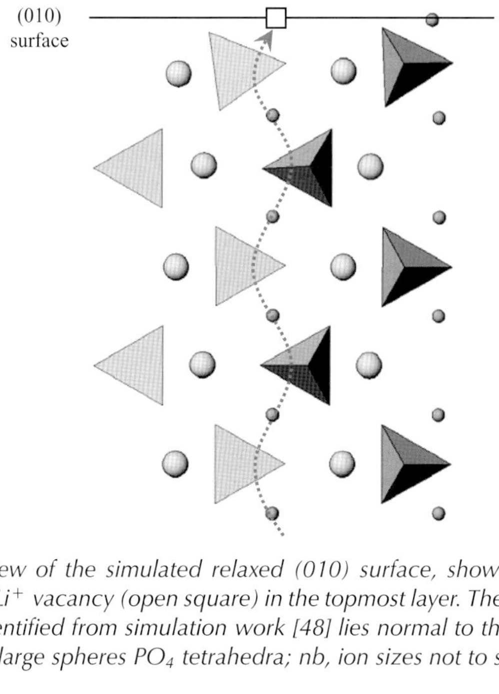

Figure 4.13 *Side view of the simulated relaxed (010) surface, showing tilting of the PO_4 tetrahedron near the Li^+ vacancy (open square) in the topmost layer. The curved Li^+ migration path (dotted line) identified from simulation work [48] lies normal to the surface plane. (Li^+, small spheres; Fe^{2+}, large spheres PO_4 tetrahedra; nb, ion sizes not to scale). (From [58].)*

computational studies will no doubt play a role in revealing how solvent molecules interact with crystal surfaces under different thermodynamic conditions to produce different final morphologies.

4.7 Current Trends and Future Directions

At present, most research into Li–ion battery materials using atomistic simulation is concerned with bulk properties, and this is set to continue as they are used to study an ever-increasing range of materials systems of greater compositional and structural complexity. In cathode materials, the cell voltage and stable phase vary with Li content (as well as dopant distribution), but lithium nonstoichiometry introduces disorder to the system, which is difficult to handle by straightforward (i.e., geometry optimization) DFT techniques. One way of addressing this problem is to combine cluster expansion methods with a statistical-mechanical treatment such as Monte Carlo calculations [40, 60]. In this case, energies derived from DFT are used to parameterize an effective Hamiltonian describing the energy of the crystal as a function of the configurational degrees of freedom arising from all possible ways of arranging Li ions and vacancies within it.

In parallel with these developments, as hinted at by recent studies on crystalline surfaces of some cathode materials, there is increased interest in nano- and micro-structural features of battery materials. Indeed, the influence of larger scale or extended crystalline defects such as grain boundaries and heterogeneous interfaces (whether solid–solid or solid–liquid) on battery performance in real systems can be better understood through systematic examination of such features. Recent studies on twin boundaries in $LiCoO_2$ [61] and growth of solid–electrolyte interphase films at a solvent–graphitic anode interface [62] are early indicators of this trend. Further recent examples of how atomistic simulation can complement experimental techniques include the use of DFT calculations to aid interpretation of scanning electron transmission microscopy images of surface regions of $LiMn_2O_4$ nanoparticles [63], and calculation of XANES and ELNES fingerprints for $LiMO_2$ ($M =$ Mn, Fe, Co and Ni) crystals using the *ab initio* multiplet method [64].

Another consequence of faster computer processor speeds and wider availability of cheap, massively parallel computing architectures will be the increasing use of first-principles molecular dynamics and hybrid DFT calculations. At present the small number of atoms and short time spans (a few picoseconds) that can be treated using first-principles MD, such as the Carr–Parrinello method, makes it difficult to extract statistically meaningful transport data. As a consequence, only a few MD studies have been carried out on Li–ion battery materials so far, and these almost exclusively on solid electrolyte materials, such as Nasicon-type $LiZr_2(PO_4)_3$ [65], lithium lanthanum titanate [66] and doped garnet-type $Li_7La_3Zr_2O_{12}$ [67], using empirical potential models because these systems are exclusively ionic conductors with no electronic conductivity or variation in charge states. Use of large-scale MD, currently limited to classical models, will be particularly useful for studying electrode nanostructures and heavily disordered systems, such as cathodes in different states of charge (Li content).

There is also increasing interest in the use of hybrid exchange functionals to treat the exchange–correlation interactions, as the accurate calculation of band gaps, which are often underestimated with pure LDA/GGA, is needed to predict cell voltages reliably. Unfortunately, hybrid functional methods require an enormous amount of computational resources when compared with standard DFT, limiting their use to simple systems with small numbers of atoms. Consequently, application to Li–ion battery cathode materials has so far been limited, but as ever-faster computing platforms continue to be developed, there is little doubt that these methods will become increasingly important in the study of complex phenomena at the atomic/electronic level. Developments in the area of computational methods are also likely to have a large impact on the search for new electrode materials in the years ahead. Already, a high-throughput *ab initio* computational approach [68] is being used to screen large numbers of candidate compounds for use as cathodes.

4.8 Concluding Remarks

This chapter has aimed to demonstrate the valuable role that advanced computational techniques now play in contemporary studies of materials for rechargeable lithium batteries. The principal aims of computer modeling can be summarized as (i) to complement and assist in the interpretation of experimental studies (e.g., phase changes, cell voltages and rate capabilities); (ii) to elucidate key features that may be difficult to extract from experiment alone (e.g., Li–ion diffusion pathways and defect chemistry); and (iii) to play a predictive role in the design of promising new electrode materials.

It is clear that for next-generation portable electronics and electric vehicles, new lithium battery materials exhibiting improved safety, higher energy density, lower cost and longer lifecycle are needed. Future computational work is likely to address these challenges, and will encompass studies of the properties of nanostructured electrodes and nanoionics, the structures and transport behavior at interfaces, and the use of high-throughput screening approaches to explore new compounds, all assisted by the continuing growth in computer power. Whichever future direction is taken, it is clear that major advances in lithium battery and energy storage technologies depend on exploring new materials and concepts, and on a greater understanding of the fundamental science of solid-state materials.

References

1. Armand, M. and Tarascon, J.M. (2008) Building better batteries, *Nature* **451**, 652.
2. Ellis, B.L., Lee, K.T. and Nazar, L.F. (2010) Positive electrode materials for Li-ion and Li-batteries, *Chem. Mater.* **22**, 691.

3. Goodenough, J.B. (2010) Challenges for rechargeable Li batteries, *Chem. Mater.* **22**, 587.
4. Bruce, P.G., Scrosati, B. and Tarascon, J.M. (2008) Nanomaterials for rechargeable lithium batteries, *Angew. Chemie* **47**, 2930.
5. Whittingham, M.S. (2004) Lithium batteries and cathode materials, *Chem. Rev.* **104**, 4271.
6. Park, M., Zhang, X., Chung, M. *et al.* (2010) A review of conduction phenomena in Li-ion batteries, *J. Power Sources* **195**, 7904.
7. Scrosati, B. and Garche, J. (2010) Lithium batteries: Status, prospects and future, *J. Power Sources* **195**, 2419.
8. Fergus, J.W. (2010) Recent developments in cathode materials for lithium ion batteries, *J. Power Sources* **195**, 939.
9. Liu, R., Duay, J. and Lee, S.B. (2011) Heterogeneous nanostructured electrode materials for electrochemical energy storage, *Chem. Commun.* **47**, 1384.
10. Palacin, M.R. (2009) Recent advances in rechargeable battery materials: a chemist's perspective, *Chem. Soc. Rev.* **38**, 2565.
11. Etacheri, V., Marom, R., Elazari, R. *et al.* (2011) Challenges in the development of advanced Li-ion batteries: a review, *Energy Environ. Sci.* **4**, 3243.
12. Bruce, P.G., Freunberger, S.A., Hardwick, L.J. and Tarascon, J.-M. (2012) Li–O_2 and Li–S batteries with high energy storage, *Nature Mater.* **11**, 19.
13. Catlow, C.R.A. (ed.) (1997) *Computer Modelling in Inorganic Crystallography*, Academic Press, San Diego, CA.
14. Harding, J.H. (1990) Computer simulation of defects in ionic solids, *Rep. Prog. Phys.* **53**, 1403.
15. Koch, W. and Holthausen, M.C. (2001) *A Chemist's Guide to Density Functional Theory*, Wiley-VCH Verlag GmbH, Weinheim.
16. Meng, Y.S. and Arroyo-de Dompablo, M.E. (2009) First principles computational materials design for energy storage materials in lithium ion batteries, *Energy Environ. Sci.* **2**, 589.
17. Ceder, G. (2011) Opportunities and challenges for first-principles materials design and applications to Li battery materials, *MRS Bull.* **35**, 693.
18. Gale, J.D. and Rohl, A.L. (2003) The General Utility Lattice Program (GULP), *Mol. Simul.* **29**, 291.
19. Kresse, G. and Furthmüller, J. (1996) Efficient iterative schemes for *ab initio* total-energy calculations using a plane-wave basis set, *Phys. Rev. B* **54**, 11169.
20. Payne, M.C., Teter, M.P., Allan, D.C. *et al.* (1992) Iterative minimization techniques for *ab initio* total-energy calculations: molecular dynamics and conjugate gradients, *Rev. Mod. Phys.* **64**, 1045.
21. Mizushima, K., Jones, P.C., Wiseman, P.J. and Goodenough, J.B. (1980) Li_xCoO_2 ($0<x\leq1$): A new cathode material for batteries of high energy density, *Mater. Res. Bull.* **15**, 783.
22. Thomas, M.G.S.R., Bruce, P.G. and Goodenough, J.B. (1985) Lithium mobility in the layered oxide $Li_{1-x}CoO_2$, *Solid State Ionics* **17**, 13.
23. Padhi, A.K., Nanjundaswamy, K.S. and Goodenough, J.B. (1997) Phospho-olivines as positive-electrode materials for rechargeable lithium batteries, *J. Electrochem. Soc.* **144**, 1188.

24. Ravet, N., Chouinard, Y., Magnan, J.F. *et al.* (2001) Phospho-olivines as positive-electrode materials for rechargeable lithium batteries, *J. Power Sources* **97–98**, 503.
25. Chung, S.Y., Bloking, J.T. and Chiang, Y.M. (2002) Electronically conductive phospho-olivines as lithium storage electrodes, *Nature Mater.* **1**, 123.
26. Subramanya Herle, P., Ellis, B., Coombs, N. and Nazar, L.F. (2004) Nano-network electronic conduction in iron and nickel olivine phosphates, *Nature Mater.* **3**, 147.
27. Delacourt, C., Laffont, L., Bouchet, R. *et al.* (2005) Toward understanding of electrical limitations (electronic, ionic) in $LiMPO_4$ (M = Fe, Mn) electrode materials, *J. Electrochem. Soc.* **152**, A913.
28. Amin, R., Balaya, P. and Maier, J. (2007) Anisotropy of electronic and ionic transport in $LiFePO_4$ single crystals, *Electrochem. Solid-State Lett.* **10**, A13.
29. Chen, J., Vacchio, M.J., Wang, S. *et al.* (2008) The hydrothermal synthesis and characterization of olivines and related compounds for electrochemical applications, *Solid State Ionics* **178**, 1676.
30. Muraliganth, T., Murugan, A.V. and Manthiram, A. (2008) Nanoscale networking of $LiFePO_4$ nanorods synthesized by a microwavesolvothermal route with carbon nanotubes for lithium ion batteries, *J. Mater. Chem.* **18**, 5661.
31. Aydinol, M.K., Kohan, A.F., Ceder, G. *et al.* (1997) *Ab initio* study of lithium intercalation in metal oxides and metal dichalcogenides, *Phys. Rev. B* **56**, 1354.
32. Ceder, G., Chiang, Y.-M., Sadoway, D.R. *et al.* (1998) Identification of cathode materials for lithium batteries guided by first-principles calculations, *Nature* **392**, 694.
33. Zhou, F., Cococcioni, M., Marianetti, C.A. *et al.* (2004) First-principles prediction of redox potentials in transition-metal compounds with LDA + U, *Phys. Rev. B* **70**, 235121.
34. Koyama, Y., Tanaka, I., Nagao, M. and Kanno, R. (2009) First-principles study on lithium removal from Li_2MnO_3, *J. Power Sources* **189**, 798.
35. Braithwaite, J.S., Catlow, C.R.A., Gale, J.D. and Harding, J.H. (1999) Lithium intercalation into vanadium pentoxide: a theoretical study, *Chem. Mater.* **11**, 1990.
36. Armstrong, A.R., Lyness, C., Panchmatia, P.M. *et al.* (2011) The lithium intercalation process in the low-voltage lithium battery anode $Li_{1+x}V_{1-x}O_2$, *Nature Mater.* **10**, 223.
37. Arroyo-de Dompablo, M.E., Dominko, R., Gallardo-Amores, J.M. *et al.* (2008) On the energetic stability and electrochemistry of Li_2MnSiO_4 polymorphs, *Chem. Mater.* **20**, 5574.
38. Eames, C., Armstrong, A.R., Bruce, P.G. and Islam, M.S. (2012) Insights into changes in voltage and structure of Li_2FeSiO_4 polymorphs for lithium-ion batteries, *Chem. Mater.* **24**, 2155.
39. Van der Ven, A., Aydinol, M.K., Ceder, G. *et al.* (1998) First-principles investigation of phase stability in Li_xCoO_2, *Phys. Rev. B* **58**, 2975.
40. Carlier, D., Van der Ven, A., Delmas, C. and Ceder, G. (2003) First-principles investigation of phase stability in the O2-$LiCoO_2$ system, *Chem. Mater.* **15**, 2651.
41. Van der Ven, A. and Ceder, G. (2000) Lithium diffusion in layered Li_xCoO_2, *Electrochem. Solid-State Lett.* **3**, 301.
42. Van der Ven, A. and Ceder, G. (2001) Lithium diffusion mechanisms in layered intercalation compounds, *J. Power Sources* **97–98**, 529.

43. Reed, J. and Ceder, G. (2002) Charge, potential, and phase stability of layered $LiNi_{0.5}Mn_{0.5}O_2$, *Electrochem. Solid-State Lett.* **5**, A145.
44. Islam, M.S., Davies, R.A. and Gale, J.D. (2003) Structural and electronic properties of the layered $LiNi_{0.5}Mn_{0.5}O_2$ lithium battery material, *Chem. Mater.* **15**, 4280.
45. Choi, S.H., Shlyakhtin, O.A., Kim, J. and Yoon, Y.S. (2005) Structural and electrochemical properties of $Li_{1+x}Ni_{0.5}Mn_{0.5}O_{2+\delta}$ ($0 \leq x \leq 0.7$) cathode materials for lithium-ion batteries, *J. Power Sources* **140**, 355.
46. Kang, K., Meng, Y.S., Bréger, J. *et al.* (2006) Electrodes with high power and high capacity for rechargeable lithium batteries, *Science* **311**, 977.
47. Morgan, D., Van der Ven, A. and Ceder, G. (2004) Li Conductivity in Li_xMPO_4 (M = Mn, Fe, Co, Ni) olivine materials, *Electrochem. Solid State Lett.* **7**, A30.
48. Islam, M.S., Driscoll, D.J., Fisher, C.A.J. and Slater, P.R. (2005) Atomic-scale investigation of defects, dopants, and lithium transport in the $LiFePO_4$ olivine-type battery material, *Chem. Mater.* **17**, 5085.
49. Fisher, C.A.J., Hart Prieto, V.M. and Islam, M.S. (2008) Lithium battery materials $LiMPO_4$ (M = Mn, Fe, Co, and Ni): insights into defect association, transport mechanisms, and doping behavior, *Chem. Mater.* **20**, 5907.
50. Nishimura, S., Kobayashi, Y., Ohayama, K. *et al.* (2008) Experimental visualization of lithium diffusion in Li_xFePO_4, *Nature Mater.* **7**, 707.
51. Malik, R., Burch, D., Bazant, M. and Ceder, G. (2010) Particle size dependence of the ionic diffusivity, *Nano Lett.* **10**, 4123.
52. Chung, S.Y., Choi, S.Y., Yamamoto, T. and Ikuhara, Y. (2008) Atomic-scale visualization of antisite defects in $LiFePO_4$, *Phys. Rev. Lett.* **100**, 125502.
53. Gardiner, G. and Islam, M.S. (2010) Anti-site defects and ion migration in the $LiFe_{0.5}Mn_{0.5}PO_4$ mixed-metal cathode material, *Chem. Mater.* **22**, 1242.
54. Ellis, B., Wang, H.K., Makahnouk, W.R.M. and Nazar, L.F. (2007) Synthesis of nanocrystals and morphology control of hydrothermally prepared $LiFePO_4$, *J. Mater. Chem.* **17**, 3248.
55. Franger, S., Le Cras, F., Bourbon, C. *et al.* (2005) Overview on the practical electrochemical behaviour of optimized lithium iron phosphate, *Recent Res. Devel. Electrochem.* **8**, 225.
56. Chen, G., Song, X. and Richardson, T.J. (2006) Electron microscopy study of the $LiFePO_4$ to $FePO_4$ phase transition, *Electrochem. Solid State Lett.* **9**, A295.
57. Benedek, R. and Thackeray, M.M. (2011) Simulation of the surface structure of lithium manganese oxide spinel, *Phys. Rev. B* **83**, 195439.
58. Fisher, C.A.J. and Islam, M.S. (2008) Surface structures and crystal morphologies of $LiFePO_4$: relevance to electrochemical behaviour, *J. Mater. Chem.* **18**, 1209.
59. Wang, L., Zhou, F., Meng, Y.S. and Ceder, G. (2007) First-principles study of surface properties of $LiFePO_4$: surface energy, structure, Wulff shape, and surface redox potential, *Phys. Rev. B* **76**, 165435.
60. Arroyo y Dompablo, M.E., Van der Ven, A. and Ceder, G. (2002) First-principles calculations of lithium ordering and phase stability on Li_xNiO_2, *Phys. Rev. B* **66**, 064112.
61. Fisher, C.A.J., Huang, R., Hitosugi, T. *et al.* (2012) A high coincidence grain boundary in lithium battery material $LiCoO_2$, *Nanosci. Nanotechnol. Lett.* **4**, 165.

62. Leung, K. and Budzien, J.L. (2010) *Ab initio* molecular dynamics simulations of the initial stages of solid–electrolyte interphase formation on lithium ion battery graphitic anodes, *Phys. Chem. Chem. Phys.* **12**, 6583.
63. Huang, R., Ikuhara, Y.H., Mizoguchi, T. *et al.* (2011) Oxygen-vacancy ordering at surfaces of lithium manganese(III,IV) oxide spinel nanoparticles, *Angew. Chem. Int. Ed.* **50**, 3053.
64. Ikeno, H., Mizoguchi, T., Koyama, Y. *et al.* (2011) Theoretical fingerprints of transition metal $L_{2,3}$ XANES and ELNES for lithium transition metal oxides by *ab initio* multiplet calculations, *J. Phys. Chem. C* **115**, 11871.
65. Kumar, P. and Yashonath, S. (2001) Lithium ion motion in $LiZr_2(PO_4)_3$, *J. Phys. Chem. B* **105**, 6785.
66. Katsumata, T., Inaguma, Y., Itoh, M. and Kawamura, K. (2002) Influence of covalent character on high Li ion conductivity in a perovskite-type Li ion conductor: prediction from a molecular dynamics simulation of $La_{0.6}Li_{0.2}TiO_3$, *Chem. Mater.* **14**, 3930.
67. Adams, S. and Prasada Rao, R. (2012) Ion transport and phase transition in $Li_{7-x}La_3(Zr_{2-x}M_x)O_{12}$ ($M = Ta^{5+}, Nb^{5+}, x = 0, 0.25$), *J. Mater. Chem.* **22**, 1426.
68. Hautier, G., Jain, A., Ong, S.P. *et al.* (2011) Phosphates as lithium-ion battery cathodes: an evaluation based on high-throughput *ab initio* calculations, *Chem. Mater.* **23**, 3495.

5

Energy Storage: Hydrogen

Viet-Duc Le and Yong-Hyun Kim
Graduate School of Nanoscience and Technology, KAIST, Daejeon, South Korea

5.1 Introduction

Various environmental issues associated with the high level of carbon dioxide emission have initiated the search for alternative energy sources. Hydrogen has great potential to replace fossil fuel since the combustion product of hydrogen is water, a harmless substance in the environment [1]. Recent advances in fuel cell technology allow the vision of a green and sustainable economy to come close to realization. However, how to store hydrogen efficiently emerged as a formidable barrier still to be overcome (Figure 5.1). As the lightest element in the periodic table, hydrogen provides excellent energy density per unit weight, but rather poor energy density per unit volume. This raised fundamental issues in constructing an optimal storage scheme aiming at high weight and volume densities for onboard application. At least 4 kg of hydrogen is required for 300 miles of driving with the current proton exchange membrane fuel cell. In 2003, the Department of Energy (DOE) launched the "Grand Challenge" to stimulate research and development in hydrogen storage technologies with onboard storage as the focal point. The ultimate target is a 2.5 kWh/kg system (7.5 wt% H_2) and 2.3 kWh/l (0.070 kg H_2/l) by 2017 [2]. If successful, the potential impact on the environment and the economy will be tremendous since a large portion of greenhouse gas emission and fossil fuel consumption originates from transportation.

It is straightforward to store hydrogen physically inside compressed gas cylinders and liquid hydrogen tanks. Great industrial efforts have been made to develop

Computational Approaches to Energy Materials, First Edition. Edited by Aron Walsh, Alexey A. Sokol and C. Richard A. Catlow.

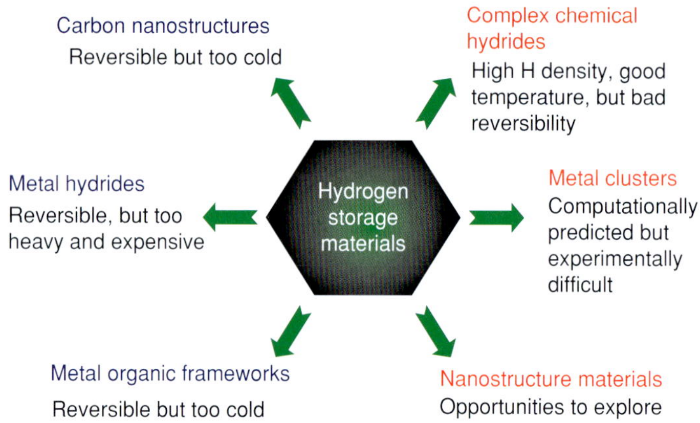

Figure 5.1 *Summary of various hydrogen storage techniques and their limitations. (Reproduced with permission from [9].)*

efficient compression and liquefaction technologies together with lightweight containers. What is not obvious is the energy penalty needed to carry out compression and liquefaction [3]. Since hydrogen is nonpolar, a typical pressure of several hundred bars is usually imposed to overcome closed-shell repulsion between molecules during the condensation stage. Taking into account the intrinsic weight of gas cylinders, the storage capacity is only within 1 wt%, but about 18% of the specific energy content is already lost for compression. Although liquid hydrogen can provide a higher volume density than compressed gas, the use of cryogenic temperatures also removes 40% of the specific energy content. In addition, hydrogen vaporization due to an unfavorable surface-to-volume ratio places great constraints on automobile design and operation. Even in the most optimistic scenario, both approaches are still inferior to petroleum on a mass and volume basis. Perhaps compressed and liquid hydrogen is more suitable for large-scale and stationary storage in which weight and volume constraints are lower.

For practical onboard application, we need to introduce suitable storage media that interact with hydrogen molecules, therefore, indirectly condensing them without high pressure and cryogenic temperatures. The nature of such interaction is extremely crucial for charge and discharge processes. If the interaction is too strong, desorption will only take place at high temperatures. In contrast, if the interaction is too weak, we again need to apply cryogenic temperatures for hydrogen uptake. Since the intrinsic weight of storage media is also included in the theoretical capacity, light elements are preferable to achieve a balance between volumetric and gravimetric densities. Other desired properties include good thermal conductivity for heat management, easy synthesis route from cheap and abundant materials, safety upon exposure to air, and recyclability. As we can see, the development of hydrogen storage materials naturally becomes an exciting research area for chemists and materials scientists. With well-established synthetic techniques in conjugation with greater understanding and control over materials properties at the atomistic level, the final solution will surely come in due time.

5.2 Computational Approach in Hydrogen Storage Research

As previously mentioned, the interaction strength between hydrogen and storage media is the most important criterion for selecting a practical hydrogen storage material. Particularly, the material should be compatible with fuel cell working conditions to deliver the best performance. This requires understanding of the material at a microscopic level so that its properties can be tuned accordingly. The physical and chemical properties of hydrogen storage materials are rather rich and diverse with many aspects still unexplored. The trial-and-error approach is deemed inappropriate and inefficient within limited time and resources. With the aid of first-principles-based simulations, candidate materials can be validated in the *virtual* laboratory before actual syntheses. This is the so-called high-throughput screening process. Due to the ability to treat electronic effects explicitly, novel interacting mechanisms as well as subtle structural effects can be easily elucidated within a simulation framework, which remain hidden inside complicated experimental phenomena. Thanks to the staggering development of theories, algorithms, and computing resources in the past decades, *ab initio* simulation has become much more popular than in the past. Collaboration between theorists and experimentalists has been shown to be fruitful on various occasions. Valuable information and bold predictions regarding hydrogen storage materials have provided a driving force for the advancement of the field.

The purpose of this chapter is not to give an exhaustive review of hydrogen storage research. Excellent reviews already exist elsewhere, and readers are encouraged to go through them for more detailed discussions according to their research interests [4–10]. Section 5.3 briefly outlines current trends and opportunities in chemisorption. Section 5.4 presents hydrogen physisorption in porous media such as a metal–organic framework (MOF), followed by current attempts to store hydrogen via spillover effects (Section 5.5). This chapter mainly focuses on the recent progress on hydrogen storage utilizing Kubas-like mechanisms in metal-decorated nanostructures and related theoretical issues.

5.3 Chemisorption Approach

The strategy of chemisorption is characterized by dissociating and storing hydrogen in the atomic form. Historically, an interstitial hydride was first observed in palladium. Hydrogen molecules first dissociate on the metallic surface, then diffuse, and finally occupy the interstitial lattice sites. The moderate binding strength between hydrogen and lattice sites offers a great advantage in terms of kinetics, thermodynamics, and reversibility in ambient conditions. Other notable interstitial hydrides including $LaNi_5$, Mg_2Ni, ZrNi, and FeTi were later developed in the quest for ways to lower both cost and total mass of the storage medium [11]. Nevertheless, the capacity remains pretty low, typically 1–2 wt%, with less room for further improvement. The paradigm shift took place in the early 1990s when scientists directed their attention toward compounds made of hydrogen and light elements in the periodic table. This combination promises the gravimetric and volumetric densities that can exceed DOE's target. Since hydrogen is stored as the reduced form in complex hydrides, the critical issues to be tackled here are high operational temperatures and slow reaction kinetics. To break the strong covalent bond and release hydrogen, an operating temperature as high as 200 °C is usually required. Moreover, this process is stepwise, accompanied by phase transition of the host material at each elementary step. First-principles simulation plays a pivotal role in providing experimentalists with fundamental understanding of the physical and chemical properties of complex hydrides.

For example, alanate (AlH_4^-) has stirred up great interest in the scientific community since the pioneer work on destabilizing sodium alanate ($NaAlH_4$) by the Ti dopant [12]. The reaction proceeds via two steps:

1. $NaAlH_4 \rightleftarrows \frac{1}{3}Na_3AlH_6 + \frac{2}{3}Al + H_2$
2. $Na_3AlH_6 \rightleftarrows 3NaH + Al + \frac{3}{2}H_2$

From the point of view of condensed matter, hydrogen release in alanate can be treated as the formation of hydrogen-charged defects and vacancies. Long-range diffusion of aluminum and hydrogen defects is also important during hydrogen release and uptake. Therefore, new insight can be obtained by extending the well-developed theoretical techniques in semiconductor physics to complex hydrides (Figure 5.2). It has been established through *ab initio* simulation that the Ti dopant prefers to occupy the Al site and weakens the strength of the Al–H bond [13, 14]. As a result, the formation energy of hydrogen-charged defects is lowered and the self-diffusion process is improved substantially [13]. Through *ab initio* molecular dynamics (MD), possible routes for hydrogen release and the dynamic origin of Ti dopants were revealed [15]. The good agreement between predicted and measured activation barriers of the rate-determining step for $NaAlH_4$ demonstrated the ability of first-principles simulation to describe phenomena that were not traceable in experiments. The microscopic level of understanding obtained here could pave the way for a more systematic design of the catalyst for alanate complexes.

The case above is not the only example of fruitful application of *ab initio* simulation. The definite advantage of a computational approach is the capability to evaluate thermodynamics of a reaction with accuracy and efficiency on the basis of total energy formalism. There are many possible ways to combine hydrogen with other light elements to make hydrides. Unfortunately, the chemistry of hydrides can be drastically changed by replacing just one component. For instance, $LiAlH_4$ and $NaAlH_4$ undergo different elementary steps with different reaction intermediates. In

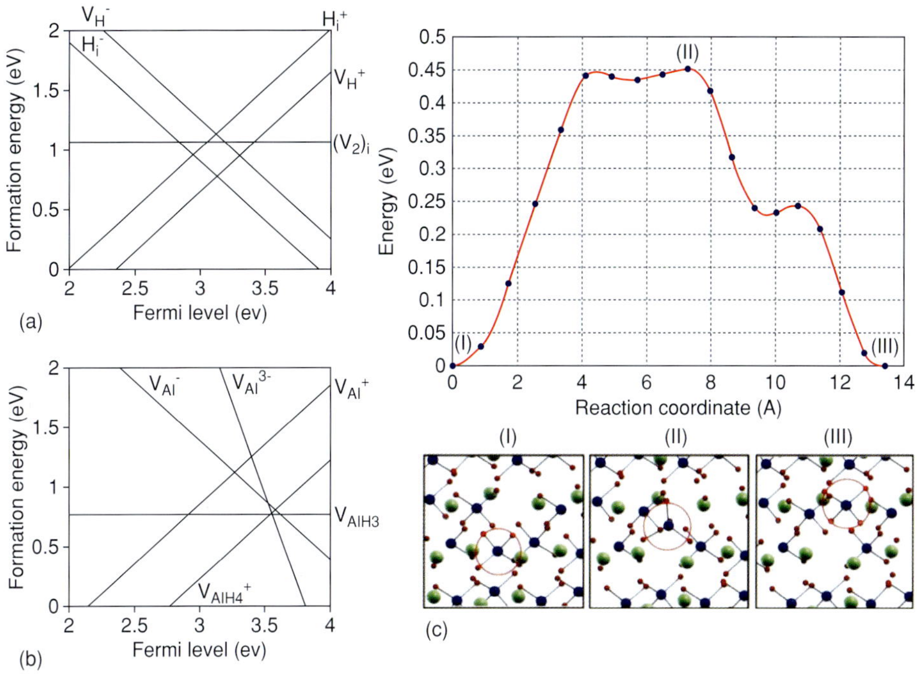

Figure 5.2 *Formation energies of (a) hydrogen-related and (b) aluminum-related defects. (c) Vacancy migration path as calculated with the nudged elastic band (NEB) method. (Reproduced with permission from [13].)*

addition, hydrides can be destabilized via alloying with other elements or hydrides to yield more desirable properties. We know from our experience that not all destabilizing schemes are useful, except a small subset among them. With the aid of theoretical study, experimentalists can avoid wasting time and effort on reactions that lie outside the thermodynamic window of interest. The benefit could be in terms of saving months or even years of experimentation.

With the rising popularity of density functional theory (DFT), as introduced in Chapter 1, many tailored functionals for thermodynamic calculation have been actively incorporated into routine calculations. Scientists must rely on their chemical intuition first to enumerate possible reaction paths as inputs [16]. This approach is biased and tedious for complicated hydride systems. By using DFT total energies, Gibbs free energy can be calculated and used for constructing a phase diagram. This grand canonical linear programming (GCLP) method predicted favorable thermodynamic reactions, stable compounds, and coexistence of phases solely based on the general chemical composition for the Li–Mg–N–H system (Figure 5.3) [17]. Last but not least, what about the compound whose crystal structure is not available in the database or not yet synthesized? It has always been the dream of computational scientists to determine crystal structure truly from first principles.

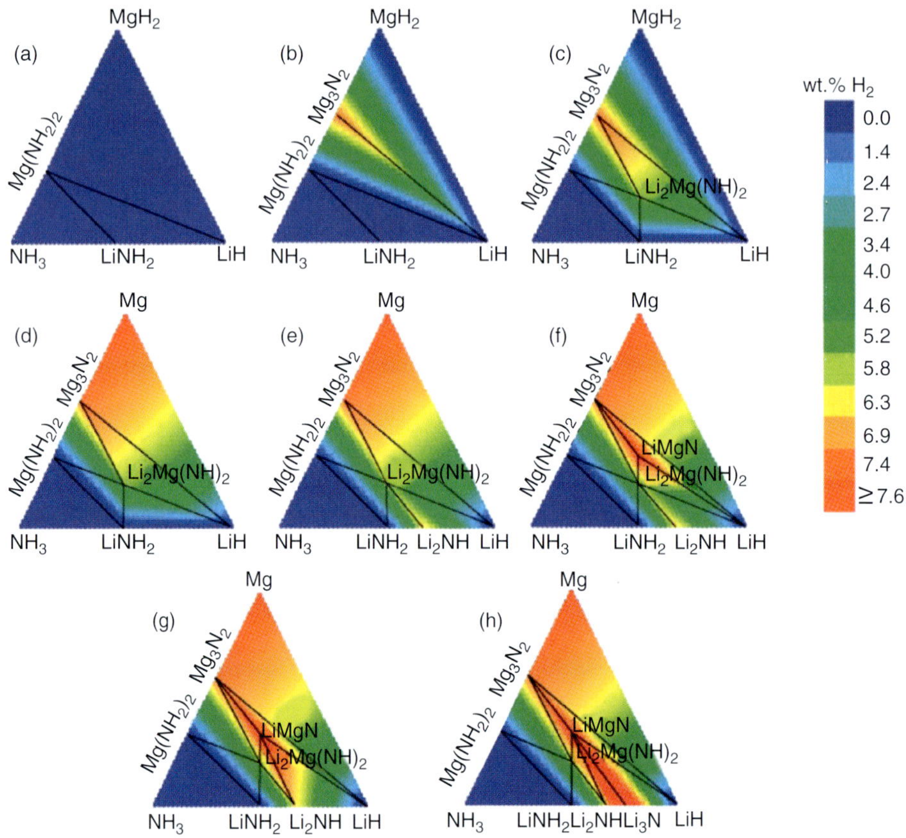

Figure 5.3 *Phase diagram of Li-Mg-N-H generated by the GCLP method. Each triangle corresponds to a different temperature: (a) $T < 130$ K, (b) $130 < T < 426$ K, (c) $426 < T < 458$ K, (d) $458 < T < 606$ K, (e) $606\ \mathrm{K} < T < 734$ K, (f) $734 < T < 780$ K, (g) $780 < T < 1077$ K and (h) $1077\ \mathrm{K} < T < 1126$ K. (Reproduced with permission from [17].)*

For hydrides, at least it is possible as demonstrated for $Mg(BH_4)_2$ using the prototype electrostatic ground state method [18]. Here, the DFT total energies are used in conjugation with global Monte Carlo (MC) optimization, assuming the dominance of electrostatic interactions within a framework of rigid complex ions. With powerful simulation techniques at their disposal, experimentalists are now able to freely explore the chemical workspace until reaching materials with desired properties. The obstacles in chemisorption will be likely overcome by combining efforts from both theory and experiment.

5.4 Physisorption Approach

Physisorption represents another extreme form of interaction in hydrogen storage, in which hydrogen adsorbs onto the surface of the substrate through weak van der

Waals (vdW) interaction. Thus, the process involves fast adsorption and desorption kinetics without activation barriers. However, high pressure and cryogenic temperatures are essential to stabilize the instantaneous dipole–dipole interaction. This means that the storage capacity quickly fades at ambient temperature. The MOF is currently the most promising candidate in this category. Since the first report about the MOF by the Yagi group [19], many research efforts have been devoted to this fascinating class of materials. As the name suggests, the MOF consists of metal ions linked by organic ligands. Due to the existence of nanosize pores and channels, MOFs possess huge surface areas. The nature of both metal ions and organic ligands is tunable, offering excellent control over materials properties such as volume, surface areas, and hydrogen affinity. The aforementioned problem of capacity fading at room temperature (RT) still remains as a key drawback for the MOF. Many adsorption isotherm studies have demonstrated the capability of the MOF to hold up 6–7 wt% hydrogen at 77 K. At RT, only 1–2 wt% is affordable. The foremost task of *ab initio* simulation for the MOF centralizes around improvement of storage behavior in ambient conditions. Similarly to chemisorption, experimentalists face many possible paths for improvement because the MOF is such a flexible class of materials. Either changing metal ions or organic linkers are likely to have an impact on storage capacity. It would be beneficial if the H_2 binding strength as well as storage capacity can be predicted beforehand. The collaboration between theory and experiment is expected to be very synergetic.

The computational approach to physisorption centralizes around three important aspects: (1) first-principles-based hydrogen interaction in the MOF, (2) grand canonical Monte Carlo (GCMC) to predict capacity, and (3) MD to describe H_2 diffusion in the MOF as in the extensive review by Han *et al.* [7]. Various strategies for improving storage capacity of the MOF are also outlined in the same review. Here, we would give some technical comments on GCMC simulation. By far, a large part of theoretical study in physisorption employs the MC method. This is a natural approach since computational time will become extremely demanding with the *ab initio* method. Nevertheless, the GCMC technique is unable to describe a subtle change in the electronic structure such as charge transfer and orbital rehybridization. However, as long as the interaction is weak, electronic effects have an insignificant contribution to the result. Empirical potential can overestimate the capacity in some cases. However, the empirical potential whose parameters are derived from high-level first-principles calculations shows better agreement with experiment. GCMC is capable of producing reasonable agreement with experiments for the MOF in many crucial aspects such as correlation between H_2 storage with heat of adsorption, surface area, and pore volume. Recently, the microscopic origin of hysteresis effect was also revealed via GCMC simulation (Figure 5.4) [20]. The foremost task of both theory and experiment is to enhance hydrogen adsorption enthalpy so that the system can operate in ambient conditions.

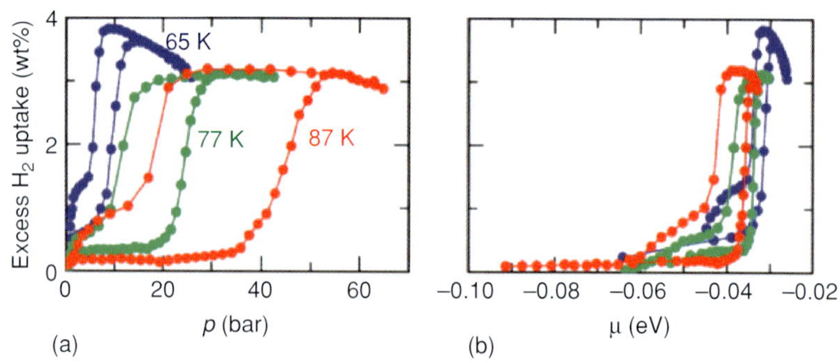

Figure 5.4 Hysteretic H_2 adsorption in Co(BDP) MOF from GCMC simulation. (Reproduced with permission from [20].)

5.5 Spillover Approach

The spillover mechanism is not a novel mechanism. It has a long history among the surface and catalyst community. Spillover can be characterized as diffusion of atomic hydrogen to the substrate surface after catalytic dissociation by precious metal such as platinum (Pt) or palladium (Pd). This is a crucial step for various surface reactions. The dissociation step of hydrogen on the catalyst is spontaneous without any energy barrier [21]. The rate-determining step is the formation of not too stable C−H bond so that H can dissociate, diffuse, and recombine to form H_2. The C−H bond formation is accompanied by sp^2 to sp^3 rehybridization of the carbon atom and energetically unfavorable. For the carbon nanotube (CNT), the curvature effect can tune the C−H bond strength. The larger the curvature, the closer the local carbon environment resembles the sp^3 configuration. The energy penalty of rehybridization will be small, which makes hydrogenation steps more facile [22]. Recently, the spillover mechanism was investigated by first-principles calculation for several systems including graphene [23, 24] and the MOF [25]. The authors have also outlined suggested design principles to improve hydrogen absorption via spillover.

5.6 Kubas-Type Approach

If chemisorption, physisorption, and spillover are at least vindicated by experiment, material for hydrogen storage using the Kubas approach is not yet realized. Therefore, many scientists are skeptic about the potential application of such approach. The groundbreaking discovery of a nonclassical complex $W(CO_3)(PR_3)_2(H_2)$ (the Kubas complex) in 1984 opened a new avenue in the hydrogen storage research [26]. Before that, scientists could not believe in the existence of a complex with elongated hydrogen bonds. The common consensus at the time was that either hydrogen stayed intact (physisorption) or completely dissociated (chemisorption).

The nonclassical two-electron three-center bond involves hybridization between vacant transition metal (TM) d and H_2 σ orbitals. Electron donation weakens and elongates the H–H bond about 20–30%. There is also a possibility of back-donation from occupied TM d to vacant H_2 σ^*. Therefore, the stability of the hydrogen complex depends on the chemical environment around TM. Donor and acceptor ligands can further tune the interaction strength. Even though the concept is well known among the chemistry community, it was physicists who recognized the potential impact of Kubas interaction in designing hydrogen storage materials.

For practical purposes, hydrogen binding energy should fall within the 0.2–0.6 eV energy window. The isolated metal atom can hold up to 12 hydrogen molecules in the form of $W(H_2)_{12}$, implying existence of a universal interaction between d orbitals and H_2 σ state [27]. Later, we will show that the Kubas interaction is not restricted to only d orbitals. From the materials point of view, it is more desirable to disperse TM atoms on the substrate to minimize potential agglomeration. One of the first studies along this line of reasoning was performed by a theory group at NREL [28]. Inspired by the ferrocene motif, they explored the possibility of multiple hydrogen adsorptions at the exposed TM site, assuming successful removal of one cyclopentadiene (Cp) ring. The maximum number of absorbed H_2 is readily calculated from the well-known 18-electron rule: $18 - N_V - 5$. N_V represents the total number of valence electrons of exposed TM and depends on the strength of metal–substrate interaction. The 18-electron rule also conveys the advantage of using early TM for the highest storage capacity. In the case of scandium (Sc), the first H_2 is stored in the form of hydride $Sc(Cp)H_2$ after one electron transfers to the Cp ring to satisfy aromaticity. Up to four H_2 can be further absorbed in the Kubas fashion with a binding energy of 0.3 eV. The 3d level becomes more stabilized across the periodic table, ensuing a more meaningful coupling and stronger Kubas binding. Therefore, from manganese (Mn) onward, the Kubas complex is formed directly without first forming hydrides. Unfortunately, the stand-alone TM–CP complex is unstable against polymerization. This undesired scenario can be avoided by locking the CP unit on the surface of the bulkyball C_{60}. With this design, roughly 7 wt% storage capacity is attainable for RT application. Lighter dopants can be introduced to lower the host mass and as a means to tune the strength of metal–substrate interaction. As an acceptor, boron (B) induces a higher degree of charge transfer between metal- and carbon-based substrates. B can pull an additional electron from TM, resulting in the formation of monohydride ScH instead of $Sc(H_2)$ for the first H_2 adsorption. Each exposed ScH site can now absorb up to five H_2 in the Kubas fashion with an estimated capacity of 8.77 wt% (Figure 5.5). Another side effect of the B dopant is the creation of an empty and strongly localized p_z state within the carbon framework. A meaningful orbital coupling exists as long as the separation between B p_z and H_2 σ is not too large. By introducing light element dopants such as B and Be, we avoid the need to use heavy TM as the adsorption center. The binding strengths are 0.39 eV/H_2 and 0.65 eV/H_2, respectively. The downside is that the number of adsorption molecules

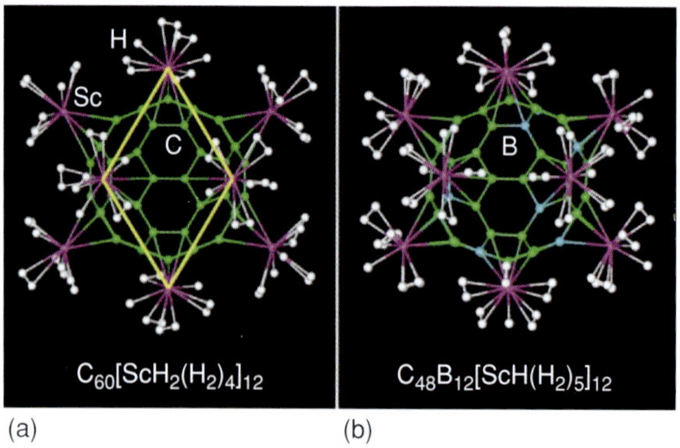

Figure 5.5 Hydrogen storage in (a) $C_{60}[ScH_2(H_2)_4]_{12}$ and (b) $C_{48}B_{12}[ScH(H_2)_4]_{12}$. (Reproduced with permission from [28].)

is limited to only one hydrogen. To maximize capacity, up to six B or Be dopants can be arranged symmetrically on the surface of C_{60} (Figure 5.6) [29].

The work on TM-decorated bulkyballs stimulated many theoretical studies to adapt this concept to different nanomaterials. However, the high cohesive energies of TMs, typically 4–5 eV, represent a big hurdle to experimental realization. The charge transfer-based dispersion scheme was proved to be insufficient to prevent TM atoms from clustering. Storage capacity will significantly decrease because of reduction of the number of available absorption sites. In addition, competitive adsorption of other gas molecules is another issue, but more engineering oriented. A system design that allows only hydrogen access to exposed metal sites would serve the purpose. We are now in a position to discuss three main strategies to overcome the aforementioned clustering problem based on the first-principles approach: (1) the MOF with exposed metal sites, (2) the replacement of TM with alkaline and alkaline earth, and (3) defect engineering for an enhanced metal–substrate binding in nanomaterials.

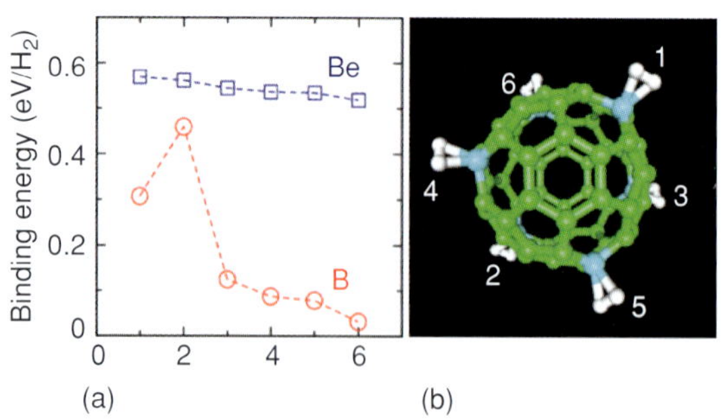

Figure 5.6 (a) Hydrogen storage in light element B and Be doped C_{60} as a function of adsorbed H_2 and (b) geometry of $C_{54}B_6$ with $6H_2$ molecules. (Reproduced with permission from [29].)

In the first approach, exposed TMs are rendered kinetically stable by the strong covalent network. Even though the largest contribution to storage capacity of the MOF comes from physisorption, the exposed metal sites can contribute to RT operation. The interaction between TM and H_2 is somewhat problematic to interpret and subjected to an elucidative debate. The measured TM–H_2 separations are 20–30% shorter than the typical 3 Å distance of vdW interaction, but the binding strengths are not strong enough to be classified as Kubas interaction [30]. In the Kubas coupling scheme, there exists an energy gaining mechanism by the downshift of the bonding level and an energy costing mechanism by the upshift of the occupied antibonding level. After H_2 adsorption, both spin-up and spin-down of Mn d_{z^2} states are pushed up. Since spin-up is occupied, the Kubas interaction is weakened. This shows the dependence of Kubas interaction strength on the occupancy of the responsive 3d orbital (Figure 5.7). The occupancy of exposed TM sites in turn depends on the crystal splitting scheme, spin configuration, and the oxidation state of TM. Realizing that d_{z^2} is the fourth highest energy in the square pyramidal crystal-field splitting scheme, we can tailor absorption energy by depopulating this level when replacing Mn with early TM atoms. Smaller binding energies were observed as expected for late TM atoms because the d_{z^2} is now completely filled.

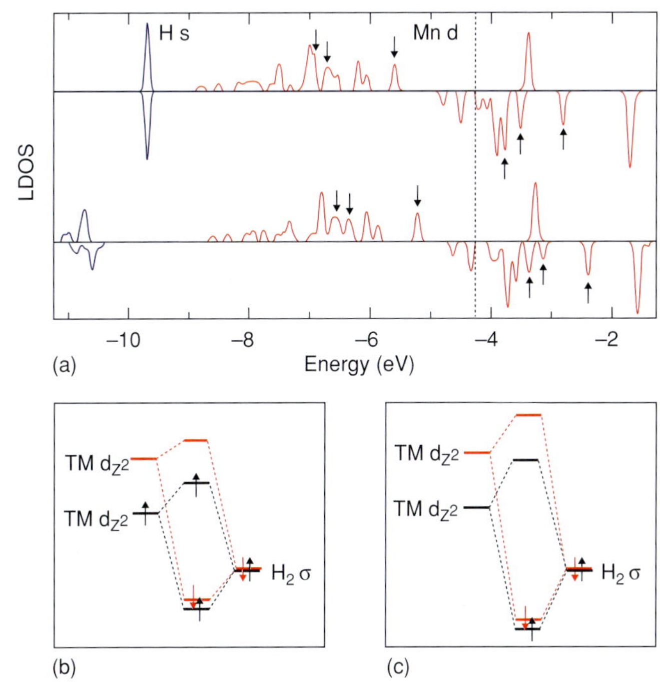

Figure 5.7 Interaction mechanism between Mn d and H_2 σ in Mn-MOF. (Reproduced with permission from [30].)

A similar square pyramidal symmetry also exists in the metal-exposed paddle wheel (TM_2-PW) [31]. Unlike Mn-MOF, multiple covalent bonds between metal atoms are the decisive factor instead of crystal splitting. Only TMs with available d and f shells can form metallic covalent bonds. Interaction between two TMs produces a set of bonding and antibonding molecular orbitals in the following energy order: $\sigma < \pi < \delta < \delta^* < \pi^* < \sigma^*$. This qualitative argument is based on the assumption that energy of the molecular orbital strongly depends on the extension of overlap between parental orbitals. The $d_{x2\text{-}y2}$ is not included in this scheme due to participation in the covalent bond with carboxylate bridging ligands. The dihydrogen binding strength shows a reverse trend compared to MOF series. Early TMs from Sc to Cr show only vdW binding due to the dominance of strong metal–metal bonds. Strong magnetic exchange impedes the absorption as in the case of Fe and Mn, leading to weak interaction as well. Only late TMs Co and Ni show noticeable Kubas behavior since the $dd\sigma^*$ is empty and half-occupied, respectively. Surprisingly, the enhanced absorption was also observed in Cu_2-PW and Zn_2-PW although there are no available d states to couple with H_2 σ. A close inspection of the electronic structure of Cu_2-PW and Zn_2-PW revealed a nontrivial s–σ coupling between localized 4s and H_2 σ [32]. The binding strength can be further tuned by incorporating elements that have a lower s state, such as Mg. Metal alloy combinations such as CuMg and CuZn show an enhanced hydrogen binding compared to Cu_2-PW. Ideally, only Mg- and Zn-based PWs should be used because they provide a desired binding strength from 0.1 eV/H_2 to 0.2 eV/H_2. However, due to the full occupation of d levels in the case of Zn, and the total absence of d levels in Mg, a local tetrahedral environment is more favorable than the octahedral environment. While distortion can dismiss the enhanced binding strength, in-plane strain can somewhat remedy this negative effect. The role of a localized state in enhanced adsorption cannot be overlooked as it has been demonstrated in the case of a light-element-doped carbon structure and post-TM PW framework. A strong localized state, regardless of s, p, or d, can couple with H_2 σ, resulting in a meaningful coupling. The MOF approach is mostly promising since the synthetic technique of the MOF is well developed. The structural flexibility of the MOF also offers opportunities to test ideas. As a final remark, we would suggest an alternative way to characterize Kubas interaction. Traditionally, we can identify Kubas interaction on the basis of the H–H bond elongation, which is a consequence of electron donation. However, if the binding strength is weak, typically less than 0.2 eV/H_2 as in TM_2-PW, the H–H distance only varies slightly. Nevertheless, the TM-H_2 can reduce up to 30%. Therefore, TM-H_2 distance serves well as an indicator of Kubas interaction.

Since TMs prefer clustering, much attention has turned to alkaline and alkaline earth metals (AEMs) since their cohesive energy is lower than that of TMs. Calcium (Ca) is the most viable candidate since its 3d orbitals are completely empty. This promises the maximum hydrogen coordination. For instance, the magic cluster with a Ca-decorated bulkyball, $C_{60}Ca_{32}$, was demonstrated to be capable of more than

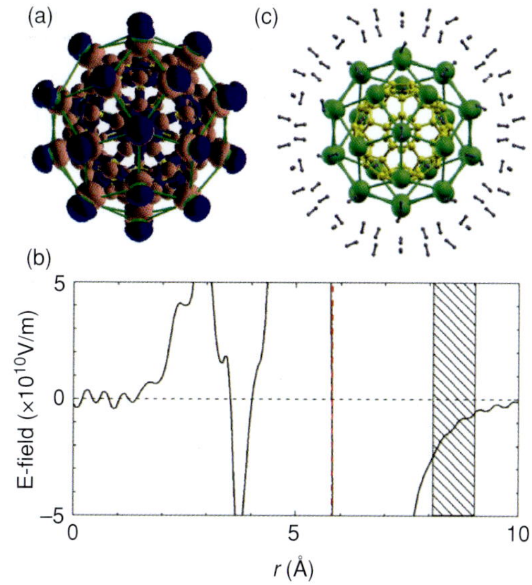

Figure 5.8 Hydrogen storage in $C_{60}Ca_{32}$ based on polarization of H_2 by charged C_{60}. (Reproduced with permission from [33].)

8.4 wt% hydrogen uptake. Yoon *et al.* associated the enhanced hydrogen binding strength solely with the induced polarization of H_2 by dispersed Ca (Figure 5.8) [33]. Since dispersed Ca forms a complete shell on the fullerene surface, it is not clear whether Kubas interaction also plays a role here. It is worth noting that fullerene is very special since the magic cluster $C_{60}Ca_{32}$ is readily observed from experiment and most likely to be realized. For other carbon structures such as CNT or graphene, we would avoid Ca clustering as much as possible. A systematic study on CNTs shows that formation of Ca dimers on pristine CNTs is favorable, leading to capacity reduction. B dopants can be again utilized to suppress dimerization. Each Ca can hold up to six H_2 molecules, providing capacity of 5 wt%. This conclusion was also shown to be insensitive with respect to curvature and chirality of CNTs [34]. For graphene nanoribbons (GNRs), Ca prefers the edge position because of enhanced electrostatic interaction by localized transferred charge. By combining the B dopant and localized GNR edge, Ca clustering can be effectively prevented [35]. The question we must address is: Why is Ca superior to other AEMs? A crucial aspect that distinguishes Ca from other AEMs is the existence of bound and localized 3d states after being oxidized due to charge transfer. Ca sandwiched between two graphene layers can attract up to eight hydrogen molecules with the adjustable oxidation state based on the distance toward the substrate. With the $\sqrt{7} \times \sqrt{7}$ packing scheme, a volumetric density higher than liquid hydrogen can be achieved [36]. However, the distance between two layers of graphene must be somehow engineered to provide the suitable channel for hydrogen desorption and uptake. The energy gain via hydrogen adsorption is not enough to initiate this process. As already stressed several times so far, to take advantage of enhanced Kubas interaction, it is crucial that the localized states can be created and controlled. In the case of AEMs, both polarization and Kubas interaction contribute to an enhanced H_2 binding. The crossover between two mechanisms depends on the ionic state of AEMs and the number of adsorbed H_2 [37].

The final approach can be considered as an extension of the first approach. Even though we can successfully lock the TM using the strong covalent network, the design of the MOF does not allow the maximum number of exposed metal sites. Thus, most of the capacity still comes from the physisorption at organic linkers. To introduce the covalent network in nanomaterials, we have borrowed the idea of metal porphyrin, which naturally exists in blood cells. These molecules play a crucial role of O_2 transportation in the human body. A systematic study of H_2 adsorption in metal porphyrin reveals that only Ti and V show a significant Kubas binding [38]. This seems confusing at first since empty d_{z^2} still exists in Cr, but as pointed out in the previously discussed Mn-MOF, the crystal splitting scheme decides positions of the responsive d_{z^2}. The abrupt change at Cr means that the energy gain via hydrogen adsorption is not large enough to compensate for the symmetry breaking process. Kubas interaction can be stabilized over vdW by introducing an axial ligand to break the square planar symmetry. Particularly, the pyridine ligand is most suitable due to a strong coupling of the N lone pair with metal d_{z^2}. This leads to the delocalization and reduction of exchange energy in d_{z^2}. Symmetry-lowered Fe-porphyrin can deliver a binding energy of 0.3 eV for RT application. We can take this idea further by implementing the Fe–N_4 unit in a 2D material such as graphene [39]. This motif is called porphyrin in graphene (PIG) (Figure 5.9). Even though Fe-PIG is thermodynamically unstable against H_2 environment, structural decomposition is prevented by a large activation barrier due to the strong covalent bond between Fe and four substitutional N defects. So, Fe-PIG is kinetically stable. This is also attributed to the fact that dispersion energy of TM is 1–4 eV larger than its cohesive energies. In terms of hydrogen binding, only Ti and V show the Kubas-type binding, similar to TM-porphyrin. The Kubas-type binding for late TM such as Fe can be stabilized by introducing in-plane strain, up to 3%.

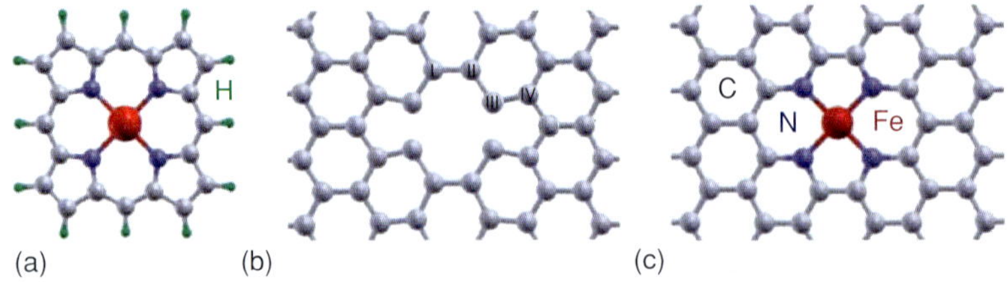

Figure 5.9 Atomic structures of (a) Fe-porphyrin molecule, (b) divacancy in graphene, and (c) Fe-porphyrin-like moiety in graphene. (Reproduced with permission from [39].)

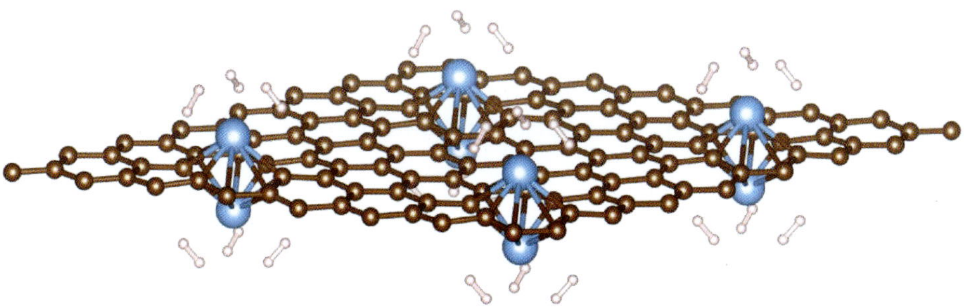

Figure 5.10 *Hydrogen storage using Ti_2 dimers incorporated in a defective graphene sheet (Le* et al.*, unpublished).*

So far, we have had a great deal of discussion on the Kubas-type approach. Even though none of the theoretical predictions have come to realization because of various experimental constraints, the Kubas approach offers a unique advantage over others in terms of operation in ambient conditions. In the case of TM-decorated nanomaterials, we learn the importance of hydrogen-accessible space around TM. To attract as many hydrogen molecules as possible, TM must maintain a *right* distance to the surface, exposing all d orbitals to H_2. Ironically, this is possible only when TM is weakly dispersed. Metal clustering is therefore unavoidable. We can make the material kinetically stable by embedding TM using strong covalent bonds just like Fe-PIG. Obviously, we do not leave enough space for TM to interact with H_2. Each TM center can only hold one H_2 with rather limited capacity. Yet, we can somewhat compromise between these two approaches, by utilizing the multiple metallic bond concepts in the TM-PW framework. By using a divacancy graphene substrate, TM can be dispersed at both sites of the sheet. The divacancy defect allows two TMs to come in contact and form the strong covalent bond, rendering the TM_2 unit kinetically stable. Unlike TM-PIG, the protrusion of TM out of plane gives it enough space to interact strongly with H_2. This prototype is the condensation of all the advantages in each approach discussed above (Figure 5.10).

5.7 Conclusion

In this chapter, we could not cover all aspects of the rapidly evolving hydrogen storage research. Each approach has its own merits. For chemisorption, although the gravimetric density is high, the kinetics needs to be further improved for practical application. In contrast, physisorption materials such as metal-organic frameworks have fast and reversible kinetics, but the required cryogenic operation temperature is far from satisfactory. The Kubas approach stands somewhere in between chemisorption and physisorption. We can say that the Kubas approach has most benefited from first-principles calculations. Many subtle and complicated aspects of the Kubas interaction have been revealed through the understanding of the electronic structure brought forward from a range of theoretical studies. We hope that the knowledge gained, combined with future computational studies, will inspire and guide experimentalists on the road toward green and sustainable economy.

References

1. Schlapbach, L. and Züttel, A. (2001) Hydrogen-storage materials for mobile applications, *Nature* **414**, 353.
2. US Department of Energy (2012) *Hydrogen, Fuel Cells and Infrastructure Technologies Program*, http://www1.eere.energy.gov/hydrogenandfuelcells/storage/ (accessed 10 December 2012).
3. Harris, R., Book, D., Anderson, P. and Edwards, P. (2004) Hydrogen storage: the grand challenge, *Fuel Cell Rev.* **1**(1), 17.
4. Jain, I.P., Jain, P. and Jain, A. (2010) Novel hydrogen storage materials: a review of lightweight complex hydrides, *J. Alloys Compd.* **503**, 303.
5. Yang, J., Sudik, A., Wolverton, C. and Siegelw, D.J. (2010) High capacity hydrogen storage materials: attributes for automotive applications and techniques for materials discovery, *Chem. Soc. Rev.* **39**, 656.
6. Paik Suh, M., Park, H.J., Prasad, T.K. and Lim, D.-W. (2012) Hydrogen storage in metal-organic frameworks, *Chem. Rev.* **112**(2), 782.
7. Han, S.S., Mendoza-Cortés, J.L. and Goddard III, W.A. (2009) Recent advances on simulation and theory of hydrogen storage in metal–organic frameworks and covalent organic frameworks, *Chem. Soc. Rev.* **38**, 1460.
8. Wang, L. and Yang, R.T. (2008) New sorbents for hydrogen storage by hydrogen spillover—a review, *Energy Environ. Sci.* **1**, 268.
9. Jena, P. (2011) Materials for hydrogen storage: past, present, and future, *J. Phys. Chem. Lett.* **2**, 206.
10. Klontzas, E. and Tylianakis, E. (2011) On the enhancement of molecular hydrogen interactions in nanoporous solids for improved hydrogen storage, *J. Phys. Chem. Lett.* **2**, 1824.
11. Osborn, W., Markmaitree, T., Shaw, L.L. *et al.* (2009) Solid-state hydrogen storage: storage capacity, thermodynamics and kinetics, *JOM* **61**, 45.
12. Bogdanovi, B. and Schwickardi, M. (1997) Ti-doped alkali metal aluminium hydrides as potential novel reversible hydrogen storage materials, *J. Alloys Compd.* **1–9**, 253.
13. Wilson-Short, G.B., Janotti, A., Hoang, K. *et al.* (2009) First-principles study of the formation and migration of native defects in $NaAlH_4$, *Phys. Rev. B* **80**, 224102.
14. Li, S., Ahuja, R., Araujo, C. M. *et al.* (2010) Dyhydrogenation associated with Ti catalyst in sodium analytes, *J. Phys. Chem. Solid* **71**, 1073.
15. Gunaydin, H., Houk, K.N. and Ozolinš, V. (2008) Vacancy mediated dehydrogenation of sodium alanate, *Proc. Natl. Acad. Sci. USA* **105**, 3673.
16. Alapati, S.V., Johnson, J.K. and Sholl, D.S. (2006) Identification of destabilized metal hydrides for hydrogen storage using first principle calculations, *J. Phys. Chem. B* **110**, 8769.

17. Akbarzadeh, A.R., Ozolinš, V. and Wolverton, C. (2007) First-principles determination of multicomponent hydride phase diagrams: application to the Li-Mg-N-H system, *Adv. Mater.* **19**, 3233.
18. Majzoub, E.H. and Ozolinš, V. (2008) Prototype electrostatic ground state approach to predicting crystal structures of ionic compounds: application to hydrogen storage materials, *J. Phys. Chem. B* **77**, 104115.
19. Rosi, N.L., Eckert, J., Eddaoudi, M. *et al.* (2003) Hydrogen storage in microporous metal-organic frameworks, *Science* **300**, 1127.
20. Kang, J., Wei, S.-H. and Kim, Y.-H. (2010) Microscopic theory of hysteretic hydrogen adsorption in nanoporous materials, *J. Am. Chem. Soc.* **11**, 1510.
21. Bhowmick, R., Rajasekaran, S., Friebel, D. *et al.* (2011) Hydrogen spillover in Pt-single-walled carbon nanotube composites: formation of stable C–H bonds, *J. Am. Chem. Soc.* **133**(14), 5580.
22. Lee, E.-C., Kim, Y.-S., Jin, Y.-G. and Chang, K.J. (2002) First-principles study of hydrogen adsorption on carbon nanotube surfaces, *Phys. Rev. B* **66**, 073415.
23. Singh, A.K., Ribas, M.A. and Yakobson, B.I. (2009) H-spillover through the catalyst saturation: an *ab initio* thermodynamics study, *ACS Nano* **3**(7), 1657.
24. Wu, H.-Y., Fan, X., Kuo, J.-L. and Deng, W.-Q. (2011) DFT study of hydrogen storage by spillover on graphene with boron substitution, *J. Phys. Chem. C* **115**(18), 9241.
25. Lee, K., Kim, Y.-H., Sun, Y.Y. *et al.* (2010) Hole-mediated hydrogen spillover mechanism in metal-organic frameworks, *Phys. Rev. Lett.* **104**, 1–4.
26. Kubas, G.J. (2007) Dihydrogen complexes as prototypes for the coordination chemistry of saturated molecules, *Proc. Natl. Acad. Sci. USA* **104**, 6901.
27. Gagliardi, L. and Pyykkö, P. (2004) How many hydrogen atoms can be bound to a metal? Predicted MH_{12} species, *J. Am. Chem. Soc.* **126**, 15014.
28. Zhao, Y., Kim, Y.-H., Dillon, A.C. *et al.* (2005) Hydrogen storage in novel organometallic buckyballs, *Phys. Rev. Lett.* **94**, 1–4.
29. Kim, Y.-H., Zhao, Y., Williamson, A. *et al.* (2006) Nondissociative adsorption of H_2 molecules in light-element-doped fullerenes, *Phys. Rev. Lett.* **96**, 1–4.
30. Sun, Y.Y., Kim, Y.-H. and Zhang, S.B. (2007) Effect of spin state on the dihydrogen binding strength to transition metal centers in metal-organic frameworks, *J. Am. Chem. Soc.* **129**, 12606–12607.
31. Bak, J.H., Le, V.-D., Kang, J. *et al.* (2012) First-principles study of electronic structure and hydrogen adsorption of 3d-transition metal exposed paddle wheel frameworks, *J. Phys. Chem. C* **116**, 7386.
32. Kim, Y.-H., Kang, J. and Wei, S.-H. (2010) Origin of enhanced dihydrogen-metal interaction in carboxylate bridged Cu_2–paddle-wheel frameworks, *Phys. Rev. Lett.* **105**, 1–4.
33. Yoon, M., Yang, S., Hicke, C. *et al.* (2008) Calcium as a superior coating metal in functionalization of carbon fullerenes for high-capacity hydrogen storage, *Phys. Rev. Lett.* **100**, 1–4.
34. Lee, H.K., Ihm, J., Cohen, M.L. and Louie, S.G. (2009) Calcium-decorated carbon nanotubes for high-capacity hydrogen storage: first-principles calculation, *Phys. Rev. B* **80**, 115412.
35. Lee, H.K., Ihm, J., Cohen, M.L. and Louie, S.G. (2010) Calcium-decorated graphene-based nanostructures for hydrogen storage, *Nano Lett.* **10**, 793.

36. Sun, Y.Y., Lee, K., Kim, Y.-H. and Zhang, S.B. (2009) *Ab initio* design of Ca-decorated organic frameworks for high capacity molecular hydrogen storage with enhanced binding, *Appl. Phys. Lett.* **95**, 033109.
37. Kim, G., Jhi, S.-H., Lim, S. and Park, N. (2009) Crossover between multipole Coulomb and Kubas interactions in hydrogen adsorption on metal-graphene, *Phys. Rev. B* **79**, 155437.
38. Kim, Y.-H., Sun, Y.Y., Choi, W.I. *et al.* (2009) Enhanced dihydrogen adsorption in symmetry-lowered metal–porphyrin-containing frameworks, *Phys. Chem. Chem. Phys.* **11**, 11400–11403.
39. Choi, W.I., Jhi, S.-H., Kim, K. and Kim, Y.-H. (2010) Divacancy-nitrogen-assisted transition metal dispersion and hydrogen adsorption in defective graphene: a first-principles study, *Phys. Rev. B* **81**, 1–5.

6

Energy Conversion: Solid Oxide Fuel Cells

First-Principles Modeling of Elementary Processes

E.A. Kotomin[1,2], *R. Merkle*[1], *Y. A. Mastrikov*[2,3], *M.M. Kuklja*[3], *and J. Maier*[1]
[1] *Max Planck Institute for Solid State Research, Stuttgart, Germany*
[2] *Institute for Solid State Physics, University of Latvia, Riga, Latvia*
[3] *Materials Science and Engineering Department, University of Maryland, College Park, USA*

6.1 Introduction

Fuel cells are electrochemical devices that directly transform the chemical free energy of combustion (e.g., H_2 + O_2 and CH_x + O_2) into electrical energy. The avoidance of a thermal detour guarantees high theoretical efficiency. As far as the temperature regimes are concerned, we distinguish between high temperature ceramic fuel cells, intermediate-temperature fuel cells, and low temperature (i.e., only slightly above room temperature) fuel cells. The high temperature fuel cells are usually based on oxide components (ternary transition metal oxides as cathodes, Ni or Cu cermets as anodes, and acceptor-doped zirconia or ceria as electrolytes). The high temperature necessary for ion conduction (though giving rise to problems concerning materials durability and compatibility) is favorable for the electrode reaction kinetics and additionally allows direct electrochemical conversion of hydrocarbons. Low temperature fuel cells basically rely on the reaction between H_2 and O_2 and typically use polymeric proton conductors (ionomers) ([1] and references cited there). A major materials challenge is the selection of the

Computational Approaches to Energy Materials, First Edition. Edited by Aron Walsh, Alexey A. Sokol and C. Richard A. Catlow.

electrocatalysts, in particular on the cathode side. Even though intense research is underway on alternative materials, noble metals are still the candidates of choice. A particularly challenging task is the development of direct methanol (or ethanol) fuel cells at these temperatures. A good compromise, as far as the dilemma between materials problems and electrochemical kinetics is concerned, is to be expected from intermediate temperature fuel cells. Here, mostly highly protonically conducting oxides, phosphates, and silicates are envisaged as electrolytes [2–4]. However, the development of properly adapted, catalytically active electrodes for this device class is still in its infancy.

A great deal of modeling work has been performed in the context of low and intermediate temperature fuel cells, in terms of understanding the conduction mechanisms of the electrolytes (e.g., see [5–9] and references therein), as well as in terms of elucidating the electrochemical reactions (e.g., see [10–15] and references cited therein). As it is impossible to take account of all of this progress in a single chapter, we concentrate on high temperature fuel cells here and lay emphasis on those issues which atomistic modeling is most helpful for.

Solid oxide fuel cells (SOFCs) have been under intensive investigation since the 1980s (for an overview see [16–20]). The detailed studies of SOFCs raise a number of questions that are interesting also from a fundamental point of view, such as the kinetics of the electrode reactions, the oxygen transport rate in solids, and the thermodynamic stability of the involved phases and interfaces. With the help of modeling one can vary parameters and handle adjusting screws that are often inaccessible by experiments. So, a major part of this chapter consists of showing how one can extract a detailed reaction mechanism for the oxygen reaction at the cathode, in particular, when the conclusions are complemented by experimental results. As the point defects are the key species, we begin with modeling them, and at the end of the chapter we will also briefly tackle transport in the electrolyte and fuel reduction at the anode.

Not least owing to the kinetic stability of the oxygen molecule, one of the main limiting contributions to the overall cell performance comes from the oxygen reduction and incorporation reaction at the cathode, which can be addressed by optimization of electrode morphology and choice of electrode material. ABO_3-type perovskites and perovskite-related oxides with adequate electronic as well as ionic (oxygen vacancy) conductivities are widely applied as cathode materials. The rate constants for oxygen incorporation into these materials vary widely, as shown in Figure 6.1 for (La,Sr)MnO_3 (LSM), (La,Sr)(Co,Fe)O_3 (LSCF), and (Ba,Sr)(Co,Fe)O_3 (BSCF) perovskites. To really reflect the intrinsic materials properties and avoid influences (e.g., from varying morphologies in porous films or gas diffusion limitations), these data were measured on dense films (prepared by pulsed laser deposition (PLD)). Without going into details, the sheer fact that the material with the highest concentration of oxygen vacancies ($V_O^{\cdot\cdot}$, Kröger–Vink notation [21]) exhibits the highest rate constant indicates that point defects play a decisive role in the electrode reactions.

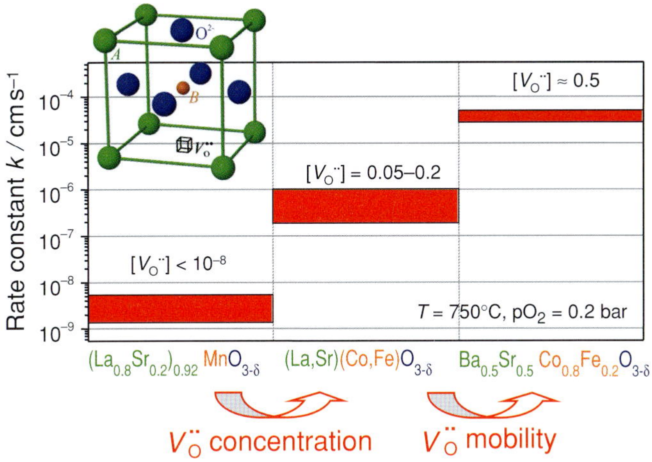

Figure 6.1 *Effective rate constants k of the oxygen surface reaction of different perovskites (measured on dense PLD films on YSZ single crystal substrates), which are investigated as fuel cell cathode materials (inset: perovskite structure). (Data taken from [22–24].)*

In recent years, quantum chemical calculations, particularly density functional theory (DFT) methods (see Chapter 1), have become a powerful tool for the understanding of transport and reactions in solids on a microscopic level. They give access to quantities that are difficult to obtain from experiments, such as the energies and mobilities of surface defects and the energies and electronic and geometrical configurations of transition states. In this way, they greatly enhance the insight into transport and reaction mechanisms, which then serves as a basis for targeted materials variation (or can help recognize fundamental limitations of a materials family).

In the context of SOFC research, several specific problems must be addressed in *ab initio* modeling. The materials under study typically have high defect or dopant concentrations that can occupy the supercells in different configurations that need to be explored. The study of effects of mutual defect interactions requires varying the defect concentrations. For the energetics of oxygen vacancy formation as well as for the reaction mechanism involving adsorbed molecules, a proper description of molecular oxygen species is important. Because the adsorbate concentrations may be small, sufficiently large surface supercells are needed.

The electrode materials typically contain transition metals with partially occupied *d* orbitals (i.e., they are typically highly correlated systems). Some efforts have to be made in order to obtain realistic band-gap values; also, the energetics of oxygen vacancy formation (involving a valence change of the transition metal) will be sensitive to this. Because the redox energetics can also affect the ion migration barriers, a good description of the band gap could be important, even if the main focus is placed on the ionic conductivity of a mixed conducting perovskite. On the other hand, different magnetic orders observed at low temperatures are often of minor importance because their energy difference is much smaller than the reaction enthalpies and reaction barriers considered, and the ion transfer processes are not directly influenced by the magnetic state. All of these issues are discussed in this chapter.

Because the SOFCs are operated at 600–1000 °C, the effect of elevated temperatures must be taken seriously (e.g., for the identification of the stable surface terminations). This is briefly discussed in Section 6.3.1. The stability of different terminations does not only depend on temperature and gas pressure, but may also be affected by the presence (or even segregation) of dopants at the respective surface. For many perovskites studied as cathode materials, the simple surface terminations are also polar.

An increase in perovskite complexity when going from $LaMnO_3$ (LMO) to solid solutions with mixed occupancy on both cation sites ((Ba,Sr)(Co,Fe)$O_{3-\delta}$ and (La,Sr)(Co,Fe)$O_{3-\delta}$) is accompanied by an increase in material instability with respect to disorder and phase transformations. The bulk (and surface) oxygen vacancy formation energy strongly depends on the actual oxidation state of the transition metals and thus on the exact materials stoichiometry (dopant content as well as oxygen deficiency), which is often not clearly addressed, not only theoretically but also in experimental studies. These issues are discussed in Section 6.3.2.

Finally, in addition to point defects, higher dimensional defects can influence the electrical properties, in particular of the electrolyte materials. Most electrolytes exhibit a more or less pronounced blocking character of the grain boundaries for ion (oxygen vacancy) transfer. *Ab initio* calculations of model grain boundaries are thus highly desirable, but owing to the many degrees of freedom (mutual crystal and grain boundary plane orientation, but also local deviations from average stoichiometry and electroneutrality), they are also extremely challenging (see Section 6.4).

In this chapter, we focus on key SOFC topics, such as stable surface terminations, bulk and surface point defect formation and migration energetics, and in particular the mechanism of the oxygen incorporation surface reaction mainly based on the results of studies obtained in our research groups. Other topics, such as the effect of strain on defect mobility and *ab initio* treatments of grain boundaries, are discussed on the basis of recent results from literature. Experimental studies will be cited when necessary for the comparison with the theoretical findings (a full coverage of experimental research would go far beyond the scope of this chapter). This chapter exemplarily illustrates the range and scope of contemporary methods based on DFT in materials research.

6.2 Computational Details

Basic computational approaches to calculations of the atomic and electronic structure of perfect and defective solids are discussed in Chapter 1 of this book (see also [25]). In most of our SOFC calculations of LSM and BSCF cathodes, we employed the *ab initio* DFT computer code VASP, which combines a plane-wave basis set and an exchange–correlation functional based on the generalized gradient approximation (GGA). Spin polarized calculations of the LMO bulk and the (001) and (110) surfaces are in good agreement with experiments (where available) [26]. In particular, for the low-temperature orthorhombic structure, the A-type antiferromagnetic configuration is the energetically most favorable, which is in agreement with experiment. The lattice constant of both the cubic (stable above 750 K) and orthorhombic phases exceeds the experimental value by only 0.5%. The calculated cohesive energy of 30.7 eV is close to the experimental value (31 eV).

The calculated surface relaxations and surface energies show a weak dependence on the magnetic configuration [26, 27]. For slabs, the ferromagnetic configuration has the lowest energy; thus, we performed all further calculations with collinear spins. We must also keep in mind that the relevant magnetic effects ($\approx$0.1 eV) are much smaller than the adsorption and migration energies under study (one to several electron volts). However, a complete neglect of spin polarization ("closed shell" calculation) results in considerable errors in material properties. We studied both orthorhombic and cubic phases for the oxygen adsorption and thermodynamic stability of surfaces with adsorbates, but focused on the orthorhombic phase in the analysis of intermediates and transition states for the kinetics treatment. Calculations for the defects/adsorbates in the high-temperature cubic phase [28] give qualitatively similar results; in particular, the atomic O adsorption energy atop Mn increases by approximately 3% only. Nevertheless, the use of symmetry restrictions may lead to artifacts from rearrangements in the bulk structure if these restrictions have to be released, for example, for low-symmetrical transition states (all results presented here were obtained without symmetry restrictions).

For surface models, convergence with respect to the number of atomic layers should always be checked. We found that seven- and eight-plane slabs are thick enough to show convergence of the main properties. The periodically repeated slabs were separated by a vacuum gap of 15.8 Å. All atomic coordinates in a slab were allowed to relax. The eight-layer slab $(LaOMnO_2)_4$ has the proper cation ratio and oxygen stoichiometry (all Mn atoms are in the formal +3 oxidation state), but exhibits a nonzero slab dipole moment and is terminated by different planes (MnO_2 and LaO). This is avoided in the symmetric seven-layer slab $MnO_2(LaOMnO_2)_3$, but this slab has a Mn excess relative to La and a higher oxygen content. This slightly affects the energy for dissociative oxygen adsorption on the $MnO_2(001)$ surface,

$$O_2 + 2Mn_{Mn}^{x} \rightleftharpoons 2O_{ad}^{-} + 2Mn_{Mn}^{\cdot}, \tag{6.1}$$

which is −2.7 eV for the $(LaOMnO_2)_4$ slab and −2.2 eV for the symmetrical $MnO_2(LaOMnO_2)_3$ slab [28]. On the other hand, the average Mn oxidation state of more than 3 implicitly mimics the situation in Sr-doped $LaMnO_3$, which is actually used in SOFCs. Due to the reduced computation time for the symmetrical configuration and the fact that the symmetrical slab allows decoupling effects of different surface terminations, all further calculations refer to the seven-layer slabs.

The choice of the $2\sqrt{2} \times 2\sqrt{2}$ (001) surface supercell for the calculation of adsorbates/surface defects corresponds to a surface coverage of 12.5%. This concentration is quite large compared to the actual adsorbate coverage at operating conditions (cf. Section 6.3.1.2), but is still acceptable for semiquantitative conclusions. Several test calculations were performed for a smaller coverage of 5%. The use of even larger defect concentrations and/or highly polar and oxygen-deficient (110) surfaces could lead to reaction enthalpies that are much too exothermic and oxygen incorporation reaction barriers that are too low (or even missing) [29–32]. The reaction and ion migration barriers (also for the other perovskites studied in Sections 6.3.2 and 6.3.3) were calculated using the nudged elastic band method [33,34]. The effective charges on the ions were determined according to the Bader (topological) analysis [35]. The effective charges both in the bulk and on the surfaces of $LaMnO_3$ are quite different from the formal charges, namely, $+2.13\,e_0$ for La, $+1.85\,e_0$ for Mn, and $-1.29\,e_0$ for O [26,27]. The considerable contribution of the covalence effects in Mn–O chemical bonding makes classical force-field calculations of these systems, and especially of diffusion-controlled processes therein, a nontrivial task.

In calculations of oxygen-exchange related processes, the proper description of molecular oxygen is important. In plane-wave calculations, the O_2 dissociation energy E_{diss} and bond length d depend considerably on both the O pseudopotential and the exchange–correlation functional used. The VASP calculations with the so-called standard pseudopotential considerably overestimate $E_{diss} = 5.9\,eV$ but give a reasonable $d = 1.23$ Å (experimental: $E_{diss} = 5.12\,eV$ and $d = 1.21$ Å [36]), whereas "soft" O pseudopotentials yield a better $E_{diss} = 5.24\,eV$ but overestimate $d = 1.29$ Å. Several procedures were suggested to correct for such DFT errors in metal oxide and O_2 calculations [28, 37, 38].

For comparison with VASP plane-wave modeling, a series of calculations were performed using the CRYSTAL computer code [39] with a linear combination of atomic orbital (LCAO) basis set. Its advantages are the efficient incorporation of the so-called hybrid exchange–correlation functionals, which permit a more accurate calculation of the band gap (see Chapter 1), and the absence of the artificial translation of slabs along the z-axis in surface calculations, which is unavoidable in plane-wave calculations. The O_2 dissociation energy and bond length are reproduced quite well: $E_{diss} = 5.30\,eV$ and $d = 1.20$ Å.

Lastly, we note that in oxygen vacancy calculations, the supercells used are always neutral (no compensating "background charge"). The charge distribution between the O vacancy and its environment adjusts itself while minimizing the total energy of the system. For reducible transition metals, the electrons are typically transferred to the nearest cations (see Section 6.3.2 and 6.3.3 on LSCF and BSCF).

6.3 Cathode Materials and Reactions

6.3.1 Surfaces: $LaMnO_3$ and $(La,Sr)MnO_3$ Perovskites

The prerequisite for investigating the oxygen incorporation reaction mechanism is to identify the stable surface terminations of the mixed conducting perovskites. Effects of doping (and potential dopant segregation to the surface) have to be considered as well as the elevated operating temperature (thermodynamic treatment, Section 6.3.1.1). Owing to the high operating temperature (allowing for rearrangements also in the cation lattice), the perovskite is expected to expose mainly the most stable surface terminations. Nevertheless, the overall reaction rate may be dominated by a contribution from a slightly less stable surface with intrinsically higher rate (see Section 6.3.1.3).

6.3.1.1 Surface Termination, Surface Point Defects

We performed a series of calculations for the structure of $LaMnO_3$ (001) and (110) surfaces using different methods: pair potentials [40], Hartree–Fock LCAO [41], hybrid B3PW LCAO [27], and GGA [42] density functional. The results for the atomic relaxations obtained using these different methods are in reasonable agreement. The next step is the thermodynamic analysis of surface stabilities based on the *ab initio* calculations [26, 43, 44]. The diagram of the LMO (001) and (110) surface stabilities in the cubic (high temperature) and orthorhombic (low temperature) phases is presented in Figure 6.2. The most stable termination depends on the oxygen chemical potential shown for different temperatures and gas pressures on the right side of the diagram. In particular, under usual SOFC operating conditions ($T = 1200$ K, oxygen pressure in the range 10^{-2}–1 atm, shown by dashed lines), the most stable cubic phase termination is MnO_2 (001) with adsorbed oxygen atoms, whereas the hypothetical orthorhombic phase (unstable at such high temperatures) would be adsorbate free. Note also that the LaO-terminated (001) surface could be stable only at very low oxygen partial pressure whereas the O-terminated (110) surface stability range lies in between these two terminations. Additionally, the stability of LMO surfaces is strongly restricted by the surface decomposition into metals and simple oxides (numbered lines in Figure 6.2). In a study of the $LaFeO_3$ (010) surface stability FeO_2 terminations were similarly predicted to be stable at high temperatures [45].

The cathode material actually used is $LaMnO_3$ typically doped with 20% Sr on the La site (LSM). This acceptor doping has a twofold effect: it increases the p-type electronic conductivity, and it brings the nominal oxygen excess (which is accommodated by cation vacancies, and can reach up to 0.2 O per formula unit) down to almost zero, thereby also increasing the oxygen vacancy concentration [46,47].

The thermodynamic stability of LSM has been analyzed using CRYSTAL and WIEN-2k (all-electron DFT) computer codes [48]. Both types of calculations

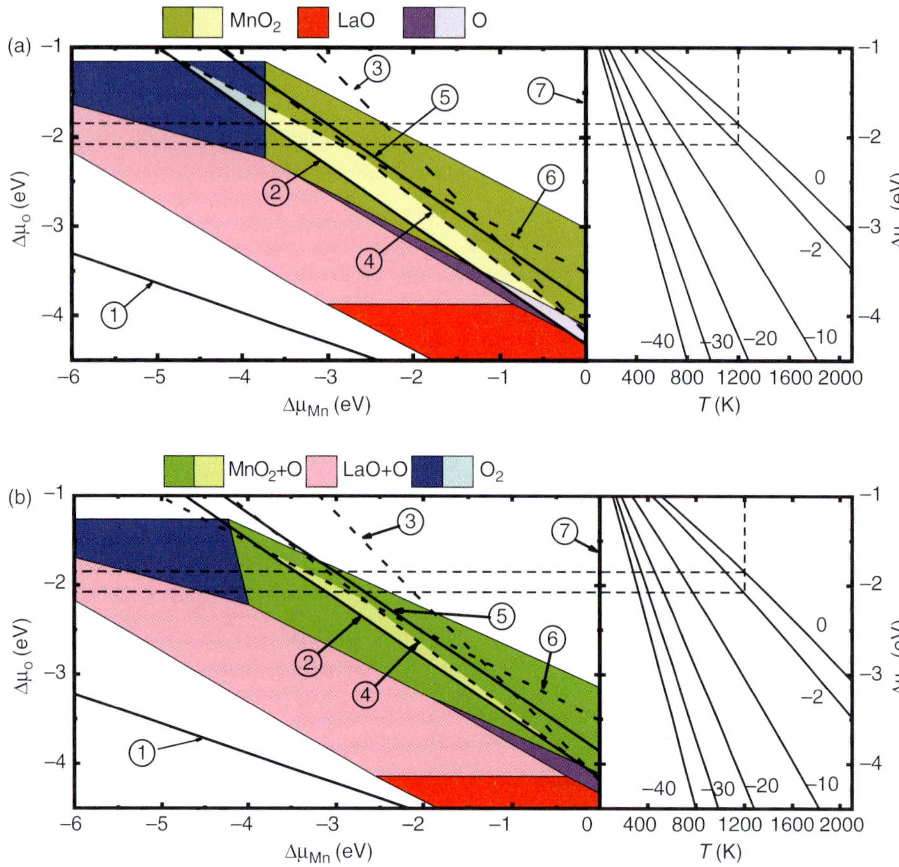

Figure 6.2 *Phase stability diagrams: The regions of stability of $LaMnO_3$ surfaces with different terminations (LaO- and MnO_2-terminated (001) surfaces with and without adsorbed O atom, O_2- and O-terminated (011) surfaces) for both phases as a function of chemical potential variations for manganese and oxygen. The numbers in circles point to lines, where metals or their oxides begin to precipitate: (1) metal La, (2) La_2O_3, (3) MnO, (4) Mn_3O_4, (5) Mn_2O_3, (6) MnO_2, and (7) metal Mn. The right side of the figures contains the chemical potential changes $\Delta\mu_O$ as a function of temperature at various oxygen gas pressures. The labels* m *on the lines represent the oxygen pressure: $pO_2 = 10^m$ bar. (a) Orthorhombic phase and (b) cubic phase. (Reprinted with permission from [38], copyright (2010) American Chemical Society.)*

demonstrate that unlike the isostructural (Sr,Ba)TiO_3 solid solution, a decomposition into $LaMnO_3$ and $SrMnO_3$ phases is energetically unfavorable for LSM. For 12% Sr, the order–disorder transition could occur only at very high temperatures (due to configurational entropy), otherwise Sr is predicted theoretically to be periodically arranged in the LMO matrix. A periodic distribution of Sr dopants was indeed observed experimentally in (La,Sr)CoO_3 using HRTEM [49].

The electronic structure and thermodynamic stability of LSM (001) surfaces was also calculated using the LCAO approach and hybrid functionals [50, 51]. Twelve percent Sr doping transforms the semiconducting LMO into a half-metal (in agreement with the experimentally observed decreased activation energy of electronic conductivity [46]). Sr exhibits a tendency to segregate to the surface, as experimentally observed for epitaxial thin films prepared by PLD [52, 53]. A related effect of Sr doping is to increase the stability of the (La,Sr)O (100) termination. At room temperature, MnO_2 and (La,Sr)O terminations are comparably stable, but at higher temperatures, the (La,Sr)O termination gradually occupies a larger portion of the stability region, and at typical SOFC operating temperatures (1100 K), this termination becomes predominant. In other words, Sr stabilizes the (La,Sr)O termination with respect to MnO_2 (100). The decrease of the surface dipole (cf. the nominal excess positive charge in an LaO layer) upon Sr doping certainly contributes to this. These developments are shown in Figure 6.3 [43] along with the predicted precipitation of La_2O_3 and MnO at high temperatures. This is consistent with experimental observations [54]. Thus, Sr doping introduced to increase the oxygen vacancy concentration (and thus the ionic conductivity through the cathode) has the side effect of changing the thermodynamic stability of various surface terminations, which will also affect cathode performance.

The calculation results predict a considerable segregation of oxygen vacancies toward the MnO_2 (001) surface because the O vacancy formation energy is reduced

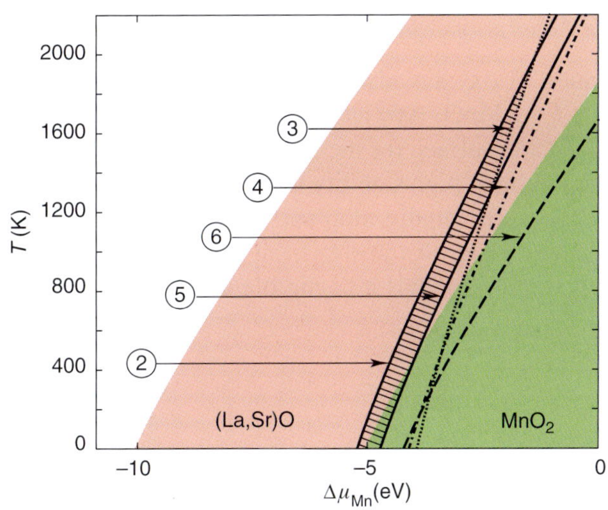

Figure 6.3 *LCAO-calculated cross-sections of the surface stability diagram for (La,Sr)MnO_3 (001) surface structures along the $LaMnO_3$ precipitation lines for an O_2 pressure of 0.2 atm. The shaded area corresponds to the perovskite stability area. Light red denotes the LaSrO termination, and green MnO_2. The numbers in the circles indicate precipitation lines for (2) La_2O_3, (3) MnO, (4) Mn_3O_4, (5) Mn_2O_3, and (6) MnO_2 (cf. Figure 6.2).*

by 0.91 eV in the surface layer compared to the bulk value for the following reaction:

$$O_O^x + 2Mn_{Mn}^x \rightleftharpoons 0.5O_2 + V_O^{\cdot\cdot} + 2Mn'_{Mn}. \quad (6.2)$$

On the other hand, there is no O vacancy accumulation on the LaO-terminated surface [28, 38] because there its formation energy is larger by 0.73 eV than in the bulk. This stems from the higher ionicity of the LaO-terminated surface as compared to the MnO_2-terminated surface. As shown in more detail in Section 6.3.1.3 (and in agreement with the experimental findings, cf. Figure 6.1), a high concentration of surface oxygen vacancies is beneficial for the fast incorporation of oxygen into the cathode and its further path to the electrolyte and finally to the anode.

Another important factor determining cathode performance is the surface oxygen vacancy migration energy (cf. Section 6.3.1.3). The GGA-DFT calculations [28] on the MnO_2 (001) surface estimate it as low as 0.67 eV, significantly smaller than in the bulk (0.95 eV, close to the typical values measured for ABO_3 perovskites such as $SrTiO_3$ [55], (La,Sr)FeO_3 and (La,Sr)CoO_3 [56] or (La,Ca)AlO_3) [57]). (The orthorhombic LMO phase has a set of barriers for bulk migration, which vary by $\approx$0.2 eV, as shown in pair potential calculations [58].) Lastly, it is not *a priori* obvious, whether or not the incorporation of the adsorbed O into a neighboring surface vacancy has a reaction barrier. Explicit calculations (Figure 6.4) indicate such a barrier to be absent. Oxygen atom incorporation into the vacancy is associated with an energy gain of 1.4 eV.

6.3.1.2 *Oxygen Adsorption and Diffusion*

For the investigation of oxygen adsorption (and O incorporation in the next section), we focused on the MnO_2 (001) termination of $LaMnO_3$, which was identified above as the most stable surface for the undoped material (and still one of the terminations with comparably high stability in case of Sr doping). Oxygen molecules adsorb on $LaMnO_3$ preferentially atop surface Mn ions. All adsorption configurations involve some electron transfer from the perovskite to the adsorbate (a physisorbed oxygen molecule has a negligible adsorption enthalpy, for example,

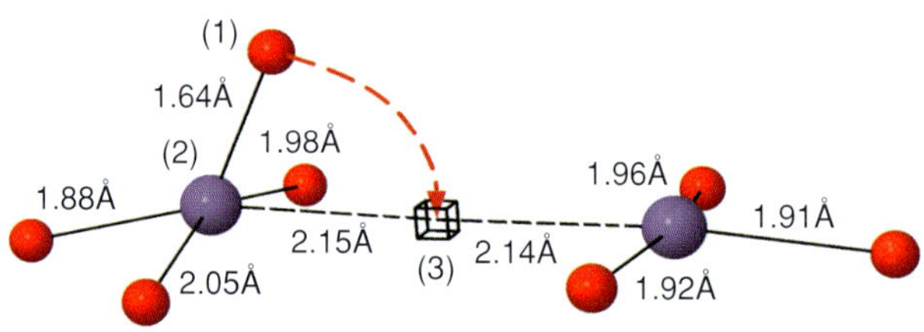

Figure 6.4 *View of the O atom (1) adsorbed atop surface Mn atom (2) on the orthorhombic MnO_2-terminated (001) surface just before dropping into oxygen vacancy (3). Some interatomic distances are indicated.*

O_2 on the $SrTiO_3$ (001) surface shows a negligible charge transfer and only a slightly exothermic adsorption energy of −0.25 eV [59]). The O−O bond length is helpful for assigning the molecular oxygen adsorbates (alternatively also the O–O stretch vibration frequencies can be used, e.g., see [29]) because the Bader atomic charges are usually smaller than the formal charges. We can identify an adsorbed superoxide O_2^- ($E_{ads} = -1.1$ eV, Bader charge = −0.42 e_0, d(O−O) = 1.36 Å, cf. HO_2 radical 1.33 Å [60]) as well as peroxide O_2^{2-} ($E_{ads} = -0.9$ eV, charge = −0.65 e_0, d(O–O) = 1.43 Å cf. H_2O_2 1.48 Å [60]), which are both attached to Mn cations. Atomically adsorbed oxygen (also bound to Mn) can be regarded as "O^-" (charge = −0.62 e_0 per O^-, $E_{ads} = -2.2$ eV for 2 O^-). This very negative chemisorption energy for atomic oxygen, typical for transition metal oxides with variable oxidation state, allows for an exothermic dissociation of O_2 molecules on the defect-free LMO surface, in contrast to large band-gap perovskites such as $SrTiO_3$. The electron redistribution, mainly from the neighboring Mn atom, is illustrated in Figure 6.5.

The adsorption enthalpies given above refer to a surface coverage of 12.5% ($2\sqrt{2} \times 2\sqrt{2}$ surface supercell). For a lower coverage of 5%, $E_{ads}(2\ O^-)$ increases to −2.5 eV, indicating some defect interaction already at these moderate degrees of coverage. To calculate adsorbate coverage under SOFC operating conditions, we used as a reasonable approximation adsorption entropies of about −200 J/mol K measured for gas adsorption on inert oxides, such as ZnO, SiO_2, and Al_2O_3. This strongly negative $\Delta_{ads}S^0$ is one reason for the resulting adsorbate coverage being comparably low. The other is that all adsorbed species are charged, leading to repulsive interactions, as it becomes obvious in the more negative E_{ads} at lower coverage (see above), and as it is described in a phenomenological model [61].

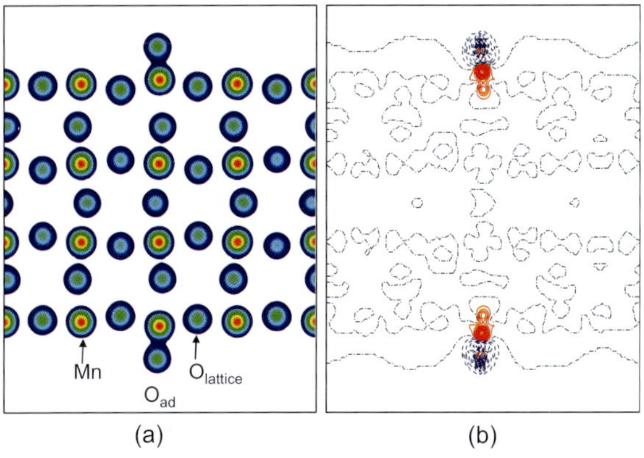

Figure 6.5 The total (a) and difference (b) electron density maps for O atoms adsorbed atop Mn on the symmetrical seven-plane slab with MnO_2 (100) termination. Red and blue indicate the electron density deficiency and excess. The density increment is 0.05 e Å^{-3}, and the black dash-dot line is the zero level. Notice the rather local character of lattice perturbation by the defect. (Reproduced from [28] by permission of the PCCP Owner Societies.)

The resulting coverage values are below 1% for the molecular adsorbates O_2^- and O_2^{2-}, and about 10% for O^- (the more exothermic E_{ads} favors higher coverage than for O_2^- and O_2^{2-}, but the mutual adsorbate repulsion leads to a leveling-off at ≈10% [38]).

The VASP modeling of atomic oxygen migration along the MnO_2 (001) surface [28, 38] shows that it occurs along the Mn–O–Mn direction. First, the O_{ad} atom moves from its position atop Mn toward the neighboring surface O, forming a tilted O–O species, which requires 1.6 eV and which can be regarded as a peroxide sticking upright in an oxygen vacancy. This dumbbell turns vertical (requiring another 0.4 eV), and then the migrating O moves toward the next surface Mn ion. Thus, the migration energy of adsorbed O atoms turns out to be quite high, 2 eV, as compared with 1 eV in $SrTiO_3$ [59]. The energy difference stems from different equilibrium positions: a strong O bonding atop surface Mn in LMO and attachment atop O in $SrTiO_3$ in the form of a less stable peroxide species. In other words, adsorbed O ions are quite immobile at the LMO surface, which will become important for the mechanism of O incorporation in the next section.

While the experimental techniques are making strong progress toward observation of surface defects, an investigation under fuel cell operating conditions remains challenging. Recently, adsorbed atomic oxygen species could be studied by scanning tunneling microscopy on the (Pr,Sr)O termination of $PrSr_2Mn_2O_7$ and even some O migration be observed [62], nevertheless the experiments are restricted to temperatures of maximum 200 K, vacuum conditions and those surfaces accessible by single crystal cleaving.

Dissociative oxygen adsorption on the LaO-terminated (001) polar surface was found to be highly exothermic (−4.57 eV) using the GGA approach [63]. However, as discussed above, a very low oxygen vacancy concentration is expected on this surface; thus, it cannot be active for oxygen incorporation. Detailed GGA + U calculations of oxygen adsorption on a series of Mn-based perovskites were performed recently [37]. The findings are similar to our calculations as discussed above, in particular, the prediction of quite a low oxygen surface coverage of the MnO_2 surface and the reduced oxygen vacancy formation energy at the MnO_2 surface as compared to the bulk. On the other hand, the choice of the optimal Hubbard U parameter is not unique. It is usually fitted to one of several experimental properties (e.g., lattice constant, band gap, and material formation enthalpy). In particular, Lee *et al.* [37] used a U_{eff} fitted to the formation energies of a series of 3d transition metal binary oxides [64], which does not necessarily ensure an optimal choice for ternary oxides such as perovskites, keeping in mind, for example, the exceptionally pronounced covalency in the transition metal–oxygen bonds.

6.3.1.3 Rate-Determining Step of the Surface Reaction

The oxygen incorporation reaction, converting oxygen molecules into regular lattice oxide ions, is a multistep chemical reaction. While adsorption on the perfect

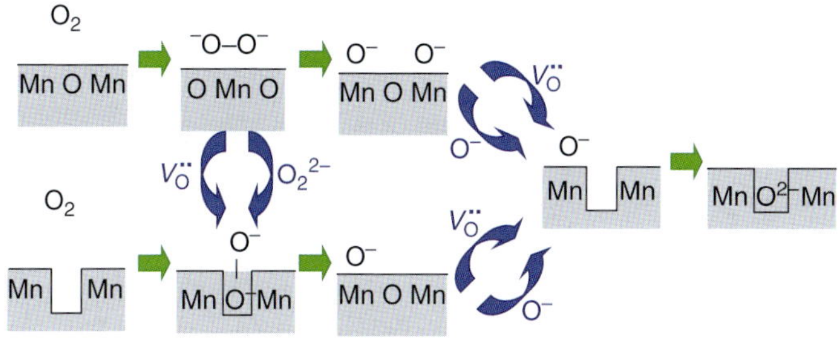

Figure 6.6 *Reaction network for oxygen incorporation on the MnO_2 (001) termination of the $LaMnO_3$ perovskite. Green arrows indicate reaction steps, blue arrows migration steps of adsorbed oxygen species or oxygen vacancies. The reaction begins with adsorption of O_2 either on the defect-free surface (top rail) or into an oxygen vacancy (bottom rail). The dissociation can occur as well with or without $V_{\ddot{O}}$ assistance, and a crossover between the rails can take place if beneficial for the overall reaction rate. For the approach of $V_{\ddot{O}}$ and adsorbed oxygen species (blue arrows) either the $V_{\ddot{O}}$ or the adsorbates migrate.*

surface (including the transfer of one or two electrons) can be assumed to be fast, the slowest of the following steps (the *rate-determining step*) will limit the overall reaction rate. Naturally, the reaction can proceed in various ways (e.g., the dissociation can occur with or without assistance of a $V_{\ddot{O}}$), and the mechanistic landscape is complex. *Ab initio* calculations offer an excellent tool to explore the reaction network (Figure 6.6) and in particular to find the slowest step in the fastest parallel branch [38].

The black lines in Figure 6.7 show the energetic profile for one of the possible reaction pathways. After exothermic adsorption the dissociation of the molecular oxygen species without assistance of $V_{\ddot{O}}$ has a barrier of 0.6 eV. For completion of the reaction, the atomic oxygen species (O^-) and oxygen vacancies have to meet. The calculations yield a barrier for migration parallel to the surface of 2.0 eV for O^-, but only 0.7 eV for the oxygen vacancies in the surface layer (see Section 6.3.1.1). Thus, it is the (very few) oxygen vacancies that approach the adsorbed O^- rather than vice versa. The subsequent incorporation of O^- into a directly neighboring $V_{\ddot{O}}$ occurs fast (without barrier). This purely energetic picture is not yet sufficient for identification of the rate-determining step and estimation of reaction rates. One has to keep in mind that the oxygen incorporation has a strongly negative overall reaction entropy (loss of translational/rotational degrees of freedom of the O_2 molecule), amounting to $T\Delta_r S^0 \approx 2\,\text{eV}$ at 1000 K. The blue lines in Figure 6.7 correspond to a (semiquantitative) estimation of the Gibbs free enthalpy profile, most of the negative $\Delta_r S^0$ occurring already in the adsorption step and resulting in comparably low adsorbate concentrations (cf. Section 6.3.1.2). In this free enthalpy profile, now the two maxima of similar height (dissociation and $V_{\ddot{O}}$ approach) represent the rate-determining steps, and it depends on the specific conditions (e.g., pO_2) which one becomes limiting.

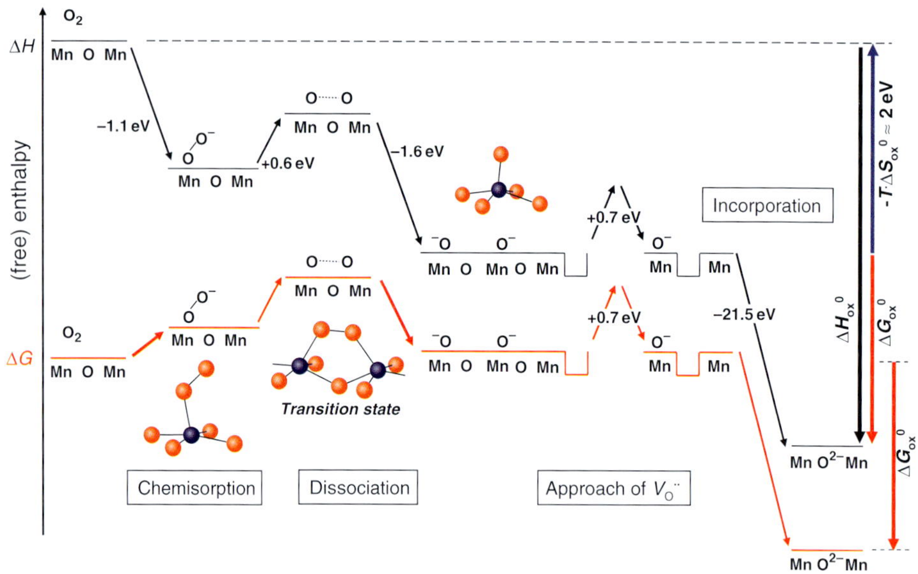

Figure 6.7 *One of the possible reaction pathways for oxygen incorporation on the* MnO_2 *(001) termination of the* $LaMnO_3$ *perovskite. The black lines are the energy profile* ($\Delta E \approx \Delta H$ *since the pV term for 1/2* O_2 *amounts to less than 0.04 eV even at 1000 K), and the red lines are a (semiquantitative) free energy profile* (ΔG) *including the effect of the overall negative reaction entropy.*

Similarly, two other reaction pathways were studied: one starting with O_2 adsorption directly into a $V_O^{\cdot\cdot}$ and the other where the dissociation occurs with assistance of a $V_O^{\cdot\cdot}$. Based on the energies of the intermediates, corresponding adsorbate coverages, and reaction barriers (combined with standard attempt frequencies), reaction rates were estimated as a function of the concentration of molecular adsorbates and surface oxygen vacancies. The O^- adsorbate concentration is assumed constant (≈0.1, cf. Section 6.3.1.2). Considering parameters typical for the $La_{0.8}Sr_{0.2}MnO_{3-\delta}$ composition actually used in SOFCs (bulk $V_O^{\cdot\cdot}$ concentration from experimental data and enhanced at surface as discussed in Section 6.3.1.1, adsorbate concentration estimated in Section 6.3.1.2, cf. [38]), as indicated by the black dotted lines in Figure 6.8, shows that the pathway depicted in Figure 6.7 does indeed yield the highest overall reaction rate. Within this pathway, at higher molecular adsorbate coverage (i.e., at high pO_2) the $V_O^{\cdot\cdot}$ approach to O^- is rate determining, while at low coverage (low pO_2) the dissociation step is the bottleneck. The mechanism with O_2 adsorption directly into a $V_O^{\cdot\cdot}$ is unfavorable for $La_{0.8}Sr_{0.2}MnO_{3-\delta}$ due to its very low bulk and surface oxygen vacancy concentration. In all of the estimates of reaction rates, we assume that electron transfer is faster than the other reaction steps, which involve either ion motion or bond breaking. This can be justified by the broad electron energy bands that are typical for perovskites.

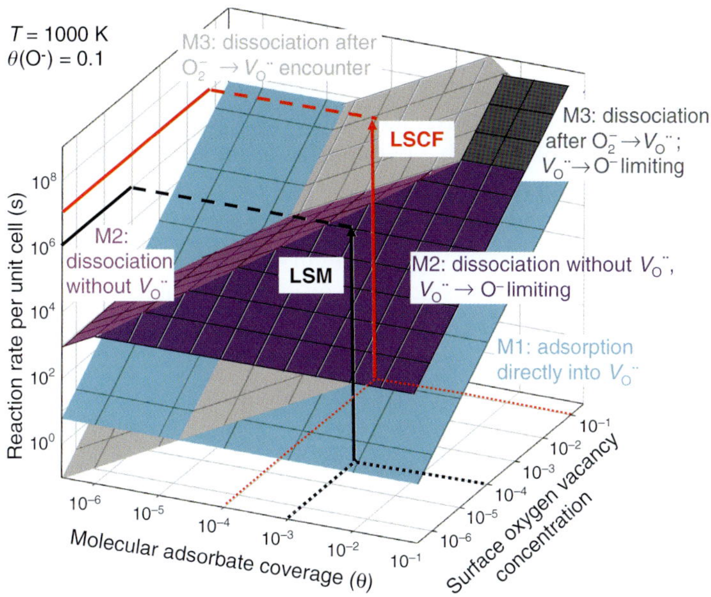

__Figure 6.8__ Reaction rates for three alternative reaction pathways of oxygen incorporation on the MnO_2 (001) surface estimated based on DFT results for energies of intermediates and barriers. Adsorbate coverage and surface oxygen vacancy concentration are given as molar fraction. The pink/magenta surfaces correspond to the pathway shown in 6.7, yielding the highest overall rate for $La_{0.8}Sr_{0.2}MnO_{3-\delta}$. (Adapted with permission from [38], copyright (2010) American Chemical Society.)

Staying within the materials family of mixed conducting perovskites and assuming that the reaction and migration barriers do not change too much, we can use Figure 6.8 also to extrapolate to related materials such as $La_{0.8}Sr_{0.2}(Fe,Co)O_{3-\delta}$ (LSCF). They have a less endothermic $V_{\ddot{O}}$ formation compared to LSM (smaller oxidation enthalpy [65,66]). This leads to lower adsorbate coverage (a less negative oxidation enthalpy is closely related to a less negative adsorption enthalpy [66]), but also to a much higher surface $V_{\ddot{O}}$ concentration. Inserting these parameters into Figure 6.8 (dotted red lines) now yields a different pathway (gray planes) as the fastest mechanism, in which the dissociation occurs with assistance of a $V_{\ddot{O}}$. In agreement with experimental data (cf. Figure 6.1), this also predicts a higher reaction rate for LSCF than for LSM.

Although oxygen adsorption is exothermic on the LaO (001) surface (Section 6.3.1.2), the oxygen incorporation rates are expected to be much lower than on MnO_2 (001). This is due to the vacancy formation being more endothermic by 0.73 eV (Section 6.3.1.1). Because the bulk oxygen vacancy concentration is very low even in Sr-doped $LaMnO_3$, the even smaller concentration on the LaO (001) termination will lead to extremely low rates for the O incorporation into the vacancies. Even if LaO and MnO_2 terminations are present in comparable amounts, the overall oxygen incorporation rate is predicted to be dominated by the kinetics on the MnO_2 termination.

In the investigation of the oxygen incorporation in the surface reaction mechanism on mixed conducting perovskites, the DFT calculations made considerable contributions, ranging from obtaining basic quantities, such as adsorption enthalpies, to elucidating the concentrations and mobilities of surface defects, and finally reaction barriers. The emerging mechanistic picture is in good agreement with the trends observed for the experimentally measured reaction rate constants. It is a fortunate point in the consideration of a realistic mechanistic pathway that the relative differences in the reaction rates of possible alternatives are greater than the uncertainties of the modeling.

Of course, one has to be aware that the calculations usually refer to an idealized surface model that neglects, for example, higher dimensional defects (e.g., steps and kinks), and may also not correctly represent dopant segregation (unless explored in very extensive calculations). Recent experiments and calculations on strained (La,Sr)$MnO_{3-\delta}$ and $LaCoO_{3-\delta}$ films have shown that moderate tensile strain favors Sr surface segregation as well as oxygen vacancy formation and oxygen chemisorption [67, 68], which may in turn change surface rate constants.

The estimate of rate constants from the *ab initio* input (energies of intermediates and transition states) can be also improved beyond the model of a single rate-determining step used here as the first approximation. The respective tools are microkinetic modeling (typically a mean field approach) and kinetic Monte Carlo simulations (as discussed in Chapters 1 and 3 of this book) [69–71]. For a fuel cell under electrical bias (current flow), one further has to consider that the relation between overpotential and electrical potential drops at the decisive solid–gas interface may be more complex than assumed for Butler–Volmer kinetics at solid/liquid electrodes [72].

6.3.2 Bulk Properties of Multicomponent Perovskites

6.3.2.1 *Oxygen Vacancy Formation in (Ba,Sr)(Co,Fe)$O_{3-\delta}$*

(Ba,Sr)(Co,Fe)$O_{3-\delta}$ (BSCF) perovskites were first optimized to have a high oxygen diffusivity for oxygen permeation membranes [73], but were soon discovered to also exhibit a fast oxygen surface reaction [74, 75]. The most studied composition is $Ba_{0.5}Sr_{0.5}Co_{0.8}Fe_{0.2}O_{3-\delta}$, which combines a high mobility of oxygen vacancies with a high vacancy concentration that can exceed $\delta = 0.5$ without formation of $V_O^{\cdot\cdot}$-ordered superstructures such as the brownmillerite structure. Before investigating the oxygen vacancy mobility in the next section, we first deal with the energetics of oxygen vacancy formation. As oxygen removal implies a gradual reduction of the material, this also requires a detailed inspection of the electronic structure.

Recently, several quantum mechanical studies on oxygen vacancy migration barriers were reported [76–79], mostly for the composition $Ba_{0.5}Sr_{0.5}Co_{0.8}Fe_{0.2}O_{3-\delta}$, which experimentally exhibits the highest diffusivity and surface exchange rate. The basis for the following discussion of the special properties of BSCF perovskites

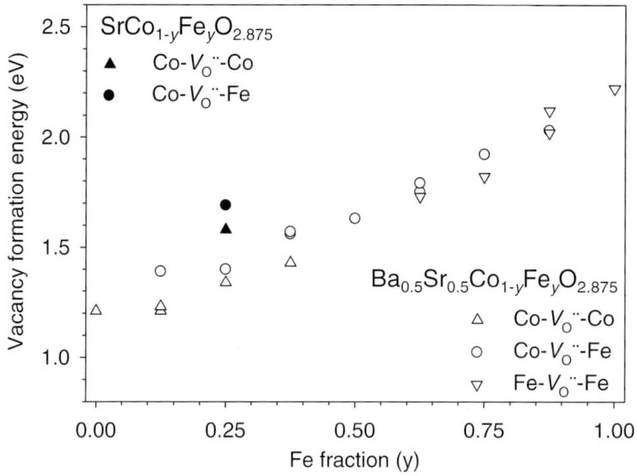

Figure 6.9 *Oxygen vacancy formation energy E_V (at 0 K) as a function of iron content y in $Ba_{0.5}Sr_{0.5}Co_{1-y}Fe_yO_{2.875}$ (open symbols) and $SrCo_{1-y}Fe_yO_{2.875}$ (solid symbols). (Reproduced with permission from [80], copyright 2012, The Electrochemical Society.)*

and their dependence on cation composition is DFT VASP calculations [80–83] with 40-atom supercells (Ba/Sr arrangement, cf. Figure 6.11a) containing one oxygen vacancy. We also modeled $Ba_{0.5}Sr_{0.5}CoO_{3-\delta}$ (BSC) and $Ba_{0.5}Sr_{0.5}FeO_{3-\delta}$ (BSF) with a cubic perovskite structure (but allowing for distortions within the supercell), although they adopt a hexagonal structure [84]. This can be justified by the fact that both can be prepared as cubic perovskites by a small additional doping (5% Ce on Fe site) or as metastable cubic perovskites by oxidation from a highly oxygen-deficient brownmillerite phase. Because the connectivity within the structure is completely different (corner-sharing vs. edge-sharing octahedra), no phase transformation will occur when relaxing the structure within the supercell.

The oxygen vacancy formation energy E_V (at 0 K) in $Ba_{0.5}Sr_{0.5}Co_{1-y}Fe_yO_{2.875}$ is shown in Figure 6.9. For a given Fe content, the E_V values vary by approximately 0.1 eV depending on the vacancy location, either between two Co ions, between Fe and Co, or between two Fe ions. Over the whole composition range, E_V increases linearly with iron content from 1.2 eV for BSC up to 2.2 eV for BSF. This smooth increase (not changing stepwise from predominant Co reduction at low y to Fe reduction at high y) reflects the pronounced electron density delocalization between Co, Fe, and O orbitals. This is in qualitative agreement with X-ray absorption spectroscopy (XAS) results [85], revealing a charge redistribution to Co and Fe as well as O ions. The vacancy formation is more endothermic (by 0.3 eV) for $SrCo_{0.75}Fe_{0.25}O_{3-\delta}$ (SCF) than for BSCF with the same iron content, in agreement with the smaller oxygen deficiency measured for SCF (in the cubic perovskite structure [86]). The effect of the barium content on the redox energetics can be related to the lattice expansion caused by the large Ba ions (SCF: $a = 3.84$ Å, BSCF: 3.90 Å) leading to an increased size mismatch for cobalt. Because a change to a lower formal oxidation state increases the ionic radius (4+ to 3+ by 0.08 Å,

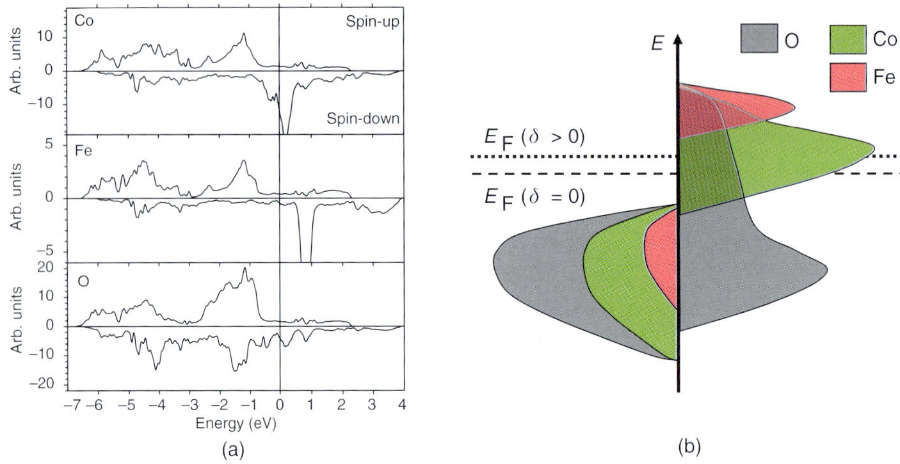

Figure 6.10 *(a) Spin-projected DOS for $Ba_{0.5}Sr_{0.5}Co_{0.75}Fe_{0.25}O_{2.875}$ derived from the GGA calculations. The oxygen vacancy is located between two Co ions. (b) Schematic DOS for oxygen-stoichiometric BSCF. The Fermi energy is indicated by the dashed line, and by the dotted line for oxygen-deficient BSCF. Spin-down DOS to the left, spin-up to the right. (Reproduced with permission from [80], copyright 2012, The Electrochemical Society.)*

3+ to 2+ by ≈0.1 Å [87]), the lattice expansion then favors the formation of lower Co oxidation states. Indeed, the highest oxygen deficiency is observed experimentally [88, 89] for Co-rich materials in the presence of a significant Ba content.

The experimental oxidation enthalpies for $Ba_{0.5}Sr_{0.5}Co_{1-y}Fe_yO_{3-\delta}$ perovskites (−58 kJ/mol for BSF at $\delta = 0.34$ and −47 kJ/ mol for $Ba_{0.5}Sr_{0.5}Co_{0.8}Fe_{0.2}O_{3-\delta}$ at $\delta = 0.52$ [90]) vary perceptibly with oxygen deficiency. This is more pronounced for Co-rich compositions, similarly to the behavior in $La_{0.6}Sr_{0.4}Co_{1-y}Fe_yO_{3-\delta}$ perovskites [91]. Although an extrapolation to $\delta = 1/8$ (used in our calculations) has quite a high numerical uncertainty, the qualitative trend of a less endothermic vacancy formation for the Co-rich materials is supported by the experimental data (larger oxygen deficiency for BSCF than for BSF).

The variation of the vacancy formation energy with Fe content can be rationalized by an analysis of the density of states (DOS) shown in Figure 6.10a, which exhibits a behavior similar to half-metals (insulator in spin-up and conductor in spin-down; note that the small but nonzero DOS at the Fermi level for spin-up electrons is due to the Fe and Co e_g orbitals [80], which correspond to metal-oxygen σ-type bonds only weakly contributing to conductivity). This is consistent with the experimental finding of a comparably modest electronic conductivity [92]. Owing to the strong covalency, Co, Fe, and O states cover a broad energy range of approximately 10 eV. This is in contrast with the DOS for typical perovskites with a large band gap (e.g., $SrTiO_3$ [93,94]), which is an insulator with an O2p bandwidth of 6 eV below the top edge of the valence band, occupied Ti states at −3 eV to −6 eV, and empty Ti and Sr states in the lower and upper parts of the conduction band. The exact nature of the two B-type ions nearest to the vacancy (Co-$V_O^{\cdot\cdot}$-Co or Co-$V_O^{\cdot\cdot}$-Fe configuration) does not noticeably affect the DOS pattern as the remaining electrons of the removed

O atom are delocalized in the whole supercell. Co exhibits a considerable DOS around the Fermi level, and thus Co^{4+} can easily transform into Co^{3+}. On the other hand, Fe shows a peak of about 1 eV higher than the Fermi level (2 eV higher than the occupied Fe states), which makes a change of its oxidation state much less energetically favorable. This is a direct consequence of the Fe high spin state and Co intermediate spin state. The resulting schematic DOS for BSCF (Figure 6.10b) is more precise than that suggested by Mueller *et al.* [85] based on semiquantitative arguments.

6.3.2.2 Oxygen Vacancy Migration in (Ba,Sr)(Co,Fe)$O_{3-\delta}$

The oxygen mobility in acceptor-doped perovskites is determined by the bottleneck in the O atom passage through the "critical triangle" formed by the B cation and the two A cations surrounding the migrating O* in the transition state (see Figure 6.11c). Allowing the O* to pass requires an outward relaxation of the A and B cations in the critical triangle, which is on the other hand limited by the other ions in the lattice, but also leads to slight rearrangements in the whole supercell (cf. Figure 6.11d). A failure to obtain these relaxations throughout the whole supercell in a consistent way for initial as well as transition states will lead to grossly wrong barriers. The calculation of the O migration barriers using the nudged elastic band

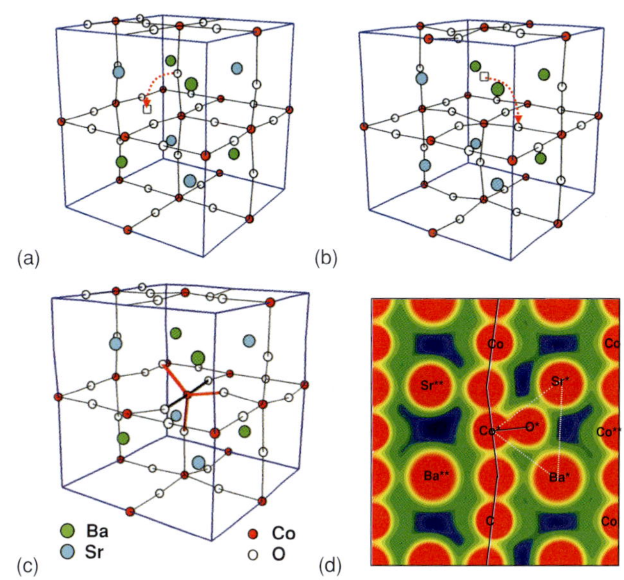

Figure 6.11 *(a) Initial state for $Ba_{0.5}Sr_{0.5}CoO_{2.875}$ with alternating arrangement of Ba and Sr; the square indicates the oxygen vacancy. (b) Initial state with modified Ba/Sr configuration. (c) Transition state for BSC; red color indicates the significantly shortened Co*–O* and Co*–O bonds. (d) Electron density map of BSC in the (110) plane for the transition state of O* migration. The black lines emphasize the structure deformations caused by O* migration; the white lines indicate the "critical triangle" of one B and two A cations around the migrating O* in the transition state. (Adapted from [83] by permission of ECS, The Electrochemical Society.)*

method for different overall cation compositions (at constant O deficiency, $\delta = 0.125$) as well as for different local configurations allows us to identify the main factors determining the barrier height.

Several factors affect the barriers only slightly. The neighboring B cations (Fe or Co) in the initial and the transition state are of minor importance (the barriers are lower by a maximum of ≈0.1 eV for a larger number of direct Co neighbors). The bond lengths to O* in the transition state are essentially identical for all BSCF compositions with the B*–O* bond being shorter by approximately 13%. Also, the decrease of the O* charge in the transition state by approximately 0.10 e_0 is similar for all (Ba,Sr)(Co,Fe)$O_{3-\delta}$.

Two factors important for the migration barrier are shown in Figure 6.12. As one might expect, the geometrical constraints in the transition state affect the barrier height. The A*–O* distance (and its change relative to r(A*O*) in BSC, which is taken as reference) proved to be a good measure. This distance changes for the

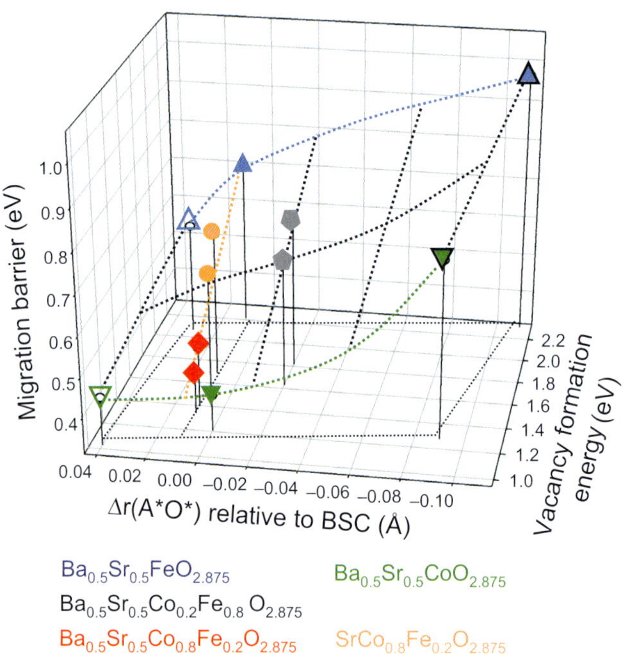

Figure 6.12 *Oxygen vacancy migration barriers E_m in $Ba_{1-x}Sr_xCo_{1-y}Fe_yO_{2.875}$ as a function of the geometric factor (deviation from A*–O* distance from the BSC values used as a reference) and the electronic factor (vacancy formation enthalpy E_V). The dotted lines are guides for the eye only. Solid symbols refer to one Ba and one Sr in the critical triangle in the transition state (TS), open symbols to a Sr,Sr, and symbols with black edge to a Ba,Ba configuration. Symbols appearing twice originate from different Co and Fe arrangements. (Adapted with permission from [80] copyright 2012, The Electrochemical Society.)*

different A-cation configurations in the critical triangle (SrSr, BaSr, or BaBa; cf. cation configurations in Figures 6.11a and 6.11b). As shown in Figure 6.12, the presence of more Ba cations enforces a decrease of this distance, in parallel with a pronounced increase of the migration barrier, from 0.4 eV to 0.75 eV in BSC and from 0.72 eV to 0.97 eV in BSF.

On the other hand, the distance $r(A^*O^*)$ is approximately constant for all O* jumps with one Ba and one Sr in the critical triangle, while the barriers increase from 0.40 eV (BSC) to 0.72 eV (BSF). In parallel, the $V_O^{\cdot\cdot}$ formation energy increases from BSC to BSF (cf. Section 6.3.2.1); thus, obviously the migration barrier and the vacancy formation energy are coupled in the BSCF perovskites. This correlation can be understood from the fact that a certain electron transfer from O* to B* occurs in the migration transition state, and the oxygen vacancy formation (involving the transfer of two electrons) gives a measure for the energetic cost of this transfer. This correlation between vacancy formation and vacancy migration energies may not naively be generalized to all mixed conducting perovskites; for example, for $(La,Sr)(Co,Fe)O_{3-\delta}$ the charge transfer in the transition state is less pronounced and thus a correlation to E_V not obvious [98]. The lowest E_V is found in the BSCF family for the perovskites with simultaneously high Ba and Co contents (cf. Section 6.3.2.1), which are also the materials with the largest size mismatch between the A and B cations. This suggests that a further improvement of the $V_O^{\cdot\cdot}$ mobility will be difficult because a too strong size mismatch, a too strong deviation of the Goldschmidt factor

$$t = \frac{r_A + r_O}{\sqrt{2}(r_B + r_O)} \tag{6.3}$$

from unity, will ultimately lead to the formation of a hexagonal instead of cubic perovskite structure. The perovskites from the $(La,Sr)(Fe,Co,Mn)O_{3-\delta}$ family [56, 57] have a higher barrier, which seems to be related to their smaller lattice parameter and probably also to a less pronounced electron redistribution in the migration transition state.

The calculated migration barriers for the Ba,Sr transition state configuration are in fair agreement with the activation energy of the vacancy diffusion coefficient $D_{V_O^{\cdot\cdot}}$ experimentally obtained for $Ba_{0.5}Sr_{0.5}Co_{0.8}Fe_{0.2}O_{3-\delta}$ (0.47 eV [95]), BSF (1.0 eV [24]), and SCF (ranging from 0.65 eV [96] to 1.0 eV [97]). Our results for barriers in $Ba_{0.5}Sr_{0.5}Co_{0.75}Fe_{0.25}O_{2.875}$ also agree with other DFT calculations [77, 79]. A comparison with calculated vacancy migration barriers in $La_{1-x}Sr_xCo_{1-y}Fe_yO_{2.875}$ perovskites is given by Mastrikov *et al.* [98]. For a complete study of ion conductivity in a multicomponent perovskite such as $(Ba,Sr)(Co,Fe)O_{3-\delta}$, the different local environments and the corresponding distribution of migration barriers (as obtained from DFT) should be considered in detail. Approaches such as the cluster expansion method (e.g., see [99, 100] for treating the spatial distribution of dopants) and kinetic Monte Carlo simulations (e.g., see [101] to extract the overall diffusivity in a landscape of varying barriers) may be applied.

6.3.2.3 Disorder and Cation Rearrangement in (Ba,Sr)(Co,Fe)$O_{3-\delta}$

The composition $Ba_{0.5}Sr_{0.5}Co_{0.8}Fe_{0.2}O_{3-\delta}$ is known to lie at the borderline of the cubic perovskite stability region, which shows up in a slow decomposition leading to the formation of an additional hexagonal perovskite observed in the temperature range of 800–1000 °C [102–104]. As the size mismatch between Co and Ba influences the oxygen vacancy formation and indirectly also the vacancy migration, as shown in the previous section, the favorable oxygen transport properties of BSCF are not independent of its limited stability in the cubic perovskite structure.

We will briefly explore this structural instability based on recent GGA VASP calculations [105, 106]. The energetics of Frenkel defect pairs (vacancy–interstitial), Schottky defects (vacancies), and cation sublattice disorder were studied along with a set of possible solid-state decomposition reactions into several phases for $Ba_{0.5}Sr_{0.5}Co_{0.75}Fe_{0.25}O_3$, which is simulated most frequently due to its high O diffusivity and exchange rate constant.

Among Frenkel defects, the oxygen vacancy–interstitial pairs have the lowest formation energy of 1.2–1.5 eV per defect, with some preference for having the oxygen vacancy between Co and Fe rather than between two Co atoms. This formation energy turns out to be much smaller than in other perovskites with small lattice constants, for example, in $SrTiO_3$ (≈10 eV [107]) with a more densely packed lattice. The dumbbell configuration of these split-oxygen interstitials is shown in Figure 6.13. The O—O distance of 1.41 Å is comparable to the bond length in adsorbed peroxide on the surface of LSM (cf. Section 6.3.1.2). While the oxygen Frenkel (vacancy–interstitial) and oxygen partial Schottky disorder (formation of vacancy + 1/2 O_2, endothermic by 1.3 eV, cf. Figure 6.9) have comparable defect formation energies in BSCF, at elevated temperatures the positive reaction entropy for the latter will favor the oxygen vacancy formation over Frenkel pairs.

In most calculations on BSCF and other perovskite solid solutions, a periodic arrangement of cations is assumed. In exploring how important the order in such a composition is, we simulated possible rearrangements in the A- and B-cation

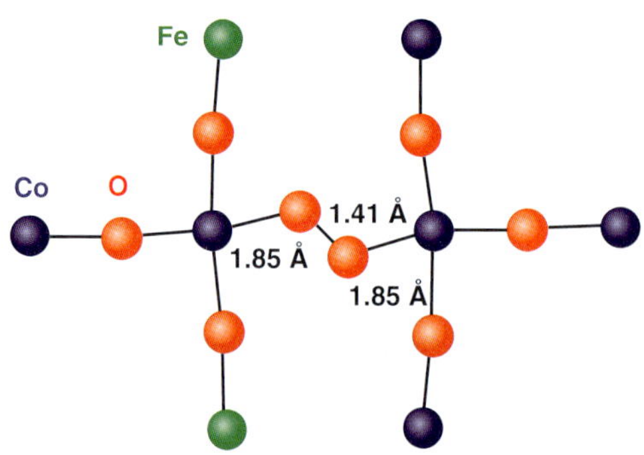

Figure 6.13 Schematic view of the split-oxygen interstitial ("peroxide") predicted in $Ba_{0.5}Sr_{0.5}Co_{0.75}Fe_{0.25}O_3$.

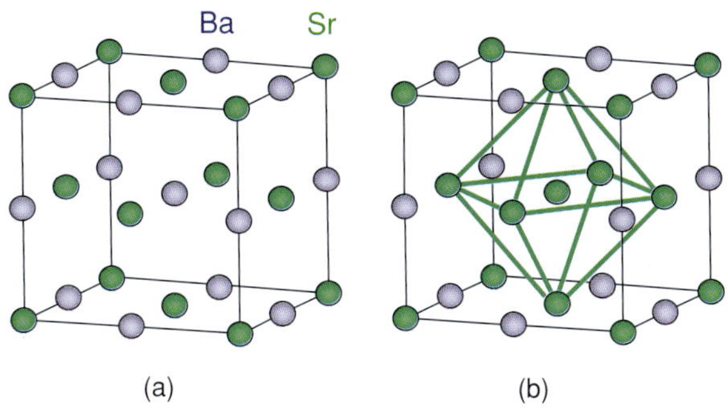

Figure 6.14 *The clustering effect caused by Sr ↔ Ba exchange in $Ba_{0.5}Sr_{0.5}Co_{0.75}Fe_{0.25}O_3$. In the original structure (a), the central A-cation site is occupied by Ba, while in the "defective" supercell (b), the site is occupied by Sr (For simplicity, only the A-cation sublattice of the ABO_3 perovskite structure is shown in a fragment of the larger 4 × 4 × 4 supercell; other parts will have a Ba excess to maintain the overall stoichiometry).*

sublattices. For example, exchanging one Ba with one Sr in the supercell causes a clustering of Sr (and Ba), as illustrated in Figure 6.14. This can be regarded as a manifestation of self-segregation in the cation sublattice.

Energies describing the cation sublattice disorder are summarized in Table 6.1. A perturbation caused by the redistribution in either the A or B-cation sublattice does not require a significant excess energy (reactions I and II). From the energetic point of view, this implies that both A metals (Ba and Sr) and B metals

Table 6.1 *Energies of cation structural disorder (within the cubic BSCF phase) and decomposition of cubic BSCF into several perovskite phases*

No.	Disorder reaction	Energy (eV)
	Cation exchange	
I	Ba ↔ Sr	−0.20
II	Co ↔ Fe	0.08
III	Ba ↔ Fe	7.87
IV	Sr ↔ Co	3.79
V	Sr ↔ Fe	5.70
VI	Ba ↔ Co	6.16
	Phase Decomposition	
VII	4 BSCF ↔ 2 $BaCo_{0.75}Fe_{0.25}O_3$ (cub) + 2 $SrCo_{0.75}Fe_{0.25}O_3$ (cub)	−0.01
VIII	4 BSCF ↔ 3 $Ba_{0.5}Sr_{0.5}CoO_3$ (hex) + $Ba_{0.5}Sr_{0.5}FeO_3$ (cub)	−0.56
IX	4 BSCF ↔ 2$BaCoO_3$ (hex) + $SrCoO_3$ (cub) + $SrFeO_3$ (cub)	−0.18

Calculations are performed for a single exchange pair of cations placed in a large 4 × 4 × 4 supercell, which consists of 64 BSCF formula units. Reactions (I) and (II) represent the clustering effect in A and B sublattices, reactions (III) and (VI) describe the formation of antisite defects, and reactions (VII) to (IX) refer to decomposition into two or three phases.

(Co and Fe) can be randomly distributed over the respective sites. This conclusion is additionally confirmed by the negative energies obtained for decomposition reactions into two phases (VII) and (VIII) described in Table 6.1. Equation (VII) may be interpreted as a complete separation of $Ba_{0.5}Sr_{0.5}Co_{0.75}Fe_{0.25}O_3$ into $BaCo_{0.75}Fe_{0.25}O_3$ and $SrCo_{0.75}Fe_{0.25}O_3$, both in the cubic phase. The decomposition of $Ba_{0.5}Sr_{0.5}Co_{0.75}Fe_{0.25}O_3$ into hexagonal $Ba_{0.5}Sr_{0.5}CoO_3$ and cubic $Ba_{0.5}Sr_{0.5}FeO_3$ perovskites according to equation (VIII) is even more exothermic due to a contribution from the cubic to hexagonal phases. Equation (IX) describes the exothermic decomposition of BSCF into perovskites with either the larger ($BaCoO_3$) or smaller ($SrFeO_3$) cations. In summary, the instability of the BSCF lattice can largely be attributed to the A- and B-site cation size mismatch. A more detailed thermodynamic analysis of phase separation is in progress.

Further, we predict that antisite substitutions (a pair of defects, in which an A metal occupies a B position of the ABO_3 lattice and the corresponding B metal fills the A position, e.g., Sr $\leftrightarrow$ Co) are in principle possible. However, as expected due to the different ion size, charge, and preferred coordination, they require significantly higher energies than the exchange of cations within the same sublattice. Table 6.1 shows that a single A $\leftrightarrow$ B cation exchange in a large supercell requires 3.79–7.87 eV. This energy corresponds to an extremely small concentration of such antisite defects and turns out to be comparable to the cation vacancy formation energies.

Strictly speaking, a direct comparison of the calculated energies to available experiments is inappropriate, although tempting, for at least three reasons. First, thermodynamic analysis at realistic temperatures is necessary, which requires quite time-consuming phonon calculations. Second, the calculated crystals are materials without O deficiency, which cannot be expected from real samples. Third, without making speculative *a priori* assumptions on compositions, an additional uncertainty complicates the analysis even more, as the relative stabilities of possible phases will strongly depend not only on defect content but also on their chemical compositions. Nevertheless, the findings are consistent with X-ray diffraction (XRD) results and transmission electron microscopy (TEM) analysis, which demonstrates that below 1073 K in air, BSCF decomposes into hexagonal and cubic perovskite phases with different cation composition [102, 103, 108]. In summary, both DFT calculations and experiments indicate a phase instability for BSCF, which will be a serious bottleneck for its application in permeation membranes and SOFC cathodes. A similar analysis of LSCF, which appears to be more stable, is in progress.

Of course, the thermodynamic treatment of phase (in)stability has to be complemented with an investigation of cation mobilities, in order to obtain the complete picture. Experimentally, cation diffusivities in perovskites are known to be much lower than oxygen diffusivity, and thus measurements are quite challenging and scarce. Even more important is the modeling of *cation* migration in perovskites preferably at the *ab initio* level of theory. A challenge is that the exact mechanism of cation migration for the two sublattices has to be resolved (e.g., the B cation can migrate along [101] or [111] or even with assistance of A-site vacancies) [109–112].

The experimental observation of comparable diffusivities for A and B cations in $BaTiO_3$ [113] and $LaMnO_{3+\delta}$ [114, 115] supports B-cation migration with the assistance of A vacancies. A combination of classical methods (exploratory studies) and quantum mechanical techniques (refinement of barriers) was employed for the modeling of Sr migration in $SrTiO_3$ [116]. Sr migration is enhanced by the presence of an oxygen vacancy, which decreases the barrier for migration of Sr to a neighboring Sr vacancy from 3.7 eV to 2.9 eV. Nevertheless, this migration energy is much higher than the typical value for oxygen migration in ABO_3 perovskites of 0.9 eV.

6.3.3 Defects in $(La,Sr)(Co,Fe)O_{3-\delta}$

Let us consider another mixed conducting SOFC cathode material (La,Sr) $(Co,Fe)O_{3-\delta}$ (LSCF). First calculations on $La_{1-x}Sr_xCo_{0.75}Fe_{0.25}O_{3-\delta}$ were performed recently using the GGA + U approach implemented into the VASP code [117]. Another method, the hybrid functional PBE0 successfully reproduces the nonmagnetic behavior of $LaCoO_3$ in contrast to the standard GGA functionals [118]. The PBE0 results suggest that $LaCoO_3$ is a semiconducting material with a band gap of approximately 2 eV. The DFT + U method confirmed the nonmagnetic nature of $LaCoO_3$ with a very narrow band gap of approximately 0.2 eV, in agreement with experiment [119] (the value of the Hubbard U parameter was optimized here as 1.5 eV [120] using the Dudarev-type of exchange–correlation functional [121]). Although the PBE0 functional overestimates the band gap, this functional reproduces the experimental phonon frequencies much better than the DFT + U [122], which is important for the calculations of thermodynamic properties.

In the next step [120], standard DFT functionals were used to calculate LSCF with and without oxygen deficiency. The distortion from the cubic structure depends significantly on the stoichiometry [123]. Thus, a particular $La_{0.875}Sr_{0.125}Co_{0.25}Fe_{0.75}O_3$ composition was studied in a 40-atom orthorhombic supercell. Different configurations for the Co/Fe and La/Sr sublattices were calculated as suggested by Fuks *et al.* [48]. The minimum perovskite formation energy was found for the superstructure with two Co atoms being direct neighbors, and Sr at largest distance from Co. This superstructure was then used for calculations with O vacancies. All degrees of freedom were relaxed, including spin magnetic moments of Co and Fe (all calculations were performed for the ferromagnetic structure). The defect formation energies were obtained using the formation enthalpy of MgO as a reference for the oxygen chemical potential [124, 125].

The free enthalpy of vacancy formation is shown in Figure 6.15 as a function of temperature for two Sr concentrations, $x = 0.125$ and 0.5, and fixed oxygen deficiency, $\delta = 0.125$, in $La_{1-x}Sr_xCo_{0.25}Fe_{0.75}O_{3-\delta}$ (one vacancy per 40-atom supercell). The temperature dependence of the gas-phase O_2 chemical potential is taken from experimental data [60]. These two dopant concentrations correspond to different transition metal oxidation states: for $x = 0.5$, Co and Fe have a formal oxidation state between 4+ and 3+ in both the oxygen-stoichiometric perovskite

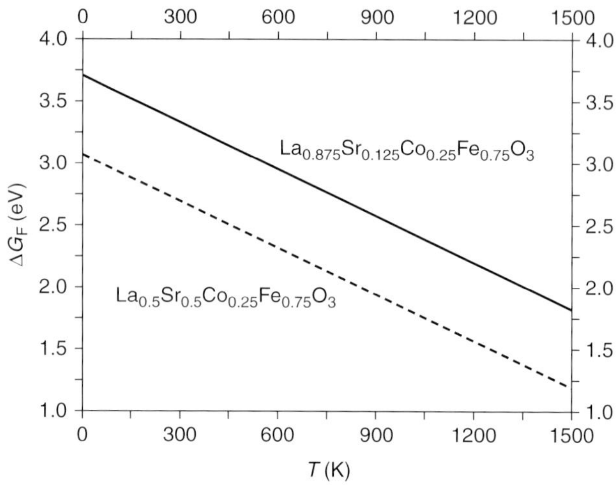

Figure 6.15 *The standard free enthalpy ΔG^0 of oxygen vacancy formation in $La_{1-x}Sr_xCo_{0.75}Fe_{0.25}O_{2.875}$ for two Sr concentrations as a function of temperature (data below 100 K are extrapolated), calculated within GGA-DFT.*

and after introducing one oxygen vacancy into the supercell. For the lower Sr doping $x = 0.125$ without oxygen vacancies ($La_{0.875}Sr_{0.125}Co_{0.25}Fe_{0.75}O_3$), the Fe/Co oxidation state remains between 4+ and 3+; however, due to introduction of vacancies ($La_{0.875}Sr_{0.125}Co_{0.25}Fe_{0.75}O_{2.875}$), it decreases to values between 3+ and 2+. Oxygen vacancy formation in the Sr-poor perovskites involves a reduction of the transition metals below the 3+ oxidation state. As one can see from Figure 6.15, the reduction in the 3+/2+ regime (Sr-poor sample) is energetically more costly than in the 4+/3+ regime. The difference in vacancy formation enthalpy is quite large (≈0.7 eV) and practically temperature independent. This is a direct result of the difference in the transition metal oxidation states, in agreement with experiments (oxidation enthalpy for $LaCoO_3$ is more negative by ≈0.8 eV compared to $La_{0.5}Sr_{0.5}CoO_{3-\delta}$ [126]; an even greater difference occurs for $La_{1-x}Sr_xFeO_{3-\delta}$ [127]). This demonstrates how carefully experimental conditions should be analyzed when comparing to theory (see also [128]).

An increase in the temperature from 300 K to 1200 K leads to a drop (≈1 eV) in the free enthalpy of vacancy formation ΔG^0 due to the considerable $T\Delta S^0$ term of the gas-phase O_2. The calculated oxygen vacancy formation energy ΔG^0 at 0 K in $La_{0.875}Sr_{0.125}Co_{0.25}Fe_{0.75}O_3$ is close to the experimental reaction enthalpy of 3.3 eV for $La_{0.9}Sr_{0.1}FeO_{3-\delta}$, averaging over the values for 4+/3+ and 3+/2+ valence change [127]. Also for the DFT + U approach [37], the vacancy formation energies for 3+/2+ valence change in $LaFeO_{3-\delta}$ and $LaCoO_{3-\delta}$ are fairly close to experiment. For $La_{0.5}Sr_{0.5}Co_{0.25}Fe_{0.75}O_3$ the agreement is less good (the experimental enthalpies are only 1.0–1.1 eV for $La_{0.5}Sr_{0.5}CoO_{3-\delta}$ and $La_{0.5}Sr_{0.5}FeO_{3-\delta}$ [126, 127]); some improvement might be achieved by including phonons (entropy contributions in the solid phase).

6.4 Ion Transport in Electrolytes: Recent Studies

Oxygen vacancy migration in SOFC electrolyte materials (mostly Y-stabilized zirconia (YSZ) but also Gd-doped Ceria (CGO) and to some degree the perovskite (La,Sr)(Ga,Mg)O_{3-z} (LSGM) and apatite-related materials) has been studied extensively mainly with interatomic potential methods but increasingly also at the *ab initio* level (e.g., see [129–133] and related work on battery electrodes in Chapter 4 of this book). Here, we want to mention some recent studies based on DFT approaches and/or focusing on interface effects. DFT calculations for acceptor-doped ceria [134] and zirconia [135] showed that depending on the actual arrangement of dopants around the migrating oxygen the barrier might change by up to 0.5 eV. For both of these fluorites, the highest barrier is found when the migrating O has to squeeze through the space between two dopants (and the barrier height increases with the dopant radius). The interaction and eventual ordering of oxygen vacancies was the focus of a molecular dynamics study using pair potentials derived from *ab initio* calculations [136]. Combining *ab initio* results with a cluster expansion approach, the dopant distribution in YSZ was investigated [137]. For zirconia doped with 8% Y_2O_3, a strong clustering of Y into nanodomains was found, which was less pronounced for higher Y content.

The effect of biaxial strain on vacancy mobility in YSZ was studied by DFT and kinetic Monte Carlo simulations in [138]. Tensile strain up to 4% was found to increase oxygen diffusivity, while larger strain leads to a severe structural distortion (Figure 6.16) and lower mobility. The increase of diffusivity at 4% strain is stronger at lower temperature (by almost 4 orders of magnitude at 400 K). While experiments also show increased diffusivity under tensile strain, the effects are much smaller (also the achievable strain was much lower [139]).

In oxide ions conducting electrolytes with fluorite as well as perovskite structures, grain boundaries typically exhibit a more or less pronounced blocking character for oxygen transport. Theoretical investigations of grain boundaries (a bicrystal being the simplest model) are challenging due to the large number of degrees of freedom to be considered (relative orientation of the crystal slabs as well as of the boundary plane, but also rigid body shifts and potential compositional variation). Particularly interesting are changes in defect formation energies in the core of the grain boundary because they can give rise to defect segregation and space charge zones [140]. For YSZ, studies were mostly performed with pair potentials (e.g., see [141]), but recently also on the *ab initio* level [142]. The obtained segregation energies depend on the dopant (more negative for Al than Y [142]), and segregation of dopant–oxygen vacancy pairs is found to be beneficial [141, 142]. Dopant segregation is also found for the surface of YSZ [143]. For YSZ films on a single-crystalline MgO substrate, a simulated amorphization and crystallization approach suggests an increased ionic conductivity in the interfacial layer [144].

For the perovskite electrolyte LSGM, *ab initio* calculations were performed for the defect formation [145] but the oxygen migration has only been studied

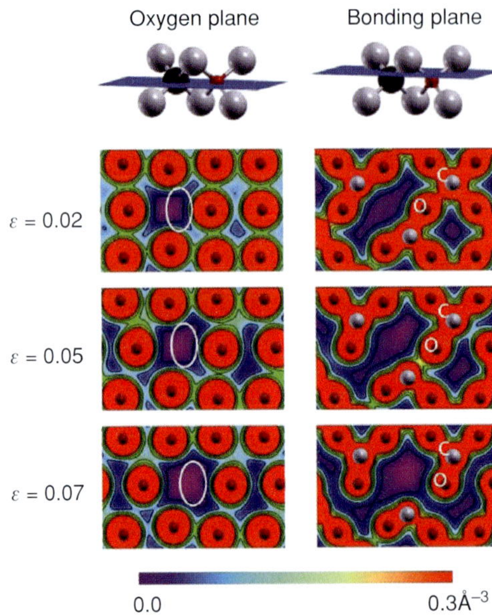

Figure 6.16 Changes in electron density in YSZ under tensile biaxial strain ε. Gray spheres, Zr; red, O; blue, oxygen vacancy. The white ellipse indicates open space for migration. (Reproduced from [138] by permission of The Royal Society of Chemistry.)

based on pair potentials so far [146]. The comparison of different cation configurations around the migrating O shows that geometrical constraints are important (as expressed by the "critical radius" [147] describing the available space for the migrating O). For Ca doping a "charge-weighted ionic radius" [146] was found to improve the correlation. While the lowest migration barriers in LSGM (0.64 eV [146]) are lower than the value of 0.8–0.9 eV found for many perovskites [56, 57], the reason for the low barriers is different than for BSCF. While for BSCF the redox properties of the B cation (related to a partial electron transfer in the transition state to Fe and Co) play a decisive role (see Section 6.3.2.2), this is expected to be less pronounced in LSGM and thus the steric constraints dominate (the "critical radius" is significantly larger in LSGM than in BSCF, facilitating fast oxygen migration).

6.5 Reactions at SOFC Anodes

Having demonstrated the case of atomistic modeling in detail for the example of cathode materials, we want to touch upon anode reactions only briefly. A complete discussion would by far exceed the length of this chapter by far. SOFC anodes can roughly be divided into two types (e.g., see [19, 148, 149]): cermets from electrolytes (doped zirconia or ceria) and metallic particles (mostly Ni, occasionally also Cu) where each phase has to form a connected network, and anodes made of electronically conducting (sometimes also mixed conducting) oxides for which a decreased sulfur poisoning and carbon deposition is expected. We will not further discuss the second case as theoretical studies for the reactions on ceramic anodes are scarce. Even for hydrogen as fuel, the experimental indications on the reaction mechanism for cermet anodes are controversial. For Ni thin films on YSZ, the reaction seems to proceed by spillover of H from Ni to the YSZ surface where water formation occurs [150, 151]. On the other hand, for porous cermets based on different metals, a volcano-shaped correlation between anode reaction rate and formation energy of the bulk metal oxide suggests that the reaction proceeds via oxygen spillover to the metal surface [152].

To confirm the O spillover mechanism, the measured reaction rates were plotted versus *ab initio* calculated adsorption energies for atomic O on the metal [153], also yielding a volcano curve, although only energies of intermediates (not transition states) were considered. Recent *ab initio* studies deal explicitly with the YSZ/Ni two-phase system and also determine some reaction and migration barriers (e.g., see [154]). Nevertheless, to get a full and reliable picture, a combination of *ab initio* determined barriers and modeling of diffusive and reaction steps is required, which also properly includes the effect of the electrical potential (the oxygen required for fuel oxidation is supplied in the form of oxide ions migrating through the electrolyte and electrons being withdrawn through the metal; and field effects on charge transfer barriers also have to be considered).

6.6 Conclusions

As demonstrated in this chapter, in recent years, first-principles materials modeling became a powerful tool for understanding atomistic details of complex processes in SOFC and permeation membranes. However, despite the great progress achieved in computer simulations, a larger number of precise large-scale calculations are required to model processes under realistic operating conditions for the relevant materials. This implies taking seriously the temperature effects and the thermochemical conditions.

The modeling results can give insight into the atomistic details of the relevant transport and reaction processes, which are difficult or impossible to address by experiments. Key examples for mixed conducting perovskites used as SOFC cathode materials are surface reaction and surface migration barriers, which reveal the important role of oxygen vacancy mobility for the actual surface reaction. For the oxygen vacancy migration in perovskites, the barriers for the different individual cation configurations can be inspected, helping to identify the parameters responsible for the low migration barriers.

Far from being able to compete with experiments for specific experimentally testable questions, modeling, though, offers an excellent complementation as it can exploit "degrees of freedom" that are not (or not readily) accessible by experimental procedures. While for some phenomena, such as conductivity in pure ion conductors, semi-empirical approaches such as pair potentials may largely cover the physics of the system, for reactions involving electron transfer—and sometimes even for ion migration in redox-active materials—*ab initio* methods are indispensable.

Acknowledgments

Authors are greatly indebted to D. Fuks, R.A. Evarestov, E. Heifets, D. Gryaznov, S. Piskunov, W. Meulenberg, J.M. Serra, R. Dronskowski, and M. Lumey for many stimulating discussions. The research leading to these results received partial funding from the German-Israeli Foundation (GIF) (grant number 1025-5.10/2009), the European Union's COST CM1104 project, and the US National Science Foundation (NSF, grant CMMI-1132451). Authors thank NSF for its support through TeraGrid resources provided by the Texas Advanced Computing Center (TACC) and the National Center for Supercomputing Applications (NCSA) under grant number TG-DMR100021. This study was also supported by a grant of computer time at the EMS Laboratory at PNNL (Project No 42498). MMK is grateful to the Office of the Director of NSF for support under the IRD Program. Any appearance of findings, conclusions, or recommendations expressed in this material are those of the authors and do not necessarily reflect the views of NSF.

References

1. Vielstich, W., Lamm, A. and Gasteiger, H.A. (eds) (2003) *Handbook of Fuel Cells*, Vol. 3 and vol. 5 (2009), John Wiley & Sons, Ltd, Chichester.
2. Kreuer, K.D. (2003) Proton-conducting oxides, *Annu. Rev. Mater. Res.* **33**, 333–359.
3. Coors, W.G. (2003) Protonic ceramic fuel cells for high-efficiency operation with methane, *J. Power Sources* **118**, 150–156.
4. Orera, A. and Slater, P.R. (2010) New chemical systems for solid oxide fuel cells, *Chem. Mater.* **22**, 675–690.
5. Münch, W., Kreuer, K.D., Seifert, G. and Maier, J. (2000) Proton diffusion in perovskites: comparison between $BaCeO_3$, $BaZrO_3$, $SrTiO_3$, and $CaTiO_3$ using quantum molecular dynamics, *Solid State Ion.* **136**, 183–189.
6. Islam, M.S., Davies, R.A. and Gale, J.D. (2001) Hop, skip or jump? Proton transport in the $CaZrO_3$ perovskite oxide, *Chem. Mater.* **13**, 2049–2055.
7. Kreuer, K.D., Paddison, S.J., Spohr, E. and Schuster, M. (2004) Transport in proton conductors for fuel-cell applications: Simulations, elementary reactions, and phenomenology, *Chem. Rev.* **104**, 4637–4678.
8. Berkelbach, T.C., Lee, H.S. and Tuckerman, M.E. (2009) Concerted hydrogen-bond dynamics in the transport mechanism of the hydrated proton: a first-principles molecular dynamics study, *Phys. Rev. Lett.* **103**, 238302.
9. Malavasi, L., Fisher, C.A.J. and Islam, M.S. (2010) Oxide-ion and proton conducting electrolyte materials for clean energy applications: structural and mechanistic features, *Chem. Soc. Rev.* **39**, 4370–4387.
10. Liu, P., Logadottir, A. and Norskov, J.K. (2003) Modeling the electro-oxidation of CO and H_2/CO on Pt, Ru, PtRu and Pt_3Sn, *Electrochim. Acta* **48**, 3731–3742.

11. Norskov, J.K., Rossmeisl, J., Logadottir, A. *et al.* (2004) Origin of the overpotential for oxygen reduction at a fuel-cell cathode, *J. Phys. Chem. B* **108**, 17886–17892.
12. Goddard, W., Merinov, B., van Duin, A. *et al.* (2006) Multi-paradigm multi-scale simulations for fuel cell catalysts and membranes, *Molec. Sim.* **32**, 251–268.
13. Gavartin, J.L., Sarwar, M., Papageorgopoulos, D.C. *et al.* (2009) Exploring fuel cell cathode materials: a High throughput calculation approach theory, anode, cathode, and oxygen evolution catalysis, *ECS Transact.* **25**(1), 1335–1344.
14. Keith, J.A., Jerkiewicz, G. and Jacob, T. (2010) Theoretical investigations of the oxygen reduction reaction on Pt(111), *Chem. Phys. Chem.* **11**, 2779–2794.
15. Mashio, T., Malek, K., Eikerling, M. *et al.* (2010) Molecular dynamics study of ionomer and water adsorption at carbon support materials, *J. Phys. Chem. C* **114**, 13739–13745.
16. Minh, N.Q. (1993) Ceramic fuel-cells, *J. Am. Ceram. Soc.* **76**, 563–588.
17. Fleig, J. (2003) Solid oxide fuel cell cathodes: polarization mechanisms and modeling of the electrochemical performance, *Annu. Rev. Mat. Res.* **33**, 361–382.
18. Adler, S.B. (2004) Factors governing oxygen reduction in solid oxide fuel cell cathodes, *Chem. Rev.* **104**, 4791–4843.
19. Jiang, S.P. and Chan, S.H. (2004) A review of anode materials development in solid oxide fuel cells, *J. Mater. Sci.* **39**, 4405–4439.
20. Jacobson, A.J. (2010) Materials for solid oxide fuel cells, *Chem. Mater.* **22**, 660–674.
21. Kröger, F.A. (1974) *The Chemistry of Imperfect Crystals*, North-Holland, Amsterdam.
22. Baumann, F.S., Fleig, J., Cristiani, G. *et al.* (2007) Quantitative comparison of mixed conducting SOFC cathode materials by means of thin film model electrodes, *J. Electrochem. Soc.* **154**, B931–B941.
23. Fleig, J., Kim, H.R., Jamnik, J. and Maier, J. (2008) Oxygen reduction kinetics of lanthanum manganite (LSM) model cathodes: partial pressure dependence and rate-limiting steps, *Fuel Cells* **8**, 330–337.
24. Wang, L., Merkle, R. and Maier, J. (2010) Surface kinetics and mechanism of oxygen incorporation into $Ba_{1-x}Sr_xCo_yFe_{1-y}O_{3-\delta}$ SOFC microelectrodes, *J. Electrochem. Soc.* **157**, B1802–B1808.
25. Catlow, C.R.A. and Kotomin, E.A. (eds) (2001) *Computational Materials Science*, IOS Publishers, Oxford, Amsterdam.
26. Mastrikov, Y.A., Heifets, E., Kotomin, E.A. and Maier, J. (2009) Atomic, electronic and thermodynamic properties of cubic and orthorhombic $LaMnO_3$ surfaces, *Surf. Sci.* **603**, 326–335.
27. Evarestov, R.A., Kotomin, E.A., Mastrikov, Y.A. *et al.* (2005) Comparative density-functional LCAO and plane-wave calculations of $LaMnO_3$ surfaces, *Phys. Rev. B* **72**, 214411.
28. Kotomin, E.A., Mastrikov, Y.A., Heifets, E. and Maier, J. (2008) Adsorption of atomic and molecular oxygen on the $LaMnO_3$(001) surface: ab initio supercell calculations and thermodynamics, *Phys. Chem. Chem. Phys.* **10**, 4644–4649.
29. Choi, Y.M., Lin, M.C. and Liu, M. (2007) Computational study on the catalytic mechanism of oxygen reduction on $La_{0.5}Sr_{0.5}MnO_3$ in solid oxide fuel cells, *Angew. Chem. Int. Ed.* **46**, 7214–7219.
30. Choi, Y.M., Mebane, D.S., Lin, M.C. and Liu, M.L. (2007) Oxygen reduction on $LaMnO_3$-based cathode materials in solid oxide fuel cells, *Chem. Mater.* **19**, 1690–1699.

31. Choi, Y.M., Lynch, D.E., Lin, M.C. and Liu, M.L. (2009) Prediction of O_2 dissociation kinetics on $LaMnO_3$-based cathode materials for solid oxide fuel cells, *J. Phys. Chem. C* **113**, 7290–7297.
32. Chen, H.-T., Raghunath, P. and Lin, M.C. (2011) Computational investigation of O_2 reduction and diffusion on 25% Sr-doped $LaMnO_3$ cathodes in solid oxide fuel cells, *Langmuir* **27**, 6787–6793.
33. Henkelman, G., Uberuaga, B.P. and Jónsson, H. (2000) A climbing image nudged elastic band method for finding saddle points and minimum energy paths, *J. Chem. Phys.* **113**, 9901–9904.
34. Henkelman, G., Johannesson, G. and Jónsson, H. (2000) *Progress on Theoretical Chemistry and Physics* (ed. S.D. Schwartz), Kluwer, New York.
35. Henkelman, G., Arnaldsson, A. and Jónsson, H. (2006) A fast and robust algorithm for Bader decomposition of charge density, *Comp. Mater. Sci.* **36**, 354–360.
36. Lyde, D.L. (1993) *CRC Handbook of Chemistry and Physics*, 74th edn, CRC Press, Boca Raton, FL.
37. Lee, Y.-L., Kleis, J., Rossmeisl, J. and Morgan, D. (2009) Ab initio energetics of $LaBO_3$ (001) (B=Mn, Fe, Co, and Ni) for solid oxide fuel cell cathodes, *Phys. Rev. B* **80**, 224101.
38. Mastrikov, Y.A., Merkle, R., Heifets, E. *et al.* (2010) Pathways for oxygen incorporation in mixed conducting perovskites: a DFT-based mechanistic analysis for (La, Sr)$MnO_{3-\delta}$, *J. Phys. Chem. C*, **114**, 3017–3027.
39. Dovesi, R., Saunders, V.R., Roetti, C. *et al.* (2009) *CRYSTAL09 User's Manual*, University of Torino, Italy.
40. Kotomin, E.A., Heifets, E., Maier, J. and Goddard, W.A. III (2003) Atomistic simulations of the $LaMnO_3$ (110) polar surface, *Phys. Chem. Chem. Phys.* **5**, 4180–4184.
41. Evarestov, R.A., Kotomin, E.A., Fuks, D. *et al.* (2004) Ab initio calculations of the $LaMnO_3$ surface properties, *Appl. Surf. Sci.* **238**, 457–463.
42. Kotomin, E.A., Evarestov, R.A., Mastrikov, Y.A. and Maier, J. (2005) DFT plane wave calculations of the atomic and electronic structure of $LaMnO_3$(001) surface, *Phys. Chem. Chem. Phys.* **7**, 2346–2350.
43. Heifets, E., Kotomin, E.A., Mastrikov, Y.A. *et al.* (2011) *Thermodynamics-Interaction Studies*, InTech Open Access Publishers, pp. 491–518.
44. Ahmad, E.A., Liborio, L., Kramer, D. *et al.* (2011) Thermodynamic stability of $LaMnO_3$ and its competing oxides: a hybrid density functional study of an alkaline fuel cell catalyst, *Phys. Rev. B* **84**, 085137.
45. Lee, C.-W., Behera, R.K., Wachsman, E.D. *et al.* (2011) Stabilization mechanisms of $LaFeO_3$ (010) surfaces determined with first principles calculations, *Phys. Rev. B* **83**, 115418.
46. Mizusaki, J., Yonemura, Y., Kamata, H. *et al.* (2000) Electronic conductivity, Seebeck coefficient, defect and electronic structure of nonstoichiometric $La_{1-x}Sr_xMnO_3$, *Solid State Ion.* **132**, 167–180.
47. Mizusaki, J., Mori, N., Takai, H. *et al.* (2000) Oxygen nonstoichiometry and defect equilibrium in the perovskite-type oxides $La_{1-x}Sr_xMnO_{3+\delta}$, *Solid State Ion.* **129**, 163–177.
48. Fuks, D., Bakaleinikov, L., Kotomin, E.A. *et al.* (2006) Thermodynamic stability and disordering in $La_cSr_{1-c}MnO_3$ solid solutions, *Solid State Ion.* **177**, 217–222.

49. Wang, Z.L. and Zhang, J. (1996) Tetragonal domain structure and magnetoresistance of $La_{1-x}Sr_xCoO_3$, *Phys. Rev. B* **54**, 1153–1158.
50. Piskunov, S., Spohr, E., Jacob, T. *et al.* (2007) Electronic and magnetic structure of $La_{0.875}Sr_{0.125}MnO_3$ calculated by means of hybrid density-functional theory, *Phys. Rev. B* **76**, 012410.
51. Piskunov, S., Heifets, E., Jacob, T. *et al.* (2008) Electronic structure and thermodynamic stability of $LaMnO_3$ and $La_{1-x}Sr_xMnO_3$ (001) surfaces: ab initio calculations, *Phys. Rev. B* **78**, 121406.
52. Fister, T.T., Fong, D.D., Eastman, J.A. *et al.* (2008) In situ characterization of strontium surface segregation in epitaxial $La_{0.7}Sr_{0.3}MnO_3$ thin films as a function of oxygen partial pressure, *Appl. Phys. Lett.* **93**, 151804.
53. Herger, R., Willmott, P.R., Schlepütz, C.M. *et al.* (2008) Structure determination of monolayer-by-monolayer grown $La_{1-x}Sr_xMnO_3$ thin films and the onset of magnetoresistance, *Phys. Rev. B* **77**, 085401.
54. Kuo, J.H., Anderson, H.U. and Sparlin, D.M. (1989) Oxygen reduction behavior of undoped and Sr-doped $LaMnO_3$ nonstoichiometry and defect structure, *J. Solid State Chem.* **83**, 52–60.
55. Denk, I., Munch, W. and Maier, J. (1995) Partial conductivities in $SrTiO_3$: bulk polarization experiments, oxygen concentration cell measurements, and defect-chemical modeling, *J. Am. Ceram. Soc.* **78**, 3265–3272.
56. Ishigaki, T., Yamauchi, S., Kishio, K. *et al.* (1988) Diffusion of oxide ion vacancies in perovskite-type oxides, *J. Solid State Chem.* **73**, 179–187.
57. Mizusaki, J., Yasuda, I., Shimoya, J. *et al.* (1993) Electrical-conductivity, defect equilibrium and oxygen vacancy diffusion-coefficient of $La_{1-x}Ca_xAlO_{3-\delta}$ single-crystals, *J. Electrochem. Soc.* **140**, 467–471.
58. Woodley, S.M., Gale, J.D., Battle, P.D. and Catlow, C.R.A. (2003) Oxygen ion migration in orthorhombic $LaMnO_{3-\delta}$, *J. Chem. Phys.* **119**, 9737–9744.
59. Alexandrov, V., Piskunov, S., Zhukovskii, Y. *et al.* (2011) First-principles modeling of oxygen interaction with $SrTiO_3$(001) surface: comparative density-functional LCAO and plane-wave study, *Integr. Ferroel.* **123**, 10–17.
60. American Chemical Society (2006) *NIST Computational Chemistry Comparison and Benchmark Database*, 14th edn, American Chemical Society, Washington, DC.
61. Fleig, J., Merkle, R. and Maier, J. (2007) The $p(O_2)$ dependence of oxygen surface coverage and exchange current density of mixed conducting oxide electrodes: model considerations, *Phys. Chem. Chem. Phys.* **9**, 2713–2723.
62. Bryant, B., Renner, C., Tokunaga, Y. *et al.* (2011) Imaging oxygen defects and their motion at a manganite surface, *Nature Comm.* **2**, 212.
63. Pilania, G. and Ramprasad, R. (2010) Adsorption of atomic oxygen on cubic $PbTiO_3$ and $LaMnO_3$ (001) surfaces: A density functional theory study, *Surf. Sci.* **604**, 1889–1893.
64. Wang, L., Maxisch, T. and Ceder, G. (2006) Oxidation energies of transition metal oxides within the GGA + U framework, *Phys. Rev. B* **73**, 195107.
65. Bakken, E., Norby, T. and Stolen, S. (2002) Redox energetics of perovskite-related oxides, *J. Mater. Chem.* **12**, 317–323.
66. Choi, Y.M., Lin, M.C. and Liu, M.L. (2010) Rational design of novel cathode materials in solid oxide fuel cells using first-principles simulations, *J. Power Sources* **195**, 1441–1445.

67. Jalili, H., Han, J.W., Kuru, Y. *et al.* (2011) New insights into the strain coupling to surface chemistry, electronic structure, and reactivity of $La_{0.7}Sr_{0.3}MnO_3$, *J. Phys. Chem. Lett.* **2**, 801–907.
68. Kushima, A., Yip, S. and Yildiz, B. (2010) Competing strain effects in reactivity of $LaCoO_3$ with oxygen, *Phys. Rev. B* **82**, 115435.
69. Stoltze, P. (2000) Microkinetic simulation of catalytic reactions, *Prog. Surf. Sci.* **65**, 65–150.
70. Hellman, A. and Honkala, K. (2007) Including lateral interactions into microkinetic models of catalytic reactions, *J. Chem. Phys.* **127**, 194704.
71. Kang, H.C. and Weinberg, W.H. (1995) Weinberg, Modeling the kinetics of heterogeneous kinetics, *Chem. Rev.* **95**, 667–676.
72. Fleig, J. (2005) On the current-voltage characteristics of charge transfer reactions at mixed conducting electrodes on solid electrolytes, *Phys. Chem. Chem. Phys.* **7**, 2027–2037.
73. Shao, Z.P., Yang, W.S., Cong, Y. *et al.* (2000) Investigation of the permeation behavior and stability of a $Ba_{0.5}Sr_{0.5}Co_{0.8}$Fe0.2$O_{3-\delta}$ oxygen membrane, *J. Membr. Sci.* **172**, 177–188.
74. Shao, Z. and Haile, S.M. (2004) A high-performance cathode for the next generation of solid-oxide fuel cells, *Nature* **431**, 170–173.
75. Baumann, F.S., Fleig, J., Habermeier, H.-U. and Maier, J. (2006) $Ba_{0.5}Sr_{0.5}Co_{0.8}Fe_{0.2}O_{3-\delta}$ thin film microelectrodes investigated by impedance spectroscopy, *Solid State Ion.* **177**, 3187–3191.
76. Ganopadhyay, S., Inerbaev, T., Masunov, A.E. *et al.* (2009) Structural characterization combined with the first principles simulations of barium/strontium cobaltite/ferrite as promising material for solid oxide fuel cells cathodes and high-temperature oxygen permeation membranes, *Appl. Mater. Interfaces* **1**, 1512–1519.
77. Ganopadhyay, S., Masunov, A.E., Inerbaev, T. *et al.* (2010) Understanding oxygen vacancy migration and clustering in barium strontium cobalt iron oxide, *Solid State Ion.* **181**, 1067–1073.
78. Stoffel, R.P., Wessel, C., Lumey, M.-W. and Dronskowski, R. (2010) Ab initio thermochemistry of solid-state materials, *Angew. Chem. Int. Ed.* **49**, 5242–5266.
79. Wessel, C., Lumey, M.-W. and Dronskowski, R. (2011) First-principles electronic-structure calculations on the stability and oxygen conductivity in $Ba_{0.5}Sr_{0.5}Co_{0.8}Fe_{0.2}O_{3-\delta}$, *J. Membr. Sci.* **366**, 92–96.
80. Merkle, R., Mastrikov, Y.A., Kotomin, E.A. *et al.* (2012) First principles calculations of oxygen vacancy formation and migration in $Ba_{1-x}Sr_xCo_{1-y}Fe_yO_{3-\delta}$ perovskites, *J. Electrochem. Soc.* **159**, B219–226.
81. Kotomin, E.A., Merkle, R., Mastrikov, Y.A. *et al.* (2011) First principles calculations of oxygen vacancy formation and migration in mixed conducting $Ba_{0.5}Sr_{0.5}Co_{1-y}Fe_yO_{3-\delta}$ perovskites, *Solid State Ion.* **188**, 1–5.
82. Mastrikov, Y.A., Kuklja, M.M., Kotomin, E.A. and Maier, J. (2010) First-principles modelling of complex perovskite $(Ba_{1-x}Sr_x)(Co_{1-y}Fe_y)O_{3-\delta}$ for solid oxide fuel cell and gas separation membrane applications, *Energy Environm. Sci.* **10**, 1544–1550.
83. Kotomin, E.A., Merkle, R., Mastrikov, Y.A. *et al.* (2010) First principles modeling of oxygen mobility in perovskite SOFC cathode and oxygen permeation membrane materials, *ECS Transact.* **35**(1), 823–830.

84. Yang, Z., Harvey, A.S., Infortuna, A. and Gauckler, L.J. (2009) Phase relations in the Ba-Sr-Co-Fe-O system at 1273 K in air, *J. Appl. Cryst.* **42**, 153–160.
85. Mueller, D.N., De Souza, R.A., Brendt, J. *et al.* (2009) Oxidation states of the transition metal cations in the highly nonstoichiometric perovskite-type oxide $Ba_{0.1}Sr_{0.9}Co_{0.8}Fe_{0.2}O_{3-\delta}$, *J. Mater. Chem.* **19**, 1960–1963.
86. Liu, L.M., Lee, T.H., Qiu, L. *et al.* (1996) A thermogravimetric study of the phase diagram of strontium cobalt iron oxide, $SrCo_{0.8}Fe_{0.2}O_{3-\delta}$, *Mat. Res. Bull.* **31**, 29–35.
87. Shannon, R.D. (1976) Revised effective ionic-radii and systematic studies of interatomic distances in halides and chalcogenides, *Acta Crystallogr. A* **32**, 751–767.
88. Harvey, A.S., Litterst, F.J., Yang, Z. *et al.* (2009) Oxidation states of Co and Fe in $Ba_{1-x}Sr_xCo_{1-y}Fe_yO_{3-\delta}$ (x, y=0.2-0.8) and oxygen desorption in the temperature range 300-1273 K, *Phys. Chem. Chem. Phys.* **11**, 3090–3098.
89. Wang, L. (2009) *PhD thesis*, University of Stuttgart, Germany.
90. Wang, L., Merkle, R., Cristiani, G. *et al.* (2008) PLD-deposited $(Ba,Sr)(Co,Fe)O_{3-d}$ thin film microelectrodes: structure aspects and oxygen incorporation kinetics, *ECS Transact.* **13**(26), 85–95.
91. Bucher, E., Sitte, W., Caramann, G.B. *et al.* (2006) Defect equilibria and partial molar properties of $(La,Sr)(Co,Fe)O_{3-\delta}$, *Solid State Ion.* **177**, 3109–3115.
92. Chen, Z.H., Ran, R., Zhou, W. *et al.* (2007) Assessment of $Ba_{0.5}Sr_{0.5}Co_{1-y}Fe_yO_{3-\delta}$ (y=0.0-1.0) for prospective application as cathode for IT-SOFCs or oxygen permeating membrane, *Electrochim. Acta* **52**, 7343–7351.
93. Kowalczyk, S.P., McFeely, F.R., Ley, L. *et al.* (1977) Electronic structure of $SrTiO_3$ and some simple related oxides (MgO. Al_2O_3, SrO, TiO_2), *Solid State Commun.* **23**, 161–169.
94. Piskunov, S., Heifets, E., Eglitis, R. and Borstel, G. (2004) Bulk properties and electronic structure of $SrTiO_3$, $BaTiO_3$, $PbTiO_3$ perovskites: an ab initio HF/DFT study, *Comput. Mater. Sci.* **29**, 165–178.
95. Wang, L., Merkle, R., Maier, J. *et al.* (2009) Oxygen tracer diffusion in dense $Ba_{0.5}Sr_{0.5}Co_{0.8}Fe_{0.2}O_{3-\delta}$ films, *Appl. Phys. Lett.* **94**, 071908.
96. Teraoka, Y., Zhang, H.M., Okamoto, K. and Yamazoe, N. (1988) Mixed ionic-electronic conductivity of $La_{1-x}Sr_xCo_{1-y}Fe_yO_{3-\delta}$ perovskite-type oxides, *Mater. Res. Bull.* **23**, 51–58.
97. Lee, T.H., Yang, Y.L., Jacobson, A.J. *et al.* (1997) The activation energy of 1.0 eV measured for D^d is expected to be also close to the $D_{V_{\ddot{O}}}$ activation energy (i.e., negligible *T*-dependence of the thermodynamic factor). *Solid State Ion.* **100**, 77–85.
98. Mastrikov, Y.A, Merkle, R., Kotomin, E.A. *et al.* (2013) Formation and migration of oxygen vacancies in $La_{1-x}Sr_xCo_{1-y}Fe_yO_{3-\delta}$ perovskites: insight from *ab initio* calculations and comparison with $Ba_{1-x}Sr_xCo_{1-y}Fe_yO_{3-\delta}$, *Phys. Chem. Chem. Phys.* **15**, 911–918.
99. Sanchez, J.M., Ducastelle, F. and Gratias, D. (1984) Generalized cluster description of multicomponent systems, *Physica* **128A**, 334–350.
100. Ducastelle, F. (1994) *Order and Phase Stability in Alloys*, Elsevier, New York.
101. Murray, A.D., Murch, G.E. and Catlow, C.R.A. (1986) A new hybrid scheme of computer-simulation based on HADES and Monte-Carlo – applications to ionic-conductivity in Y^{3+} doped CeO_2, *Solid State Ion.* **18**, 196–202.

102. Svarcova, S., Wiik, K., Tolchard, J. *et al.* (2008) Structural instability of cubic perovskite $Ba_xSr_{1-x}Co_{1-y}Fe_yO_{3-\delta}$, *Solid State Ion.* **178**, 1787–1791.
103. Mueller, D.N., De Souza, R.A., Weirich, T.E. *et al.* (2010) Mayer and M. Martin, A kinetic study of the decomposition of the cubic perovskite-type oxide $Ba_xSr_{1-x}Co_{0.8}Fe_{0.2}O_{3-\delta}$ (BSCF) (x=0.1 and 0.5), *Phys. Chem. Chem. Phys.* **12**, 10320–10328.
104. Efimov, K., Xu, Q. and Feldhoff, A. (2010) Transmission electron microscopy study of $Ba_{0.5}Sr_{0.5}Co_{0.8}Fe_{0.2}O_{3-\delta}$ perovskite decomposition at intermediate temperatures, *Chem. Mater.* **22**, 5866–5875.
105. Kuklja, M.M., Mastrikov, Y.A., Rashkeev, S.N. and Kotomin, E.A. (2011) The Structural disorder and lattice stability of (Ba,Sr)(Co,Fe)O_3 complex perovskites cathode materials, processing and performance, *ECS Transact.* **35**(1), 2077–2084.
106. Kulkja, M.M., Mastrikov, Y.A., Jansang, B. and Kotomin, E.A. (2012) The intrinsic defects, disordering, and structural stability of $Ba_xSr_{1-x}Co_yFe_{1-y}O_{3-\delta}$ perovskite solid solutions, *J. Phys. Chem. C* **116**, 18605–18611.
107. Thomas, B.S., Marks, N.A. and Begg, B.D. (2007) Defects and threshold displacement energies in $SrTiO_3$ perovskite using atomistic computer simulations, *Nucl. Instrum. Meth. B* **254**, 211–218.
108. Mueller, D.N., de Souza, R.A., Yoo, H.-I. and Martin, M. (2012) phase stability and oxygen nonstoichiometry of highly oxygen-deficient perovskite-type oxides: a case study of (Ba,Sr)(Co,Fe)$O_{3-\delta}$, *Chem. Mater.* **24**, 269–274.
109. De Souza, R.A., Islam, M.S. and Ivers-Tiffee, E. (1999) Formation and migration of cation defects in the perovskite oxide $LaMnO_3$, *J. Mater. Chem.* **9**, 1621–1627.
110. De Souza, R.A. and Maier, J. (2003) A computational study of cation defects in $LaGaO_3$, *Phys. Chem. Chem. Phys.* **5**, 740–748.
111. Jones, A. and Islam, M.S. (2008) Atomic-scale insight into $LaFeO_3$ perovskite: defect nanoclusters and ion migration, *J. Phys. Chem. C* **112**, 4455–4462.
112. Kilo, M., Taylor, M.A., Argirusis, C. *et al.* (2004) Modeling of cation diffusion in oxygen ion conductors using molecular dynamics, *Solid State Ion.* **175**, 823–827.
113. Koerfer, S., De Souza, R.A., Yoo, H.-I. and Martin, M. (2008) Diffusion of Sr and Zr in $BaTiO_3$ single crystals, *Solid State Sci.* **10**, 725–734.
114. Palcut, M., Christensen, J.S., Wiik, K. and Grande, T. (2008) Impurity diffusion of Pr-141 in $LaMnO_3$, $LaCoO_3$ and $LaFeO_3$ materials, *Phys. Chem. Chem. Phys.* **10**, 6544–6552.
115. Miyoshi, S. and Martin, M. (2009) B-site cation diffusivity of Mn and Cr in perovskite-type $LaMnO_3$ with cation-deficit nonstoichiometry, *Phys. Chem. Chem. Phys.*, **11**, 3063–3070.
116. Walsh, A., Catlow, C.R.A., Smith, A.G.H. *et al.* (2011) Strontium migration assisted by oxygen vacancies in $SrTiO_3$ from classical and quantum mechanical simulations, *Phys. Rev. B* **83**, 220301.
117. Lee, Y.L., Kleis, J., Rossmeisl, J. *et al.* (2011) Prediction of solid oxide fuel cell cathode activity with first-principles descriptors, *Energy Environm. Sci.* **4**, 3966–3970.
118. Ernzerhof, M. and Scuseria, G.E. (1999) Assessment of the Perdew-Burke-Ernzerhof exchange-correlation functional, *J. Chem. Phys.* **110**, 5029–5036.

119. Arima, T., Tokura, Y. and Torrance, J.B. (1993) Variation of optical gaps in perovskite-type 3d transition-metal oxides, *Phys. Rev. B*, **48**, 17006–17009.
120. Gryaznov, D. and Finnis, M.W. (2010) *HPC-Europa2 Research Highlights*, www.hpc-europa.eu/files/SSCinEurope/CD2010/index.html (accessed 12 December 2012).
121. Dudarev, S.L., Botton, G.A., Savrasov, S.Y. *et al.* (1998) Electronic structure and elastic properties of strongly correlated metal oxides from first principles: LSDA + U, SIC-LSDA and EELS study of UO_2 and NiO, *Phys. Status Solidi A* **166**, 429–443.
122. Gryaznov, D., Evarestov, R.A. and Maier, J. (2010) Hybrid density-functional calculations of phonons in $LaCoO_3$, *Phys. Rev. B* **82**, 224501.
123. Świerczek, K., Dabrowksi, B., Suescun, L. and Kolesnik, S. (2009) Crystal structure and magnetic properties of high-oxygen pressure annealed $Sr_{1-x}La_xCo_{0.5}Fe_{0.5}O_{3-\delta}$ (0 <= x <= 0.5), *J. Solid State Chem.* **182**, 280–288.
124. Finnis, M.W., Lozovoi, A.Y. and Alavi, A. (2005) The oxidation of NiAl: what can we learn from ab initio calculations?, *Annu. Rev. Mater. Res.* **35**, 167–207.
125. He, J., Behera, R.K., Finnis, M.W., Li, X., Dickey, E.C. and Phillpot, S.R. (2007) Prediction of high-temperature point defect formation in TiO_2 from combined ab initio and thermodynamic calculations *Acta Mater.* **55**, 4325–4337.
126. Mizusaki, J., Mima, Y., Yamauchi, S. *et al.* (1989) Nonstoichiometry of the perovskite-type oxides $La_{1-x}Sr_xCoO_{3-\delta}$, *J. Solid State Chem.* **80**, 102–111.
127. Mizusaki, J., Yoshihiro, M., Yamauchi, S. and Fueki, K. (1987) Thermodynamic quantities and defect equilibrium in the perovskite-type oxide solid-solution $La_{1-x}Sr_xFeO_{3-\delta}$, *J. Solid State Chem.* **67**, 1–8.
128. Gryaznov, D., Finnis, M., Evarestov, R.A. and Maier, J. (2013) Oxygen vacancy formation energies in Sr-doped complex perovskites: ab initio thermodynamic study. Submitted to Solid State Ion.
129. Khan, M.S., Islam, M.S. and Bates, D.R. (1998) Cation doping and oxygen diffusion in zirconia: a combined atomistic simulation and molecular dynamics study, *J. Mater. Chem.* **8**, 2299–2307.
130. Wilde, P.J. and Catlow, C.R.A. (1998) Defects and diffusion in pyrochlore structured oxides, *Solid State Ion.* **112**, 173–183.
131. Islam, M.S. (2000) Ionic transport in ABO_3 perovskite oxides: a computer modelling tour, *J. Mater. Chem.* **10**, 1027–1038.
132. Zacate, M.O., Minervini, L., Bradfield, D.J. *et al.* (2000) Defect cluster formation in M_2O_3-doped cubic ZrO_2, *Solid State Ion.* **128**, 243–254.
133. Chroneos, A., Yildiz, B., Tarancon, A. and Kilner, J.A. (2011) Oxygen diffusion in solid oxide fuel cell cathode and electrolyte materials: mechanistic insights from atomistic simulations, *Energy Environm. Sci.* **4**, 2774–2789.
134. Nakayama, M. and Martin, M. (2009) First-principles study on defect chemistry and migration of oxide ions in ceria doped with rare-earth cations, *Phys. Chem. Chem. Phys.* **11**, 3241–3249.
135. Pornprasertsuk, R., Ramanarayanan, P., Musgrave, C.B. and Prinz, F.B. (2005) Predicting ionic conductivity of solid oxide fuel cell electrolyte from first principles, *J. Appl. Phys.* **98**, 103513.
136. Marrocchelli, D., Madden, P.A., Norberg, S.T. and Hull, S. (2011) Structural disorder in doped zirconias, Part II: vacancy ordering effects and the conductivity maximum, *Chem. Mater.* **23**, 1365–1373.

137. Dalach, P., Ellis, D.E. and van der Walle, A. (2010) First-principles thermodynamic modeling of atomic ordering in yttria-stabilized zirconia, *Phys. Rev. B* **82**, 144117.
138. Kushima, A. and Yildiz, B. (2010) Oxygen ion diffusivity in strained yttria stabilized zirconia: where is the fastest strain?, *J. Mater. Chem.* **20**, 4809–4819.
139. Araki, W., Kuribara, M. and Arai, Y. (2011) Effect of uniaxial stress on ionic conductivity of 14 mol%-yttria-stabilized zirconia single crystal, *Solid State Ion.* **193**, 5–10.
140. Jamnik, J., Maier, J. and and Pejovnik, S. (1995) Interfaces in solid state ionic conductors – equilibrium and small-signal picture, *Solid State Ion.* **75**, 51–58.
141. Yoshiya, M. and Oyama, T. (2011) Impurity and vacancy segregation at symmetric tilt grain boundaries in Y_2O_3-doped ZrO_2, *J. Mater. Sci.* **46**, 4176–4190.
142. Marinopoulos, A.G. (2011) First principles study of segregation to the $\Sigma5(310)$ grain boundary of cubic zirconia, *J. Phys.: Condens. Matter.* **23**, 085005.
143. Lee, H.B., Prinz, F.B. and Cai, W. (2010) Atomistic simulations of surface segregation of defects in solid oxide electrolytes, *Acta Mater.* **58**, 2197–2206.
144. Sankaranarayanan, S.K.R. and Ramanathan, S. (2011) Interface proximity effects on ionic conductivity in nanoscale oxide-ion conducting yttria stabilized zirconia: an atomistic simulation study, *J. Chem. Phys.* **134**, 064703.
145. Kuwabara, A. and Tanaka, I. (2004) First principles calculation of defect formation energies in Sr- and Mg-doped $LaGaO_3$, *J. Phys. Chem. B* **108**, 9168–9172.
146. De Souza, R.A. and Martin, M. (2009) An atomistic simulation study of oxygen-vacancy migration in perovskite electrolytes based on $LaGaO_3$, *Monatsh. Chemie* **140**, 1011–1015.
147. Kilner, J.A. and Brook, R.J. (1982) A study of oxygen ion conductivity in doped nonstoichiometric oxides, *Solid State Ion.* **6**, 237–252.
148. Gorte, R.J. and Vohs, J.M. (2009) Nanostructured anodes for solid oxide fuel cells, *Curr. Opin. Colloid Interface Sci.* **14**, 236–244.
149. Cowin, P.I., Petit, C.T.G., Lan, R. *et al.* (2011) Recent progress in the development of anode materials for solid oxide fuel cells, *Adv. Energy Mater.* **1**, 314–322.
150. Vogler, M., Bieberle-Hütter, A., Gauckler, L.J. *et al.* (2009) Modelling study of surface reactions, diffusion, and spillover at a Ni/YSZ patterned anode, *J. Electrochem. Soc.* **156**, B663–672.
151. Goodwin, D.G., Zhu, H., Colclasure, A.M. and Kee, R.J. (2009) Modeling electrochemical oxidation of hydrogen on Ni-YSZ pattern anodes, *J. Electrochem. Soc.* **156**, B1004–B1021.
152. Setoguchi, T., Okamoto, K., Eguchi, K. and Arai, H. (1992) Effects of anode material and fuel on anodic reaction of solid oxide fuel-cells, *J. Electrochem. Soc.* **139**, 2875–2880.
153. Rossmeisl, J. and Bessler, W.G. (2008) Trends in catalytic activity for SOFC anode materials, *Solid State Ion.* **178**, 1694–1700.
154. Shishkin, M. and Ziegler, T. (2009) Oxidation of H_2, CH_4, and CO molecules at the interface between nickel and yttria-stabilized zirconia: a theoretical study based on DFT, *J. Phys. Chem. C* **113**, 21667–21678.

7

Energy Conversion: Heterogeneous Catalysis

Rutger A. van Santen, Evgeny A. Pidko, and Emiel J.M. Hensen
Department of Chemical Engineering and Chemistry, Eindhoven University of Technology, Eindhoven, The Netherlands

7.1 Introduction

Chemical reactions are essential to life, to the manufacture of products, and to energy technologies including fuel generation, storage, transport, and conversion. Without the utilization of catalysts to enhance the rates of these reactions, many of these applications would not be possible.

The topic of heterogeneous catalysis deals with the reactivity of surfaces of solid materials. According to the classical Berzelius definition, a catalyst is a substrate that directs and facilitates a reaction to specific products, but is not consumed itself during the reaction [1]. The science of heterogeneous catalysis relates chemical properties of the surface of a catalytically active solid material with its activity and selectivity in a particular chemical reaction. In contrast to the original Berzelius definition, it has become clear that actually during a catalytic reaction the state of the catalyst often changes. It has been realized that in addition to the fundamental factors determining the activity and selectivity characteristics, the relationship between the properties of the catalyst and its stability under the reaction conditions represents a key question for the science of heterogeneous catalysis.

The activation of reacting substances and the formation of new molecular fragments that by recombination give the desired product molecules is only possible

Computational Approaches to Energy Materials, First Edition. Edited by Aron Walsh, Alexey A. Sokol and C. Richard A. Catlow.

when there are strong chemical interactions between the solid surface and the reacting molecule. This interaction comes at the cost of the bonding forces between the surface atoms. The associated bond weakening may be so large that surface atoms become displaced from their equilibrium positions to form intermediate surface structures. This surface instability, induced by reaction, implies that catalysts do not necessarily remain unaltered during the catalytic reaction but may change [2]. Often these changes lead to the long-term catalyst deactivation due to the formation of more stable but less reactive surface structures. More important is the possibility of the reaction-driven formation of transient surface configurations that are much more reactive than the equilibrium structures and ultimately determine the catalytic properties of the solid catalyst. Thus, in addition to the need to model the reactivity of static catalyst surfaces, an important challenge is to understand the dynamic behavior of the surface structures and the influence of the reaction-induced surface modifications on their catalytic reactivity [2, 3].

Fundamental research in catalysis begins with the need to know the structure and chemical composition of the catalytically reactive centers, which are also relevant to many of the other technologies discussed in this book. It is often very difficult to experimentally obtain direct information on the state of the catalyst surface under the reaction conditions. Therefore, many studies in computational catalysis make use of various assumptions and simplifications related to the structure and chemical composition, which are often based on indirect experimental data. At the early stage of the development of the field of computational catalysis, when the first-principle studies using quantum chemical programs became possible, such choices were mostly determined by the computational convenience, rather than solid experimental evidence. With the current advances in computational software and hardware, the much more computationally demanding but at the same time much more realistic structures are becoming accessible for *ab initio* calculation, and thus, for computational reactivity studies. As a result there is now an increasing understanding of the structure sensitivity of many catalytic reactions [4]. This is an important topic of this chapter.

In addition to the recent progress in bridging the structural gap between modeling and experiment, there is great progress in bringing the simulations of surface structures to the catalytically relevant conditions. Thermodynamic techniques have been developed that allow predicting the actual chemical composition of the catalytic surface under chosen conditions [5, 6]. Since the state of the catalyst surface becomes predictable, studies of the reactivity of heterogeneous catalysts at the reaction condition become more and more realistic.

Thus, the state-of-the-art computational techniques allow a realistic description of reactive surface structures under the catalytically relevant conditions. However, in order to make a prediction of the macroscopic reaction rates, one has to unravel the overall mechanism of the complex sequence of elementary transformations on the catalyst surface that are involved in the catalytic reaction. A detailed model of the mechanism implies that a succession of consecutive or parallel elementary reaction steps has to be constructed. Since the mechanism of most reactions, except for very simple ones, is generally not known, the first goal of computational catalysis is to establish the very reaction mechanism at the molecular level and the energetics of the elementary reaction steps involved. Once this highly complex information is available, the relevant elementary rate constants can be calculated. The experimental validation of the computational predictions is also challenging. To validate the proposed reaction mechanism, the predicted reaction intermediates, and the catalytically relevant state of the surface, advanced *in situ* characterization tools should be applied to probe the system at a molecular level. Predicted kinetic behavior can be compared with measured kinetics of the reaction. Since often neither the catalytic site nor the mechanism is well established, the computational approach implies an iterative process of proposing the structural model for the reaction center, predicting an alternative reaction mechanism and comparing the computational results with the measured kinetics. This process will lead to the least ambiguous result when the computed data are consistent with as much molecular information one can obtain on the catalytic system as possible. Agreement with the overall catalysis without additional verification can never be convincing, because many reaction mechanistic models lead to the same overall kinetics even though they differ substantially in the chosen parameters. Clearly, for many reactions this process has not yet been completed and the mechanism of many reactions is currently subject of intensive research, which makes rapid progress. The most advanced in this iterative cycle is the ammonia synthesis reaction [2], which has been extensively discussed elsewhere.

A characteristic time scale of a reaction catalyzed by a heterogeneous catalyst is 1 s. This is to be compared with the time scale of a molecular frequency of 10^{-12} s or that of electronic motion of 10^{-16} s. Elementary reaction steps on the surface usually occur at a time scale of 10^{-6} s. The overall rate of a catalytic reaction depends on phenomena that occur at very different length and time scales. As a result a comprehensive theoretical description of a catalytic process necessitates a complementary utilization of a wide range of computational techniques. This is illustrated in Figure 7.1.

This implies that to make a first-principle comprehensive description and therefore to predict a catalytic reaction, a multiscale problem has to be solved. The codes for electronic structure calculations have evolved such that structural models can be constructed, which simulate proposed catalytic centers in a chemically realistic way [7]. Elementary reaction steps that correspond to the mechanism of a particular reaction can be studied with high accuracy. This results in the prediction of the structures and (free) energies of reaction intermediates as well as the corresponding transition states. Since entropies as well as energies can be calculated, Eyring's transition state reaction rate expressions can be deduced that predict rates at finite temperature. The key assumption to be made in the use of the transition state reaction rate expression is that the energy transfer between the reacting system and the environment is fast [8]. For a condensed system, a liquid or high-pressure gas,

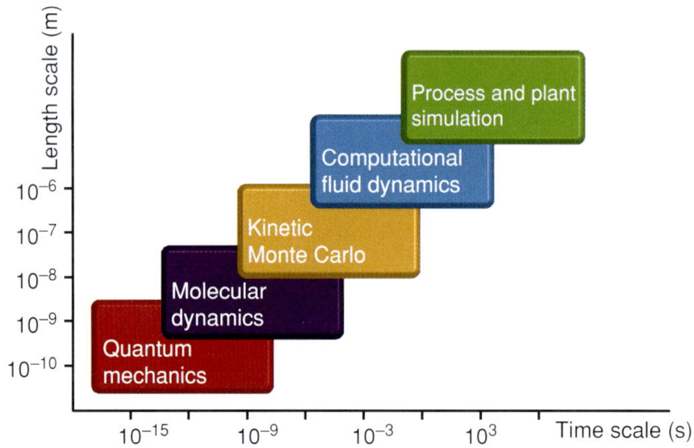

Figure 7.1 *Time and length scales of different computational techniques.*

this implies that activation energies of the reaction should be higher than 5 kJ/mol. Faster reactions such as diffusion processes with lower barriers can be studied using molecular dynamics [9]. Once elementary reaction rates are known, they can be combined into kinetic schemes by using such methods as microkinetic modeling [10] or kinetic Monte Carle techniques, which ultimately allow computing macroscopic rate constant of the overall catalytic reaction [11].

Processes that apparently occur on an even slower time scale can be practically very important. The examples of such processes can serve slow chemical processes on the catalyst surface leading to the catalyst deactivation. These processes can be very different. One should think of deactivation by coke formation, change of the surface phase by oxidation, or even deactivation of the catalyst by reconstruction of the reactive surface. Dispersed particles may slowly disintegrate and take part in sintering processes leading to the catalyst deactivation. Most of these "secondary" reactions involve mobility of the catalyst surface atoms. Methods are becoming available not only to predict the altered steady state (i.e., equilibrated), structure, and composition of the catalyst surface during reaction but also to actually study the slower kinetics of the deactivating processes [12]. The computational approaches to the investigation of these "secondary" processes are still in their early beginnings and represent a very promising avenue for future progress.

Through application and further development, computational approach in heterogeneous catalysis has become essential to formulate the conceptual basis to catalysis. It builds on the physical–chemical understanding developed in the last century. The major new methodological step is that many of the previous ideas became accessible to computational quantification and are becoming accessible to experimental validation on a molecular level.

The fundamental principle basic to any catalytic reaction is the Sabatier principle, which is already known for over a century [13]. It leads to the extensively studied phenomenon of the volcano curve dependence of the rates of catalytic reactions on the catalyst composition. A volcano-shaped curve is observed when the overall rates are plotted as a function of the substrate–surface interaction parameter. Currently, quantitative computational studies of such reactivity trends have become possible [14]. The recent deep understanding of chemical bonding on the solid surfaces applicable to many different systems has already led to new chemical insights into the fundamental origin of these phenomena. The state-of-the-art methodologies do not only enable the prediction of such interactions with great accuracy but also allow one to relate this information systematically to chemical structure and composition of the catalytic substrate. This resulted in the development of theories of reaction rates once it became realized that linear activation energy–reaction energy relationships could be derived for a wide range of surface reactions with well-understood values of the respective proportionality constants. For detailed discussions on these advances, we refer to available comprehensive reviews [14,15] and relevant books [4].

Herein, we will highlight these advances and new concepts in computational heterogeneous catalysis by applying them to our current understanding of a few classical questions in heterogeneous catalysis, which are introduced below.

7.1.1 Particle Size Dependence of Catalytic Reactivity

One of the most intriguing and important questions in heterogeneous catalysis is the dependency of the activity and selectivity of catalytic reactions on the size of catalytic particles. Figure 7.2 illustrates the experimentally observed dependencies of catalytic performance on the size of metal nanoparticles in three distinct examples of heterogeneously catalyzed reactions.

It is clear from this figure that the reactions involving cleavage of CH bonds exhibit a rapid increase of the reaction rate with decreasing particle size, whereas when CH bonds are formed in the course of the catalytic reaction, the overall rate

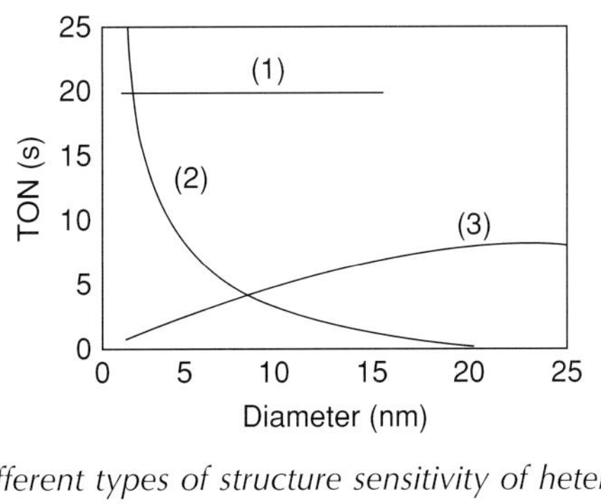

__Figure 7.2__ The three different types of structure sensitivity of heterogeneous catalytic reactions: (1) hydrogenation of benzene by Pt/SiO_2, (2) hydrogenolysis of ethane on Pt/SiO_2, and (3) CO hydrogenation to methane on Ru/Al_2O_3. (After [16].)

does not depend on the particle size. On the other hand, the rate of a reaction, in which a CO triple bond is broken, decreases along with the particle size. We will discuss the results of quantum chemical calculations that indicate that these three types of particle size dependences are associated with the differences in the optimum structure of the active site required for promoting chemical reactions, in which σ-bonds are broken or formed versus those involving a cleavage of a π-bond at the rate-controlling step [17]. These theoretical insights will be illustrated with the computational results on the mechanism of the steam methane reforming (SMR) reaction, by which methane is converted into synthesis gas [18], and the Fischer–Tropsch (FT) reaction, in which synthesis gas is converted into hydrocarbons [19, 20]. The application of microkinetic modeling and the kinetic Monte Carlo method to calculate macroscopic reaction rates and ultimately predict the catalytic activity from the molecular-level information will be discussed for the SMR and FT processes, respectively. These two reactions are important to energy conversion technology. Methane reforming is a crucial step in the overall process that converts natural gas into liquid hydrocarbon-based energy carriers. The synthesis gas produced by the reforming is further converted into diesel-type fuel by means of the FT process.

7.1.2 Activity and Selectivity as a Function of the Metal Type

Another very important fundamental question is how the selectivity and reactivity of catalytic reactions depend on the nature of the catalytic material. Figure 7.3 shows how the selectivity of the synthesis gas conversion reactions changes when different transition metals are used as catalysts.

The main differences between these reaction products can be simplified to the presence or absence of the C–O bond and the length of the hydrocarbon part in the formed molecules. This implies that the differences in actual product selectivity should be associated with the reactivity of a particular catalyst toward CO activation. The relation between the surface reactivity and the surface chemical bonding will be discussed here. Brønsted–Evans–Polanyi linear activation energy–reaction energy relationships [15] can be used to correlate the adsorption properties and chemical reactivity of catalyst surfaces.

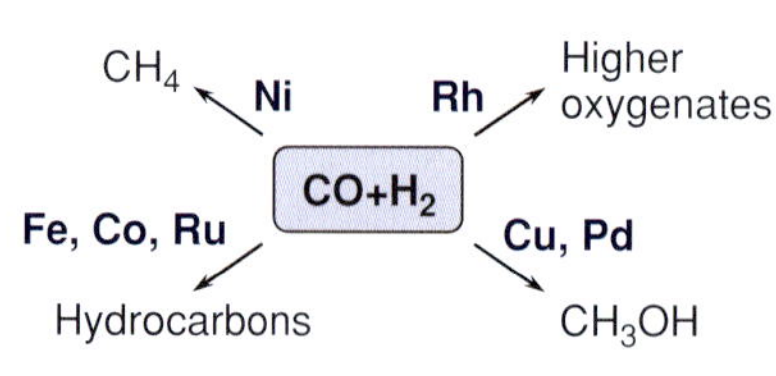

Figure 7.3 *The major products of synthesis gas conversions over different metal catalysts.*

7.1.3 Reactivity as a Function of State of the Surface

Besides the structural properties and the intrinsic nature, the chemical composition of the reactive surface is crucial for its catalytic reactivity. The role of adsorption over layers on reactivity of catalytic surfaces will be illustrated with an example of selective oxidation of ethylene by silver. This reaction is industrially very important. The epoxide product serves as an intermediate in the production of ethylene glycol, which is used as antifreeze. Depending on the oxygen coverage, the nature of the surface oxygen atoms changes from strongly nucleophilic to electrophilic [21], thus altering the predominant reaction channel from nonselective total oxidation to highly selective ethylene epoxydation.

Another example that will be considered is the dependence of the rate of the SMR reaction on the type of metal. It will be shown that the catalytic reactivity in this case depends strongly on the concentration of surface oxygen during the reaction. The activity is maximal for the metals that are most reactive in CH_4 activation and show optimum M–O bond energy [18].

7.1.4 Mechanism of Acid Catalysis: Single Site versus Dual Site

To illustrate this problem we will discuss the mechanistic aspects of chemical transformations promoted by Brønsted and Lewis acid sites in heterogeneous catalysts. Here, the question is whether the so-called single-site active components resembling well-defined catalytic species in homogeneous catalysis are responsible for the enhanced reactivity of a particular solid, or some more complex species and/or factors (e.g., self-organization of catalytic sites and confinement-induced reactivity) arising from the intrinsic heterogeneity of the latter catalysts lead to their specific chemical properties.

Protons in a zeolite are examples of single-site systems (Figure 7.4). Upon contact with organic molecules, chemical transformations occur that indicate the acidic properties of these protons [24]. The transition states of reaction intermediates are similar to the carbonium or carbenium ions observed in classical Brønsted acid catalysis. We will illustrate this with catalytic reactions involving C–C and C–H bond activation. Such reactions form the basis to the applications of zeolites in oil reforming processes. Overall catalytic reactivity of these systems relates not only

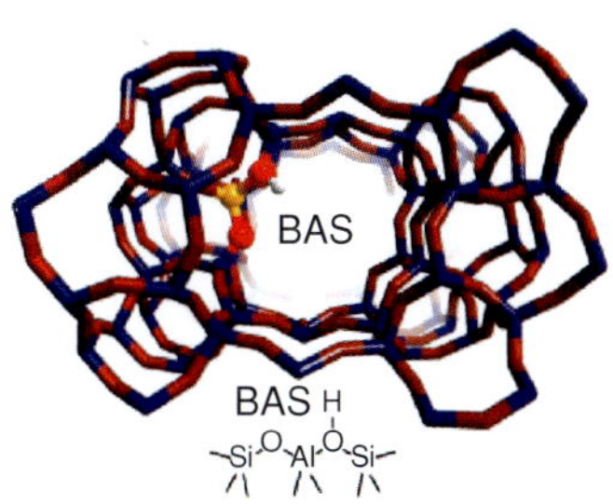

Figure 7.4 Brønsted acid proton in zeolite.

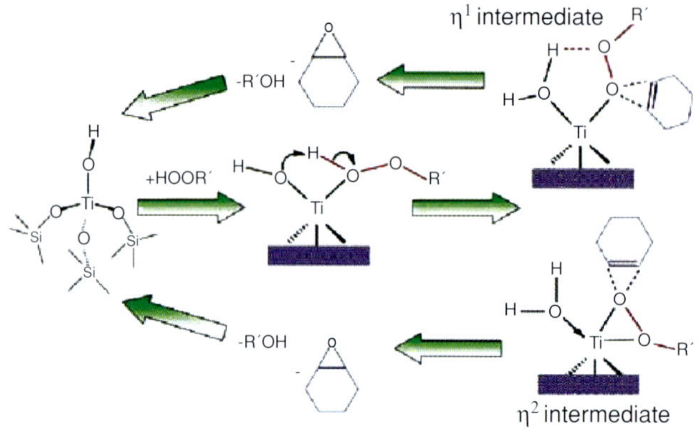

Figure 7.5 Mechanism of cyclohexene oxidation by H_2O_2/organic peroxides over a single-site Lewis acid Ti catalyst supported on mesoporous silica surface. Adapted with permission from J.M. Thomas, C.R.A. Catlow, G. Sankar., Chem. Comms, 24, p. 2921–2925 © 2010 The Royal Society of Chemistry.

to the intrinsic activity of the protons but also to the free energies of adsorption of reactants and products. These free energies can be obtained from Monte Carlo simulations [9]. It appears that the reactivity descriptors depend very strongly on the shape and size of the zeolitic pores where the reactions take place.

The prototype of a single-site Lewis acid catalyst is Ti-substituted zeolites. In such materials Ti atoms isomorphically substitute lattice silicons, thus introducing Lewis acid sites into the zeolite framework. Such materials are able to very selectively catalyze epoxydation of olefins with hydrogen peroxide. A detailed spectroscopic and quantum chemical study has been carried out on a related system of immobilized titanocene on a mesoporous silica surface [25, 26]. Quantum chemical calculations complemented with extended X-ray absorption fine structure (EXAFS) spectroscopy have allowed to elucidate the mechanism of this reaction (Figure 7.5).

This reaction proceeds via a nonmedical mechanism. During the reaction the cyclopentadienyl ligand on Ti site is replaced by a hydroxyl group producing, thus, a complex that can be converted into the actual oxidizing species by coordinating an organic peroxide molecule. Subsequent coordination of an alkene to the peroxo moiety leads to the selective oxidation involving a proton transfer to the OH ligand. This activates the oxidizing oxygen center that is introduced then into the π-bond of the olefin. This is followed by a back transfer of a proton to the alkoxy group resulting, thus, in the regeneration of the initial active complex.

Shape selective radical-type reactivity by single-site centers in oxidation reactions with molecular oxygen can be used to selectively activate alkanes [27]. The replacement of oil-derived olefins by alkanes originating from natural gas for the production of bulk chemicals is of increasing interest. Here, the catalytic systems are the zeolitic materials with the $AlPO_4$ framework composition. Reducible Mn or cocations substituted at Al positions are catalyze hydroperoxide formation from activated O_2 and the hydrocarbon as elegantly illustrated in a computational study by Catlow *et al.* [28]. The specific confinement properties of the zeolitic cavity allow for stereochemical control of the reaction.

Oxygenated cationic species stabilized at the extraframework positions within the microporous cavities of zeolites often show unique chemical reactivity. Depending on the type of the metal center in these complexes, specific Lewis acid or redox properties can be introduced into the zeolite. Chemical properties and catalytic reactivity of such sites appear to depend strongly on the nuclearity of such species. In other words, it is important whether these sites are present as isolated single-site species or self-organize in the zeolitic matrix into dual- and multi-site aggregates.

These phenomena will be illustrated by discussing the Lewis acid–base behavior of Ga-containing extraframework sites in high-silica zeolites. The difference in basicity of the oxygen ligands, which are the part of the oxycationic cluster and those which are the part of the zeolite lattice, will be demonstrated to be the main reason for the lower reactivity of the single-site systems. In addition, a relevant example of oxidation catalysis by extraframework Fe sites in high-silica zeolites will be given. In this case, the Panov reaction upon which benzene is selectively converted into phenol with N_2O will be considered. A long debate has been held on the oxidation state and nuclearity of the catalytic iron sites responsible for the selective benzene oxidation. By using fully periodic density functional theory (DFT) calculations on realistic zeolite models, these questions can be finally convincingly addressed. It appears that in this redox reaction formation of reactive Lewis basic oxygen atoms has to be suppressed.

7.2 Basic Concepts of Heterogeneous Catalysis

Theoretical catalysis has developed several important concepts that we will present in this section as a necessary background knowledge that will be used in the sections that follow. One of the most crucial concepts for understanding the relationship between the surface structure and the surface interaction energies is the bond-order conservation principle (BOCP) [29].

The interaction with an atom or a molecule with transition metal changes the electronic structure of both the adsorbed species and the metal atoms involved in the surface chemical bond. Upon the formation of the chemical bond with adsorbate, the interaction between the metal atoms within the surface weakens. The essential assumption made according to the BOCP [30] is that the valency of a metal atom is constant. The sum of the bond orders (x) of all bonds with a surface metal atom should be equal to its valency. This implies that the total bond strength can be described as a sum of two-body interactions and that the valency or the total bond order of an atom does not depend on the number and the type of the neighbors. The latter actually means that the total bond order or valency is conserved. The strength of the chemical bonds can then be expressed via the bond orders as proposed by Shustorovich in the following formula:

$$Q_n^{\mathrm{t}} = Q_{\mathrm{O}} \sum_{i=1}^{n} \left(2x_i - x_i^2\right) = Q_{\mathrm{O}} \left(2 - \frac{1}{n}\right), \tag{7.1}$$

where Q_n^{t} is the total bond strength between an atom and all its neighbors, which are assumed to be identical. There are n neighbors. Q_{O} is the individual bond strength between two atoms at equilibrium.

By dividing Q_n^{t} by n, one deduces that the bond strength of an individual chemical bond decreases when the number of neighbors of a metal atom increases. However, this decrease becomes less pronounced when the number of neighbors increases.

By using BOCP one can readily explain why the interaction energies with a surface atom increase when its coordination number to neighboring metal atoms decreases (see also [4, 15]). This principle can also be used for a qualitative description of the strength of bonding of adsorbates with surface atoms as a function of the adsorbate structure. When the number of substrate atoms coordinated to a surface atom increases, the interaction within the adsorbate–surface complex decreases.

The second important concept is a linear activation energy–reaction energy relationship that can be deduced for elementary surface reactions [15]. In heterogeneous catalysis, this concept is usually called the Brønsted–Evans–Polanyi (BEP) relationship and is described by the following formal equations:

$$\delta E^{\#} = \alpha \delta E_{\mathrm{R}} (0 < \alpha < 1). \tag{7.2a}$$

with the BEP coefficient

$$\alpha = \frac{1}{2}\left(1 + \frac{E_{\mathrm{R}}}{E_{\mathrm{O}}^{\#}}\right). \tag{7.2b}$$

For the reverse reaction the BEP relationship is then

$$\delta E^{\#} = -(1 - \alpha)\delta E_{\mathrm{R}}. \tag{7.2c}$$

$E^{\#}$ is the activation energy of a reaction, and E_{R} is the reaction energy. $E_{\mathrm{O}}^{\#}$ is the activation energy for the system for which the reaction energy $E_{\mathrm{R}} = 0$.

For many surface elementary reaction steps, these relations have been validated. The relation applies as long as there is no major change in the transition state structure of the reactions that are compared. For dissociation reactions of CO or N_2 into surface adatoms, the value of α is typically 0.9. The activation energy depends very strongly on the type of metal catalyst. Note that equation (7.2c) implies that the reverse recombination reaction of the surface adatoms to produce the molecule is independent of the nature of the catalyst surface. The value of α close to 1 implies that the transition state of dissociation is close in structure to the dissociated product state. Its mobility is low and hence the transition state entropy is close to zero. Thus, the values of activation enthalpies and activation free energies for elementary reaction steps on a catalyst surface are usually similar. This is generally the case for reactions that take place on surfaces of solids. Note, however, that for adsorption or desorption processes the values of the entropy contributions become very substantial because of the decreasing or increasing, respectively, degrees of freedom in the species involved. Therefore, entropical contributions to reaction free energies have to be explicitly taken into account when describing these reaction steps.

The third important catalytic concept is the Sabatier principle. This principle gives rise to the frequently observed bell-shaped dependences when the rate of a catalytic reaction is plotted as a function of the interaction energy of reagent molecules with the catalyst surface [31]. It essentially relates to the kinetic consequence of the cyclic nature of a catalytic reaction event. When the catalytic cycle is completed, the active site must become again available for reaction. The rate of the catalytic reaction is maximum for an optimum value of the interaction energies between the surface and the reagent. For example, if the affinity between the reagents and the catalyst surface is low, there is only a small chance that they collide and the reaction event occurs. In this case, the low rate of the initial activation of the reagents will lead to a decrease of the overall rate of the catalytic process. Similarly, when the interaction between the catalyst surface and the converting molecules is too strong, the regeneration of the active surface sites will require too high energies and again the overall catalytic reaction will be slow. Only when the elementary rates of the initial reagent activation and the final product desorption balance each other, the overall reaction rate is maximal. For a two-step reaction such as the dissociation of a diatomic molecule followed by hydrogenation of the resulting atoms to produce product molecules, one finds that the rate can be formulated as a function of the parameter λ:

$$\lambda = \frac{r_{\mathrm{H}}}{r_{\mathrm{diss}}}. \tag{7.3a}$$

Here, λ is related to the surface coverage θ:

$$\theta = \frac{1}{1+\lambda}. \tag{7.3b}$$

The overall rate of the reaction is given by

$$r = \frac{r_{\mathrm{H}}}{1+\lambda}. \tag{7.3c}$$

As illustrated in Figure 7.6, the rate maximum is found for a value of λ_{max}, which depends on the ratio of the two BEP constants of the rate of the initial activation of the reagents (α_1) and that of the final product desorption and the active site regeneration step (α_2) [15]. This value determines the conditions at which rates of dissociation and product formation are in balance and the overall rate of the catalytic reaction is maximum.

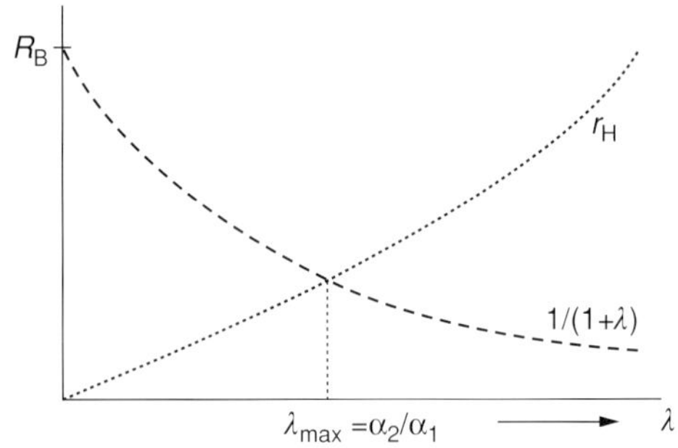

Figure 7.6 The plot of rate r of diatomic dissociation atoms. The overall rate r is given by the crossing of normalized rate of surface atom hydrogenation r_H and $1/(1 + \lambda)$.

7.3 Surface Sensitivity in CH Activation

As it has been introduced in the beginning of this chapter, understanding of how catalytic reactivity depends on the nature, structural properties, and size of catalytic species represents one of the grand challenges for the science of heterogeneous catalysis. In this section, we will discuss the phenomenon of structure sensitivity in metal catalysis with the examples of reactions involving activation of CH bonds and formation of CO bonds.

7.3.1 Homolytic Activation of CH Bonds

A characteristic of the structure sensitivity of reactions, in which the overall rate is determined by the rate of the initial activation of CH bonds in saturated hydrocarbons, is the uniform increase of the rate of alkane conversion along with the decreasing particle size. Such a behavior corresponds to the situation (2) in the schematic graph in Figure 7.2 and is further illustrated by the experimental results for the SMR reaction catalyzed by different metals summarized in Figure 7.7. During this reaction methane is converted with steam into CO and H_2. Because the

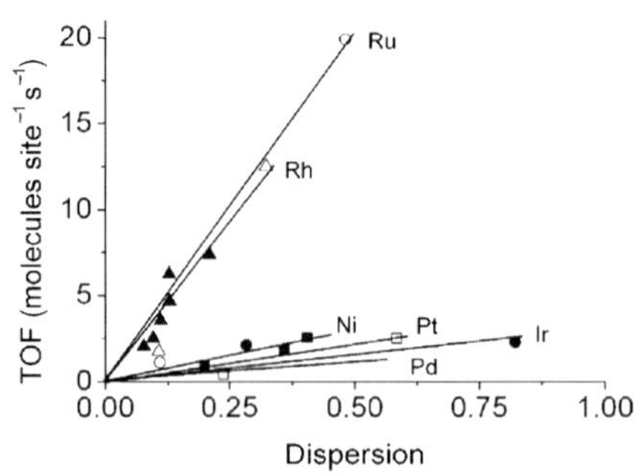

Figure 7.7 Particle size dependency of metal catalysts in SMR reaction. Reprinted with permission from G. Jones et al Journal of Catalysis, 259(1): p. 147–160 © 2008, Elsevier.

water–gas shift reaction that converts CO with H_2O into CO_2 and H_2 is fast, the outcome of the SMR process is a mixture of CO_2, CO, and H_2.

The experimentally observed rate enhancement is associated with the increasing ratio of unsaturated corner and edge atoms over terrace atoms on the particle surface when its size decreases [17]. The associated increasing coordinative unsaturation of the exposed metal atoms leads to a stronger activating interaction with the CH bonds of an alkane (e.g., the formation of σ-complexes, in which the bonding and antibonding CH orbitals are involved in strong donor–acceptor interactions with the surface). These interactions lower the corresponding activation energies as can be deduced from the results of DFT at the level of the generalized gradient approximation summarized in Table 7.1.

Such a behavior is not only associated with the increased strength of the interaction with the coordinatively unsaturated surface atoms. It is also due to the particular structure of the transition state of the CH activation. It closely resembles the transition states known for oxidative addition reactions within organometallic complexes. The homolytic CH bond dissociation is a reaction that proceeds over a single metal center. In the transition state, the dissociating C and H atoms are interacting with a single metal atom from the catalyst surface (Figure 7.8). Thus, according to the BOCP, the lower the number of neighbors a catalytic site has, the more efficient the bonding will be with the newly formed CH dissociation products.

Similar to the oxidative addition reactions of organometallic complexes, the CH bond cleavage over a metal surface implies that the two new bonds are formed with the formally negatively charged hydrogen and the alkyl fragments. Thus, the reactivity of this reaction step should strongly depend on the affinity of a particular metal surface to hydride and alkyl species. The CH dissociation activity dependency

Table 7.1 *Comparison of CH_4 activation energies on different metal surfaces [15]*

	$E^{\#}$ (kJ/mol)
Ru(0001)[a]	76
Ru(1120)[b]	56
Rh(111)[c]	67
Rh step[c]	32
Rh kink[c]	20
Pd(111)[c]	66
Pd step[c]	38
Pd kink[c]	41
Pd atom[d]	5

[a]Ciobîca *et al.* [33].
[b]Ciobîca *et al.* [34].
[c]Liu and Hu [35].
[d]Diefenbach *et al.* [36].

Figure 7.8 Transition state for methane activation on a dense metal surface (the figure is the result of a periodic structure DFT calculation). (After [33].)

on the nature of the transition metal is illustrated by the DFT-computed activation barriers for methane dissociation on dense surfaces of different metals (Table 7.2). The highest reactivity is found for such metals as Ru, Rh, Ir, and Pt, whereas Co and Ni are less reactive toward CH activation.

During the SMR process, a range of surface-catalyzed processes take place in a consecutive as well as in a parallel manner. At the initial stage of the reaction, methane is activated on the surface resulting in the adsorbed CH_x fragments. Besides that, water also dissociatively adsorbs on the surface. To produce CO, the adsorbed CH_x fragments have to react with oxygen adatoms resulting from the dissociative water adsorption. DFT calculations evidence a pronounced surface structure dependency of the activation barriers for this recombination reaction. The dependency on the type of metal is only weak though. This is in line with the BEP relationship in equation (7.2c) for the recombination reactions. On dense surfaces the barriers for the CO recombination are in the order of 160–180 kJ/mol [18]. On more open stepped surfaces, these values decrease to below 100 kJ/mol. H_2O dissociation proceeds with low barrier when preadsorbed oxygen atoms are present at the surface. Such an O_{ads}-assisted water splitting yields adsorbed OH species. This reaction also shows only a weak dependence on the metal type because it involves a simultaneous formation of a new metal oxygen bond and the weakening of another one. These two elementary processes effectively compensate each other. Oxygen adatoms are generated by further splitting of adsorbed OH species. The activation barriers for this reaction depend strongly on the strength of the M–O

Table 7.2 Activation energies for methane dissociation over different metals (kJ/mol)

	100	118
	Co(0001) [37]	Ni(111) [37]
77	67	100
Ru(0001) [38]	Rh(111) [35]	Pd(111) [35, 39]
	~40	75
	Ir(111) [40]	Pt(111) [38,39]

surface bond, in other words, on the affinity of a particular metal toward oxygen. The weaker this bond, the more difficult the OH bond cleavage [41].

Clearly, the strong structure dependence of the $CH_{ads}-O_{ads}$ recombination implies that based on a comparison of the calculated activation energies, catalytic nanoparticles that expose more open stepped surfaces will exhibit higher reactivity in this reaction. Because these surfaces are characterized by higher surface energies, they are expected to be present with a reasonable probability only for larger particles. When particle size decreases, only the dense nonreactive surfaces will be exposed resulting in the increase of the activation energies for the recombination reaction. This expected particle size dependence is inconsistent with the observed increase of the overall rate of SMR reaction with increasing dispersion of the transition metal particles. This further supports the conclusion that the initial methane activation is the rate-controlling step because of the high activation free energy of this reaction.

Since the activation of methane is a reaction between a gas phase molecule and a surface, there is a large entropy loss associated with the loss of translational and rotational degrees of freedom upon the surface reaction. The activation free energy for methane dissociation, that for thermodynamic reasons has to occur at high temperature, is approximately 200 kJ/mol at a temperature of 700 K. This is indeed dramatically different from the respective activation enthalpies, which are generally in the order of 30–70 kJ/mol, discussed above.

The difference in reactivity of the metals only partially relates to the differences in their ability to activate methane molecules. To produce CO surface, CH_x fragments have to recombine with adsorbed oxygen atoms, which are generated upon water dissociation. It has been established that the minimal adsorption energy of atomic oxygen that is sufficient to compensate for the energy losses due to water dissociation must be equal to 480 kJ/mol. Table 7.3 summarizes the DFT-computed energies of adsorption of atomic oxygen on dense surfaces of different metals. From the data presented, one can conclude that the dense metal surfaces of such metals as Pt and Ir are unreactive. Rh is borderline, and Ru is reactive toward water dissociation. Thus, one can conclude that ruthenium is a superior catalyst for the SMR reaction because it exhibits low barriers for the initial CH activation of methane and shows a very high reactivity in water dissociation.

Table 7.3 *Computed O adsorption energies on (111) metal surfaces*

Fe (hcp)	**Co**	**Ni**	*Cu*
−714	**−550**	**−496**	*−429*
Ru	Rh	*Pd*	*Ag*
−557	−469	*−382*	*−321*
Os	*Ir*	*Pt*	*Au*
−533	*−428*	*−354*	*−270*

Italic and **bold** refer to strong and weak binding, respectively.

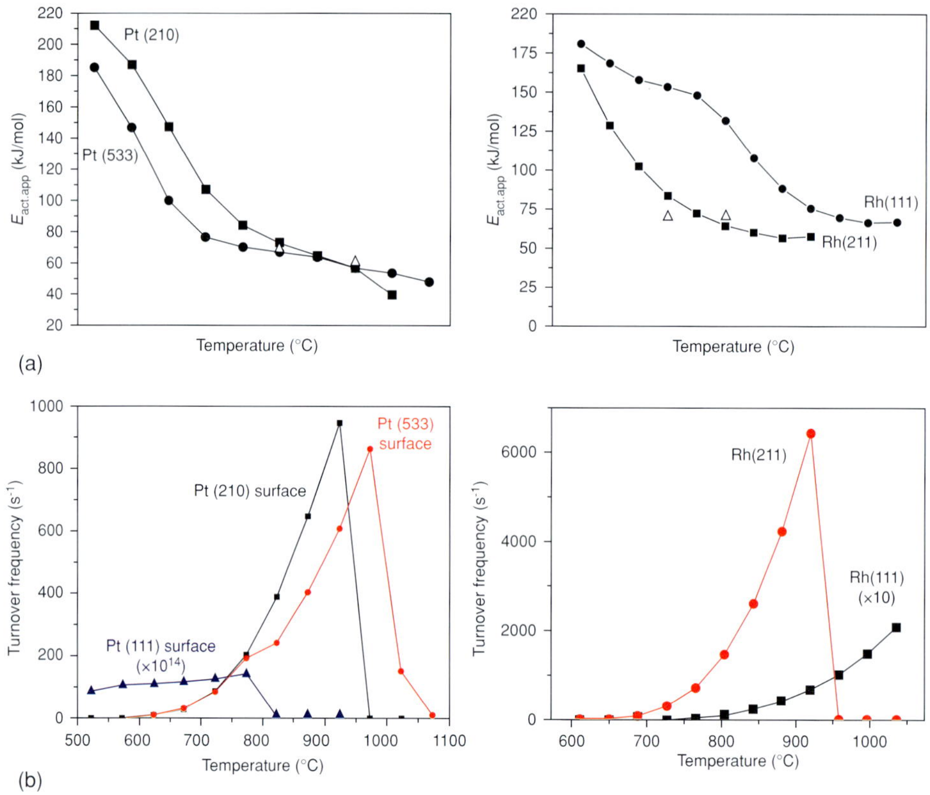

Figure 7.9 *(a) Activation energies (top) and (b) turnover frequencies (bottom) for SMR reaction on Rh (111), Rh (211), Pt (210), and Pt (533) surfaces calculated by using a microkinetic modeling approach. The triangle blocks indicate the experimental data (P.W. van Grootel, T.W. Zhu, and E.J.M. Hensen, unpublished results).*

One can see that by using quantum chemical calculations, one can unravel the detailed picture of the mechanism of surface-catalyzed reactions as well as compute individual energy parameters for elementary reaction steps. However, this information is not enough to directly compare the mechanistic details derived from quantum chemical modeling with the actual experimentally determined rates of catalytic reactions. Thus, we will conclude this section by illustrating the consequences of differences in metal surface reactivity for the actual kinetics of Pt and Rh-catalyzed SMR reactions (Figure 7.7).

In microkinetic modeling, rate constants of elementary reaction steps are used in the mean field differential equations that describe the kinetics of the reaction [10]. The output of such a simulation is production rate and surface concentration. The results presented in Figure 7.9 show large differences in the production rate of CO and H_2 (expressed in turnover frequencies) for different surfaces as well as for different metals. The differences found for Pt and Rh agree with the experimentally

Figure 7.10 The stepped surface site with transition state of CO recombination.

observed trend of Figure 7.8. This is essentially due to the different reactivity of the two metals toward H_2O.

In the case of platinum, the overall reaction is limited by the dissociation of water. The large difference between the reaction rates on the open versus the dense surfaces relates to the very different rates of atomic C or CH recombination with atomic oxygen to give CO. The recombination of adsorbed C and O to CO has a very different surface topological requirement than the formation or cleavage of CH bonds. A transition state corresponding to the low barrier dissociation or recombination of CO is shown in Figure 7.10.

The difference in the activation barriers of CO recombination on a stepped surface (as the (211) surface) and a planar surface (as the (111) surface) can be of the order of 100 kJ/mol. This structural dependence of the CO activation barrier also has important consequences for the FT reaction as we will see later. At very high temperature, the methane reforming reaction becomes poisoned by carbon deposition.

7.3.2 Heterolytic Activation of CH Bonds

Very different compared to the activation of CH bonds by transition metals are the mechanism and therefore the fundamental trends observed for catalytic reactions involving the activation of CH bonds by nonreducible oxygen-containing cationic complexes or even by protons. Reactions over such sites proceed through a heterolytic bond cleavage. In this section, we first discuss reactions involving activation of alkanes by protons in zeolites. This is followed by a discussion of CH activation by oxycationic clusters confined within the zeolite micropores. The main difference in the catalysis by these two systems is that activation of CC bonds and CH bonds in alkanes occurs with comparable activation energies by zeolitic protons. In contrast, the oxycationic clusters with Ga or Zn will only catalyze CH bond activation, without competitive CC bond activation [42–44].

7.3.2.1 Brønsted Acid Catalysis

Alkanes in contact with acidic protons will react through the intermediate formation of carbonium ions sketched in Figure 7.11. If the CH bond of an alkane is attacked by the acidic proton of the zeolite, an H_2 molecule is formed and an alkyl cation that readily attaches to the negatively charged basic oxygen atom of the zeolite lattice resulting in the formation of surface alkoxyls. Alternatively in the transition state, the proton attack can take place by the C–C bond. This splits the alkane molecule into a shorter alkane and a shorter grafted alkoxy group. Activation barriers for the reactions involving *n*-alkanes are typically of the order of 200 kJ/mol. The activation energy for H_2 formation is generally lower by 10–20 kJ/mol compared to the C–C bond cleavage reaction. These values are in line with the weaker CH bond energy compared to that of the sigma C–C bond. The relatively high activation energies are the reason for the need of highly reactive (acidic) protons for this reaction. It also explains the high temperature necessary for the reaction.

The reaction is to be considered a Brønsted acid–Lewis base catalyzed reaction because it involves the stabilization of the positively charged carbocationic intermediate by the negatively charged Lewis basic lattice oxygen atom. This stabilization is essential for the reaction to have an acceptable activation barrier. These positively charged intermediates form alkoxy species covalently bonded to the lattice oxygen atoms. A proton is back donated to the zeolite lattice by a β-H transfer reaction from the alkoxy intermediate resulting in an olefin molecule.

Activation of branched alkanes shows a related chemistry. The activation barrier for the dehydrogenation follows the stability trend of carbocations. The barrier is the lowest when a tertiary C–H bond is activated and the highest in the case of the activation of the primary CH moiety with the secondary showing the intermediate reactivity. This is due to the decreasing stability of the corresponding carbenium ions that are generated as a part of the transition states. Sometimes due to the steric repulsion effects exerted by the zeolitic walls, these carbenium ions can be considered true intermediates and no alkoxy species are formed [45].

To compute the relative free energy of a free carbenium ion versus alkoxy intermediate, methods based on a combination of DFT calculations and statistical thermodynamics have to be used that properly estimate the corresponding entropy

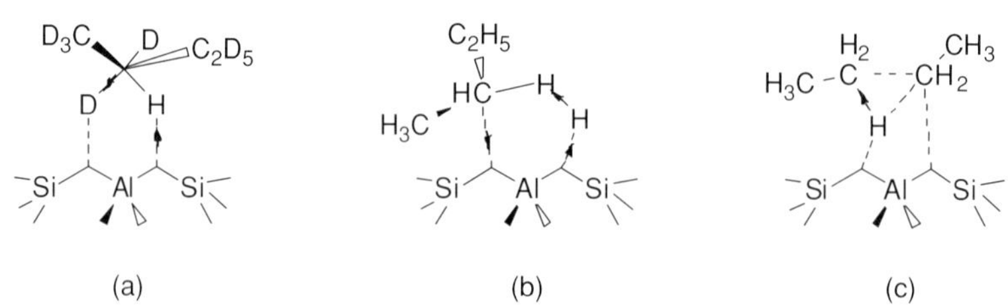

Figure 7.11 Schematic representation of the transition states for (a) H/D exchange, (b) butane dehydrogenation, and (c) cracking catalyzed by a zeolitic Brønsted acid site.

contributions. An interesting procedure to estimate these has been developed by de Moor *et al.* [46]. In essence, they consider the carbenium ions to move freely in the zeolite, so the corresponding partition functions can be easily calculated. The vibrational contributions to the partition functions can be deduced from the vibrational frequencies calculated by quantum chemical (usually DFT) calculations.

For a long time an important open problem in computational zeolite catalysis has been associated with the inability of most of the quantum chemical methods to correctly account for the weak dispersive interactions. Indeed, for studying chemical reactivity of heterogeneous catalysts including zeolites, DFT is the method of choice. However, most of the DFT methods are not capable of even qualitative description of these types of interactions. Although van der Waals or dispersive forces are individually very weak, their additive nature may result in very substantial contributions to the overall stabilization energies of hydrocarbon species confined within the zeolitic channels. For an alkane the adsorption energy is dominated by the van der Waals interactions and therefore depends strongly on the micropore dimensions that determine the number of effective van der Waals CH−O contacts between the confined alkane molecule and the zeolite walls. When the chain of the alkane contains six to eight carbon atoms, the adsorption energy varies in the range 30–100 kJ/mol [47, 48].

Based on force fields B. Smit used configurationally biased equilibrium Monte Carlo methods that enable the prediction of adsorption isotherms based on computed differences of the adsorption free energies. These simulations have convincingly shown that for hydrocarbon conversion reactions in zeolites differences in selectivity relate to the zeolite structure that determines the differences in the internal concentration equilibrium of particular reaction intermediates in the micropores [49]. This implies that zeolitic Brønsted acid catalysis represents a special case of the structure-sensitive reactions, in which the reaction selectivities and sometimes even the overall catalytic reactivity are controlled by the geometrical properties of the zeolitic confinement matrix, but not by the intrinsic properties of the active site.

Including such specific structure sensitivity in the catalytic reactivity studies on realistic zeolite models is not trivial. One has to include the van der Waals stabilizing contributions in the description of local minima states as well as transition states along the reaction paths. Tuma and Sauer [50–52] developed an interesting approach to deduce these contributions from high-level *ab initio* quantum chemical cluster calculations embedded into fully periodic zeolite structure described overly by DFT. In this approach the high level computational methods include electron correlation effects explicitly and are applied to a relevant portion of the zeolite where the dispersion interactions make the largest impact. More practical are the methods based on addition of semi-empirical correction terms to the potentials derived from electronic structure DFT calculations [53, 54]. These methodologies represent a robust and at the same time accurate enough approach for computing reaction intermediates and transition states for such extensive models as zeolites with inclusions of the van der Waals correction.

7.3.2.2 Lewis Acid Catalysis

The mechanism of alkane activation by Lewis acid sites in zeolitic catalysts is very different. Generally speaking, whereas in the case of the Brønsted acid catalysis, the substrate hydrocarbon molecule plays a role of a base that accepts a proton from the zeolite and thereby becomes activated, in the case of the catalysis by Lewis acid, an alkane molecule eliminates a proton resulting in an anionic alkyl species that is stabilized by the zeolitic Lewis acid. This mechanism implies that the cracking route is no longer the dominant in the conversion of an alkane, whereas the dehydrogenation is. To illustrate the structure sensitivity of such Lewis acid-catalyzed reactions, we will discuss the mechanism of ethane dehydrogenation by extraframework Ga species stabilized within the zeolite micropores [55, 56]. By means of periodic DFT calculations, not only the relative stability of such clusters in zeolite channels but also the molecular details of their catalytic reactivity can be investigated.

The results of DFT calculations convincingly demonstrate that binuclear oxygenated Ga clusters are much more stable compared to the respective monomeric species that were generally proposed in the experimental literature [55, 56]. More important is that the observed preference for the binuclear cations does not depend on the distribution of framework Al species in the zeolite. The very high basicity of the terminal O ligand in the mononuclear GaO^+ species induces their self-organization into binuclear clusters, in which the excessive basicity is largely compensated by the interaction with the additional Ga^{3+} Lewis acid center (see Figure 7.12).

Nevertheless, one can still speculate that Ga can be present in the microporous space in a wide range of different cationic forms. It has been demonstrated that a zeolitic material containing predominantly isolated Ga^+ species shows a stable although rather low dehydrogenation catalytic activity. The reactivity of such a material can be greatly improved by reacting the extraframework Ga^+ sites with N_2O stoichiometrically. Originally it was proposed that such a reaction would lead to a very selective formation of isolated GaO^+ species containing Ga^{3+} centers. The subsequent DFT studies raised serious doubts with regards to such an interpretation.

For all the extraframework Ga systems, the initial CH bond activation proceeds via its heterolytic dissociation resulting in a formally negatively charged alkyl intermediate that adsorbs onto the cationic Ga species. Even in the case of the Ga^+ sites where the alkane activation results in the product of alkane oxidative addition, the activation mechanism still involves the heterolytic CH bond dissociation. In this case, the extraframework Ga^+ plays a role of the Lewis acid that polarizes the activating CH bond and stabilizes the forming anionic alkyl group, while the lattice basic oxygens of the zeolite are the proton-accepting sites. Obviously, the univalent Ga species show rather low Lewis acidity, and similarly the basicity of the high-zeolite framework is also quite low. As a result, the computed activation

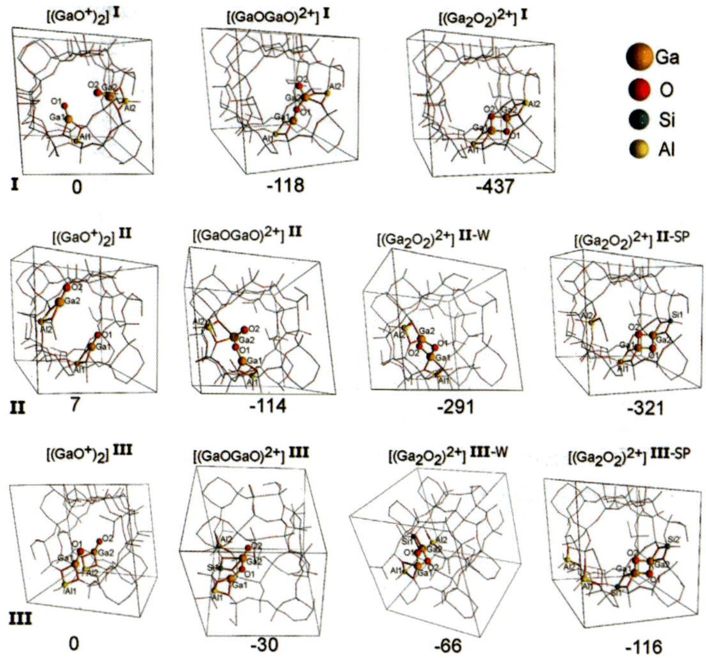

Figure 7.12 *The relative stability (kJ/mol) of oxygenated Ga species in the mordenite unit cell with different distribution of framework Al. It can be seen that mononuclear gallyl (GaO^+) species tend to self-organize in the zeolite matrix into binuclear clusters. This process is so favorable that the associated energy gain is sufficient to compensate for the destabilization of the system due to the distant location of the cationic extraframework Ga complex and the anionic sites on the lattice that compensate for its positive charge [55, 56].*

energy for this initial CH activation is rather high (210 kJ/mol; Figure 7.13). When the dissociation of an alkane over the isolated $Ga{=}O^+$ species takes place, the double Ga=O bond becomes singular, Ga accepts the anionic alkyl species, while the terminal oxygen ligand accepts a proton and becomes an OH group. Obviously, Ga^{3+} species is a much stronger Lewis acid compared to the univalent gallium ions. At the same time, the terminal oxygen atoms show an extreme basicity. This results in a very facile initial alkane activation with an activation barrier of only 69 kJ/mol. As it has been mentioned above, the formation of a binuclear complex with bridging oxygens substantially reduces the basicity of the terminal oxygen ligand, and at the same time the positive charge on the Ga centers becomes more efficiently compensated leading to the decreased effective Lewis acidity. This is reflected in a slightly higher barriers computed for the initial CH activation over such binuclear $Ga_2O_2{}^{2+}$ clusters. However, the dehydrogenation reaction involves not only the activation of CH bonds but also the regeneration of the initial active site. Because of the extreme reactivity of the mononuclear gallyl ions, the initial CH activation leads to the formation of very stable species that cannot be subsequently

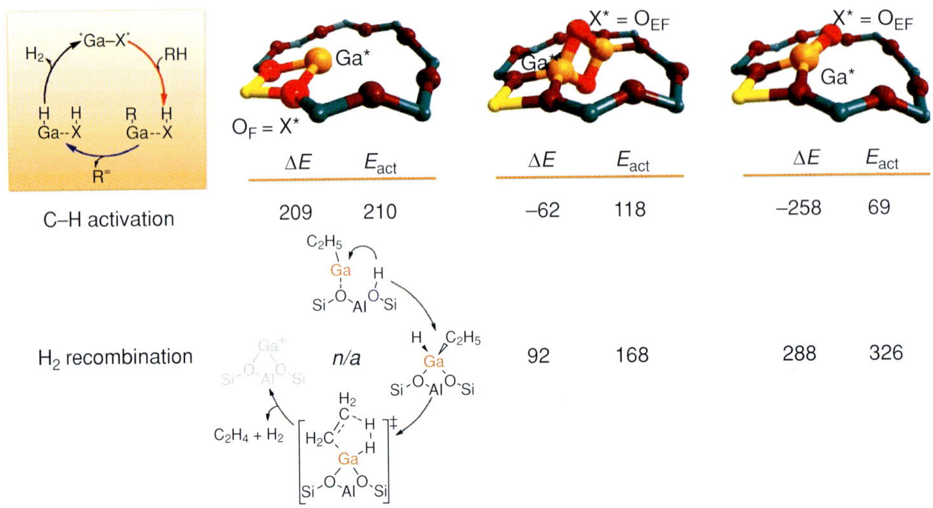

Figure 7.13 *The reactivity of ethane catalyzed by Ga complexes [55, 57].*

decomposed to regenerate the initial active site. The DFT-computed barrier for the regeneration of GaO^+ species at the last step of the catalytic dehydrogenation cycle exceeds 300 kJ/mol (Figure 7.13). The modest reactivity of the binuclear Ga-oxo cluster pays back the price that has been paid for the initiation of the catalytic cycle at the stage when the active site should be regenerated. Indeed, for these more stable reactive sites, the regeneration of the active complex via H_2 desorption proceeds with a barrier of only 168 kJ/mol. As a result, the overall barrier for alkane dehydrogenation is reduced by approximately 50 kJ/mol when Ga^+ sites are converted into binuclear oxogallyl species. This situation illustrates the Sabatier principle, according to which the maximum rate of catalysis is found for an optimum interaction energy of reactants. In the binuclear complex, the increased barrier of activation of the CH bond is balanced by the decreased barrier for product formation. These theoretical insights allowed establishing a detailed structure–reactivity relationship for a wider range of intrazeolitic Ga-containing complexes for alkane dehydrogenation [58].

Besides gallium, oxygenated complexes of other metal cations stabilized in zeolite show high dehydrogenation activity associated with a very similar molecular mechanism. A relevant example is Zn-modified ZSM-5 catalysts. It has been demonstrated that binuclear ZnO/Zn^{2+} cationic clusters show a reactivity toward CH activation comparable to that of the Ga-based systems. Again in this case, because of the enhanced basicity of the extraframework bridging oxygen ligand, the regeneration of the active site at the end of the catalytic reaction is problematic and isolated Zn^{2+} sites complemented with framework basic sites of the zeolite are the active sties for alkane dehydrogenation in the actual zinc-modified zeolitic catalyst [59]. In addition, the Bi_2O_3 component of the Bi_2O_3/MoO_6 oxidation catalyst for propylene to acrolein oxidation has a similar role. It activates the CH bond in propylene to form the allyl intermediate that is further converted in consecutive reaction steps with oxygen.

7.4 Surface Sensitivity for the C–C Bond Formation

7.4.1 Transition Metal Catalyzed FT Reaction

In the FT reaction, CO is reacted with H_2 yielding linear alkane or alkene molecules. In some systems, hydrocarbon oxygenates may also be formed. The elementary reaction steps that correspond to this reaction have been extensively studied by periodic DFT calculations involving modeling of various sites on many different metal surfaces. Iron, cobalt, and ruthenium are the mostly used catalysts for the FT reaction. The main reason for the uniqueness of these metals is that they are able to catalyze CO dissociation with relatively low barriers and the ease of the subsequent transformations of the resulting adsorbed C and O species. Many reaction mechanisms for this reaction have been proposed so far [20]. Here, we will only discuss the carbide mechanism because currently it constitutes the generally accepted reaction route. According to the carbide mechanism, the FT reaction is initiated by the decomposition of CO into adsorbed CH_x intermediates and O_{ads}. O_{ads} then rapidly reacts with CO to give CO_2 (on Fe) or with H_2 (on Co and Ru) to give H_2O. During the reaction Co and Ru catalytic surfaces remain in the metallic state, whereas Fe is converted into an iron carbide phase. As a consequence several studies are available also on CO dissociation on the surfaces of iron carbide [60].

The chain-growth reaction yielding longer hydrocarbons will only have a high probability if the rates of the elementary steps involved in the catalytic reaction satisfy a particular condition. The chain-growth reaction proceeds by the sequential incorporation of CH_x species into the growing hydrocarbon chain attached to the catalyst surface. The rate of termination of the chain via hydrogenation or CO insertion has to be slow as compared to the chain-growth reaction. The latter is controlled by the relative concentration of the surface CH_x species. Their formation therefore has to be fast compared to the termination reaction.

From theoretical calculations, a typical number of 100 kJ/mol can be deduced for the activation barrier of the hydrocarbon termination step. The chain-growth reaction has been found to occur with a lower barrier. It is sensitive to the details of the hydrocarbon chain-growth reaction and the type of the CH_x species that are being inserted. The CH_x species can differ in the number of hydrogen atoms attached to the carbon. The reaction is also structure sensitive.

A high chain-growth rate requires a high concentration of surface CH_x species. To ensure the large concentration of the initial CH_x species necessary as building blocks, the initial CO dissociation has to be very fast and to occur with a barrier lower than that of the methanation reaction that is estimated to be typically of the order of 100 kJ/mol. As mentioned in the section dealing with methane reforming, CO dissociation is a highly structure-sensitive reaction. The necessary low activation barrier is only possible when unique surface step-edge-type sites are available (see Figure 7.10). Because of the relatively high energy of formation of such specific surface structures, this conclusion implies that there are only a few reactive centers present in a practical catalyst.

CO activation may occur directly or through the intermediate formation of an activated formyl intermediate, that is, with the assistance of adsorbed H atoms [61]. It appears that the surface sensitivity of the latter reaction path is only weak. Typical activation barriers for the CH formation through the H-assisted CO dissociation route are in the order of 130 kJ/mol.

A second important requirement for this reaction center is that the growing hydrocarbon chain does not hamper the dissociation of CO. Indeed, if the same sites are required for the hydrocarbon chain growth and the initial CO dissociation, the latter can be completely eliminated in the course of the reaction, resulting in the very low overall catalytic performance. Interestingly, quantum chemical computations clearly demonstrate that the preference of the site position is quite different for the formation of C1 (C_{ads} or CH_{ads}) species and their subsequent insertion into the growing hydrocarbon chain [62]. For instance, when CO dissociation occurs via a direct mechanism resulting in the C_{ads} formation on a fourfold site, the subsequent hydrogenation resulting in the CH_{ads} intermediate is accompanied with the migration of this species with very low diffusion barrier to a threefold site on the step edge of the reaction center. The subsequent formation of C$-$C bonds occurs on this step-edge site, while the fourfold site required for the CO dissociation remains unhindered.

To provide input to kinetics simulations, a reaction energy scheme for the complete catalytic cycle has to be devised that corresponds to a particular proposed mechanism. For the formation of ethylene from synthesis gas such a reaction energy scheme is shown in Figure 7.14 [19].

To estimate rate constants for the elementary reaction steps, these DFT-computed energies of activation have to be complemented by activation entropy estimates. A reasonable approximation is to assume transition states of surface elementary reaction steps to be immobile. This implies that the activation entropies of the surface reactions can usually be ignored, but the entropy losses and gains during the adsorption and desorption processes from the gas phase have to be properly accounted for. These steps mainly involve the changes in the translational and rotational degrees of freedom, and therefore the associated entropy changes can be well estimated from statistical thermodynamics.

One should note that even after the complete reaction mechanism has been computed and all energy parameters for the involved elementary reaction steps have been estimated, the overall reaction kinetics cannot be deduced. To analyze

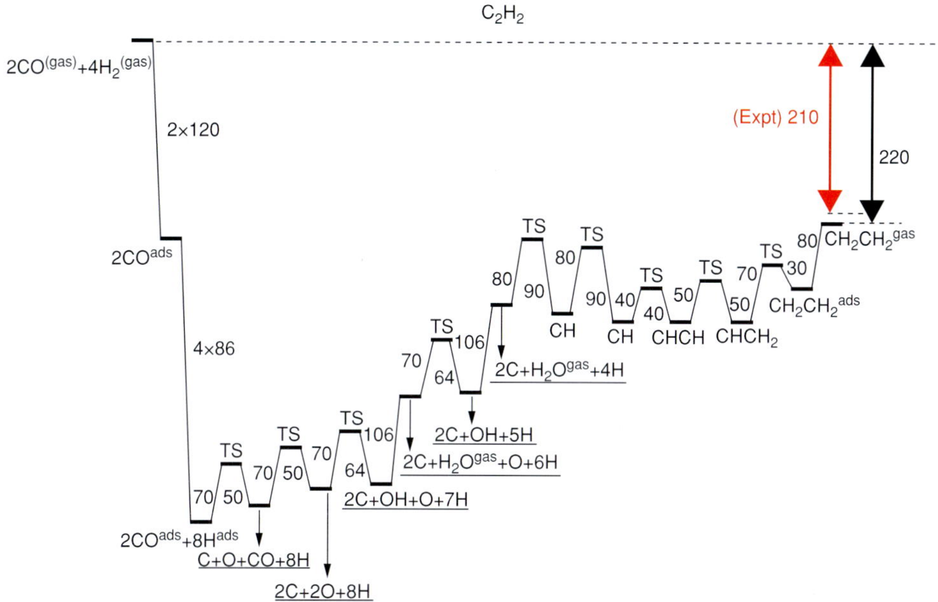

Figure 7.14 *Reaction energy diagram for ethylene formation as a part of the FT reaction. Reprinted with permission from R.A. van Santen, M.M. Ghouri, S. Shetty, E.J.M. Hensen, Catalysis Science & Technology, 1(6): p. 891–911 © 2011 Royal Society of Chemistry.*

such complex reaction networks correctly, additional computational approaches have to be involved. In this case, kinetic Monte Carlo simulations represent a suitable computational technique to compute overall kinetics. In this approach, a surface grid has to be chosen on which reaction intermediates are located that can react and diffuse. The reaction probabilities are deduced from the DFT-computed elementary rate constants. These grids can be of varying complexity simulating the differences in the coordination of reaction intermediates and allowing for surface reconstruction effects.

For a simplified grid containing only one type of sites, this approach has been used to investigate the dependence of the selectivity of the FT reaction as a function of varying CO activation energy. The kinetic Monte Carlo results for a reaction producing C1 to C3 linear hydrocarbon species are shown in Figure 7.15.

One notes the increase in methane selectivity when the activation energy of CO dissociation is larger. In Figure 7.16, the dependence of selectivity on M–C bond energy is sketched and the experimental results on methane selectivity as a function of catalyst with a decreasing M–C interaction are summarized that confirm this prediction.

Whereas low-barrier CO dissociation requires the specific step-edge B5-type surface topology and strong interaction of atomic C_{ads} as well as atomic O_{ads} with the metal surface, the chemistry that controls the CH_x insertion into the growing

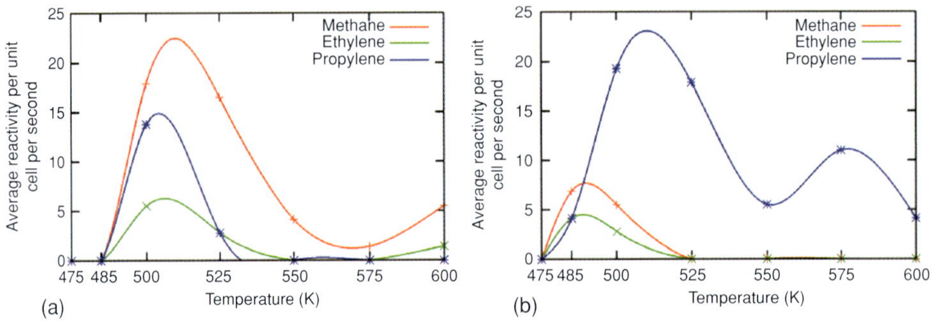

Figure 7.15 Kinetic Monte Carlo simulations of the formation of C1, C2, and C3 hydrocarbons as a function of temperature [19]. All rate parameters are the same, except for the activation energy of CO dissociation: (a) activation energy (CO) = 50 kJ/mol; (b) activation energy (CO) = 70 kJ/mol. Reprinted with permission from R.A. van Santen, M.M. Ghouri, S. Shetty, E.J.M. Hensen, Catalysis Science & Technology, 1(6): p. 891–911 © 2011 Royal Society of Chemistry.

hydrocarbon chain shows a different structure sensitivity. The main reason is that the part of the M–C interaction remains intact when a new C–C bond is formed. In the limiting cases, two reactions can be dominating; namely, when the interaction between the carbon species and the metal surface is strong, the carbide formation is favored, or when it is weak, graphene formation takes place in the absence of hydrogen. As indicated in Figure 7.16 in the presence of hydrogen, the surface carbide phase will be mainly converted. When the M–C interaction is weak, in the presence of hydrogen, graphene formation is prevented. With increasing M–C interaction, methane formation becomes suppressed in favor of the chain-growth reaction.

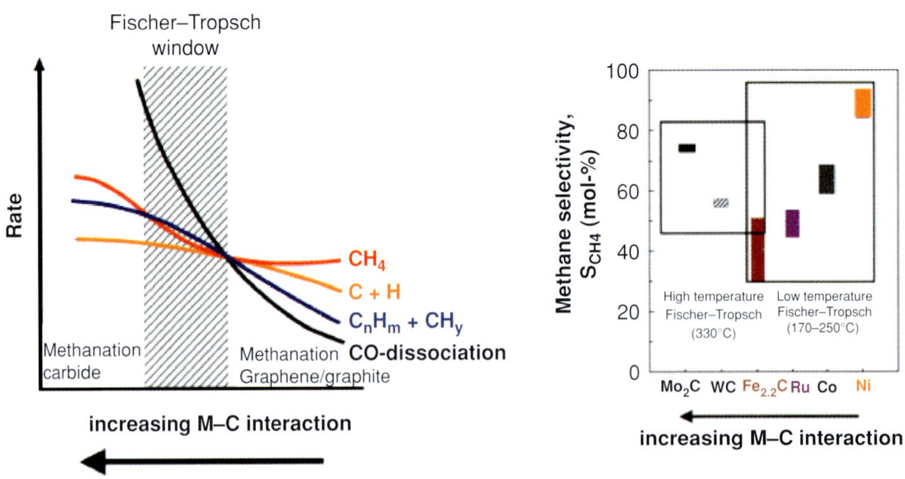

Figure 7.16 Selectivity of the Fischer-Tropsch process as a function of M–C interaction [20].

7.4.2 C–C Bond Formation Catalyzed by Zeolitic Brønsted Acids

In the FT reaction, hydrocarbon chain growth is initiated by the activation of CO by metal surfaces. The formation of C–C bonds can also be observed in the case of Brønsted acid-catalyzed processes occurring in zeolites. In this case, the formation of aromatics or short olefins proceeds by activation of methanol by the zeolitic acid sites. Both reactions are important processes that can be used to convert natural gas or coal-derived synthesis gas into liquid energy carriers and chemicals. Methanol is formed from synthesis gas by a catalytic process involving Cu/ZnO as a catalyst. In this process CO is hydrogenated, while the dissociation of CO is suppressed because of the low reactivity of copper. The thus formed methanol can then be converted into a wide range of useful products by using zeolitic catalysts. The product distribution of the zeolite-catalyzed methanol conversion strongly depends on the type of the microporous catalyst. When ZSM-5 material is used, which contains rather narrow micropores formed by 10 interconnected framework silicon or aluminium-occupied oxygen tetrahedra, the product distribution is complex (Figure 7.17).

On the other hand, when a SAPO-34 zeolitic material with a chabazite topology representing a cage structure is connected through the narrower eight-ring windows, the reaction products are mainly ethylene or propylene. The smaller channels between the microcavities prevent the passage of the larger hydrocarbons resulting in a dramatic shift of the product profile to lower olefins.

The mechanism of C–C bond formation has been a subject of extensive experimental and theoretical studies [63]. It has been shown that whereas protic weak acid sites in the zeolite catalyst will cause the initial dehydration of methanol and the formation of dimethyl ether, strong acid sites (e.g., weakly bonded protons) are required for the C–C bond formation reactions. Initial computational works were

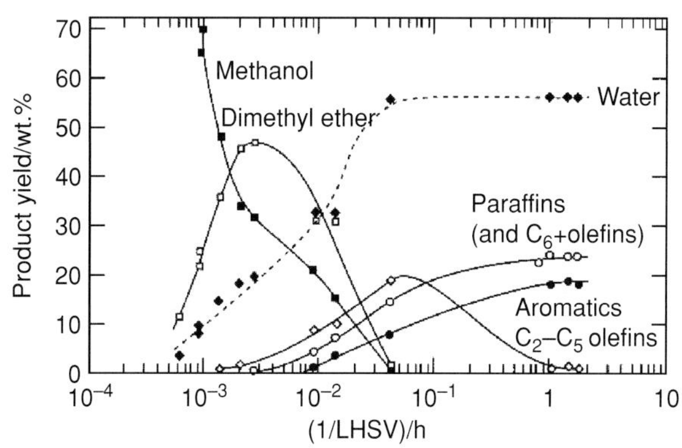

Figure 7.17 *Product selectivities for methanol to hydrocarbon reaction over H-ZSM-5 catalyst [63]. LHSV, liquid hourly space velocity. Reprinted with permission from C.D. Chang, A.J. Silvestri., Journal of Catalysis, 47(2): p. 249–259 © 1977 Elsevier.*

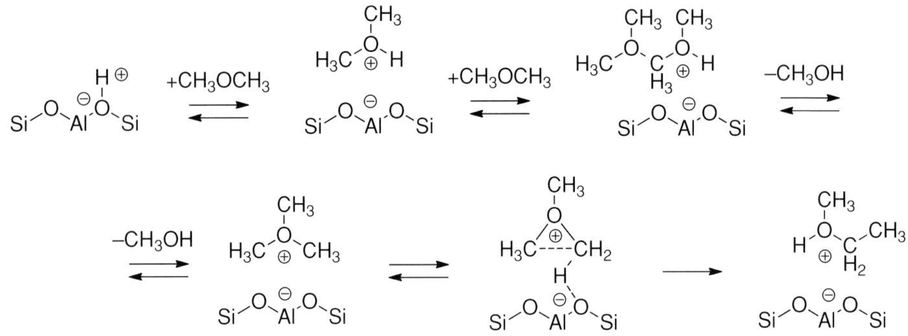

Figure 7.18 *Stevens rearrangement of a trimethyloxonium ion.*

mainly focused on the mechanism of the formation of the first C–C bond. van den Berg *et al.* [64] as well as Olah *et al.* [65] suggested the intermediate formation of a trimethyloxonium ion, formed via a series of conversions of two molecules of dimethyl ether over a zeolitic Brønsted acid site involving an intermediate elimination of methanol. Formation of the positively charged trimethyloxonium ion leads to the generation of a weakly compensated negative charge on the zeolite framework. The resulting enhanced basicity is the driving force for the subsequent Stevens rearrangement (Figure 7.18), which leads to the formation of the protonated ethyl methyl ether. More than a decade ago, Blaszkowski *et al.* [66] explored this proposition by DFT calculations using small clusters as models for zeolitic acid sites. They found that the formation of the first C–C bond had to overcome a prohibitively large activation barrier of 280 kJ/mol. This value is to be compared with an activation energy of the formation of the methoxy group (210 kJ/mol [66]) calculated by using the same cluster modeling approach. The key intermediate turns out to be a complex of dimethyl ether and methanol adsorbed to the zeolite proton that after the reaction yield ethyl methyl ether (Figure 7.19).

A decade later the mechanism underlying the methanol transformations in zeolites was reinvestigated with more advanced methods. Large cluster representatives of the zeolite cavity were used within a combined approach, in which the catalytically reactive site and the actual chemical transformations were described by using high-level *ab initio* DFT methodology, whereas the zeolite confinement environment was modeled by using cheaper semi-empirical methods [67, 68]. These calculations confirmed the very high activation barrier for the direct C–C formation step. These later results suggest that the actual direct C–C bond formation is unlikely during the catalytic methanol to hydrocarbon conversion [67]. Other highly activated processes compete with such a direct C–C coupling. These processes include the formation of methane and formaldehyde from reaction of an adsorbed methoxy species with methanol.

Experimental indications are very strong [63]—at the initial stage of the catalytic reactions, aromatic intermediates are formed within the zeolitic nanopores.

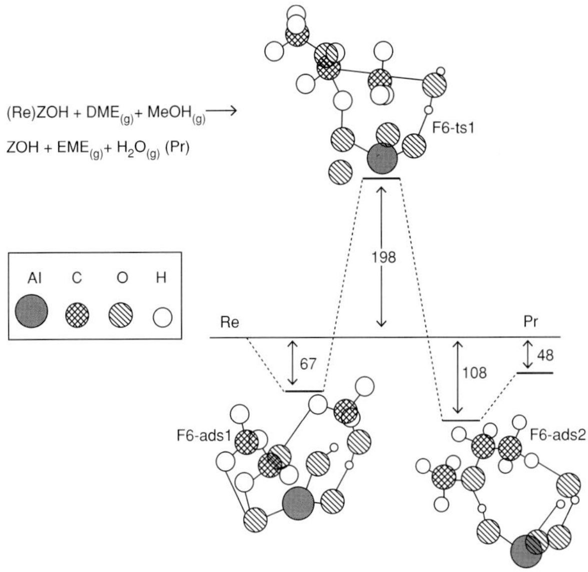

Figure 7.19 *Ethyl methyl ether formation via associative mechanism (energies are in kJ/mol). Reprinted with permission from S.R. Blaszkowski, R.A. van Santen., J. of the Am. Chem. Soc., 119(21): p. 5020–5027 © 1997, American Chemical Society.*

These confined species act then during the methanol conversion as an organocatalyst for the C–C bond formation reactions, resulting thus in short hydrocarbon molecules that desorb from the zeolite. This corresponds to the so-called hydrocarbon pool mechanism. Computational studies demonstrated that the activation energies involved in such a mechanism could be less than 150 kJ/mol [68]. If one excludes the possibility that the aromatic organocatalytic intermediates are formed by the rare, highly activated reactions of methanol, these organocatalytic molecules result from small olefinic impurities in the methanol stream.

According to the hydrocarbon pool mechanism [69], the key reaction that initiates the formation of C–C bonds is the methylation of the aromatic ring of the organocatalytic molecule. This can occur via the exocyclic methylation mechanism (Figure 7.20) as well as via the paring reaction (Figure 7.21). Both mechanisms lead to the formation of higher hydrocarbons. Calculations by Lesthaeghe *et al.* [68] provide activation energies for the methylation reaction of different aromatic molecules that are potential organocatalysts. The results obtained allow a molecular level rationalization of the observed reactivity and selectivity trends in methanol conversion by zeolitic catalysts with different topology.

When the reaction (Figure 7.22a) is studied on a small cluster with complete neglect of the zeolite environment, the rather high activation energies are found and the methanation reaction is seen to be highly endothermic (Figure 7.22b). Figure 7.22c illustrates the large promoting effect of the zeolite environment on this reaction. Depending on the zeolite topology, the steric effects influencing the

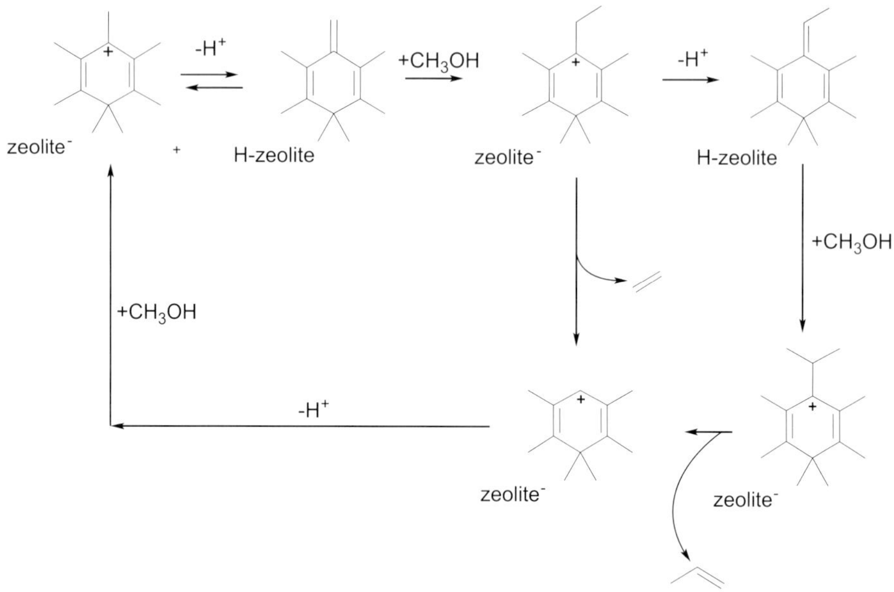

Figure 7.20 The exocyclic methylation reaction.

stability and reactivity of the confined organocatalyst are very different (Figure 7.22c). Different zeolite topologies offer a very different stabilizing environment (Figures 7.22d–f) to the confined bulky aromatic species, influencing thereby their catalytic reactivity.

The methanol conversion assisted by methylated aromatic compounds confined in a zeolite matrix would lead to the production of ethylene and propylene via both the exocyclic and pairing reaction sequences. Calculations by Hemelsoet *et al.* [71]

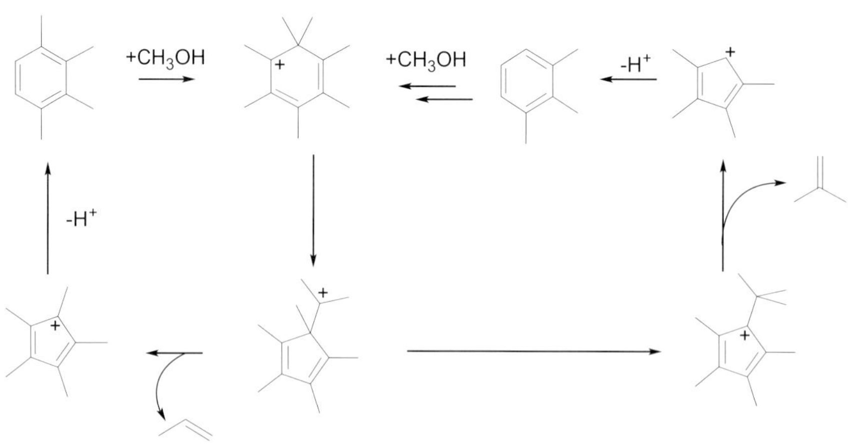

Figure 7.21 The pairing reaction.

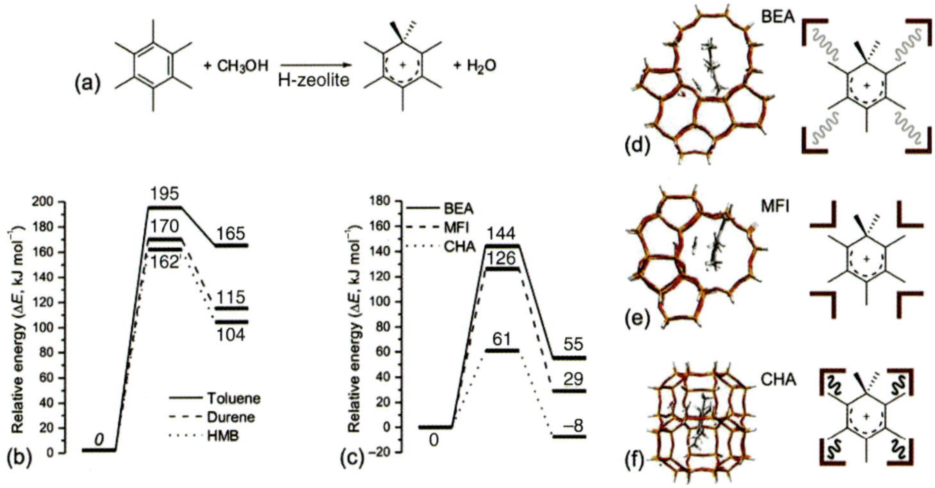

Figure 7.22 Methylation of polymethylbenzenes in acidic zeolites (a) and the computed activation barriers and reaction enthalpies depending on the nature of the organic molecule (b) and on the zeolite topology (c). The molecular recognition effects are illustrated with the schematic representation of the resulting carbenium ions in the confined space and with the structures of the transition states for methylation of hexamethylbenzene (HMB) in zeolite beta (BEA) (d), ZSM-5 (MFI), (e) and chabazite (CHA) (f). Adapted with permission from D. Lesthaeghe, B. De Sterck, V. Van Speybroeck, G.B. Marin, M. Waroquier, Angewandte Chemie International Edition, 46(8): p. 1311–1314 © 2007 WILEY-VCH Verlag GmbH & Co. KGaA, Weinheim.

have demonstrated that the activation barriers for the elementary reaction steps within the exocyclic mechanism are excessively high so that this route toward C–C bond formation can be excluded.

7.5 Structure and Surface Composition Sensitivity: Oxygen Insertion versus CH Bond Cleavage

7.5.1 Silver-Catalyzed Ethylene Epoxidation

The chemical composition of the catalyst surface can greatly affect the accessibility and electronic properties of catalytic sites, thus altering their reactivity and selectivity. Obviously, depending on the surface coverage and on the type of species adsorbed, the properties of the catalyst surface may be very substantially altered. In this section, we will discuss the situation when the steady-state chemical composition of the catalyst surface determines its reactivity and the selectivity of the reaction catalyzed by it. We will use the example of silver-catalyzed ethylene epoxidation to illustrate how surface coverage with specific species can alter its catalytic reactivity. The reaction of ethylene with oxygen adatoms on silver yields ethylene oxide (EO) with high selectivity. This is an important reaction intermediate for

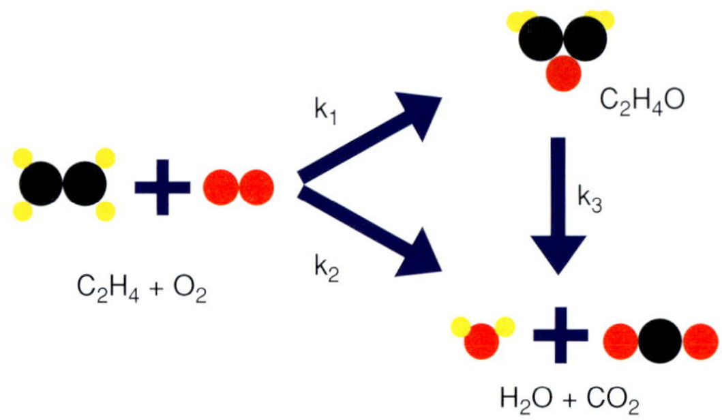

Figure 7.23 *Kinetic scheme for ethylene epoxidation with O_2 catalyzed by the Ag-based catalyst [21].*

the production of ethylene glycol. Besides EO, this catalytic process yields only the products of total oxidation, namely CO_2 and H_2O. In view of the fact that EO molecules themselves can react over the catalyst resulting in their total oxidation, a kinetic scheme shown in Figure 7.23 has been proposed to describe the ethylene oxidation reaction.

For a long time the mechanistic details of the processes involved in this reaction scheme were not understood. The actual mechanism of the total combustion of ethylene (k_3) as well as that of EO (k_3) represented open questions. It was suspected that an intermediate formation of acetaldehyde (AA) played a crucial role in these nonselective pathways because of the experimental observations of the rapid combustion of AA over Ag. Since neutralization of the α-alumina support by alkali metal salts leads to a substantial suppression of the k_3 reaction path, the protons of this support were proposed to catalyze the isomerization of EO to AA, which is then rapidly oxidized by silver particles distributed over the alumina support.

Computational support for the hypothesis of the AA formation leading to nonselective conversion path during the ethylene epoxydation reaction appeared a decade ago from the cluster DFT calculations by Barteau and Linic [72, 73], which were later confirmed with more advanced periodic DFT calculations preformed using slab models of silver surfaces [22,23, 74–76]. It has been shown that ethylene interacts with atomically adsorbed oxygen on silver surface to give an oxametallocycle (OMC) intermediate (Figure 7.24). The OMC species is then decomposed through two competing reaction channels, namely the selective formation of EO and the formation of AA representing a nonselective route leading to the total combustion. The respective DFT-computed reaction pathways are shown in Figure 7.25. The discovery of this OMC implies that AA is indeed formed as an intermediate upon the ethylene oxidation. Its allylic hydrogen makes it reactive toward the total combustion. However, for the decomposition of the OMC complex, the DFT-computed activation energies for the two competing reaction paths clearly show that the decomposition of the OMC intermediate will lead to the formation of EO with very low selectivity.

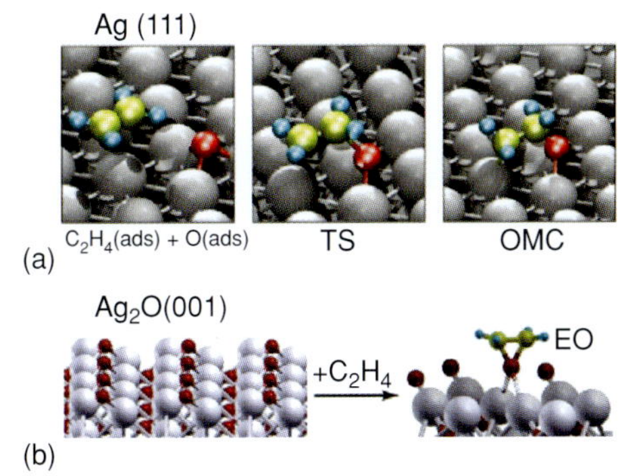

Figure 7.24 *(a) The OMC intermediate of ethylene–aldehyde conversion on metallic silver; (b) the direct epoxidation intermediate, the twofold adsorbed O issue. Reprinted with permission from M.O. Özbek, I. Önal, R.A. van Santen., ChemCatChem, 3(1): p. 150–153 © 2011 WILEY-VCH Verlag GmbH & Co. KGaA, Weinheim.*

In contrast, the DFT-computed reaction energy diagrams convincingly show that in the case of the reaction over the silver oxide surface, the AA formation is much less favored kinetically as compared to the desired EO production (Figure 7.26). At the low O coverage, the oxygen adatoms are strongly bound to the surface and are nucleophilic in character. The preactivation of the π-system by coordination to Ag sites is necessary to favor the subsequent OMC formation.

A route involving the direct formation of EO has been identified that shows an activation barrier 200 kJ/mol lower compared to that of the reaction path producing AA. The binding of the oxygen adatoms on the oxide surface is weaker and moreover the O atoms in this case are more electrophilic compared to those on the metallic surface. This opens the possibility of the direct interaction between the surface oxygen and the π-system of the ethylene molecule that ultimately results in the facile oxygen insertion into the C–C π-bond and the direct formation of

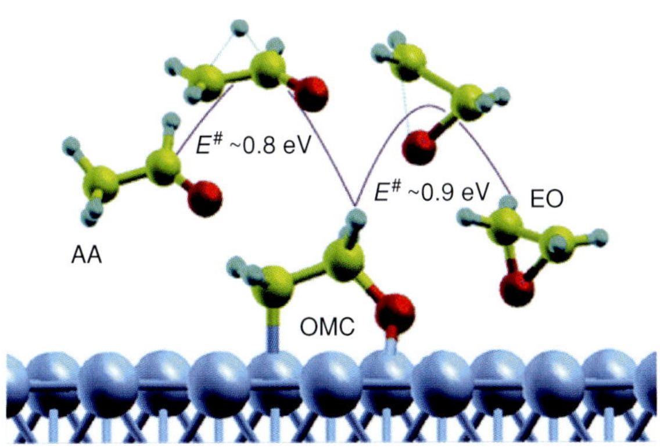

Figure 7.25 *OMC intermediate on Ag(111) surface and the two competing reaction paths for its decomposition [74].*

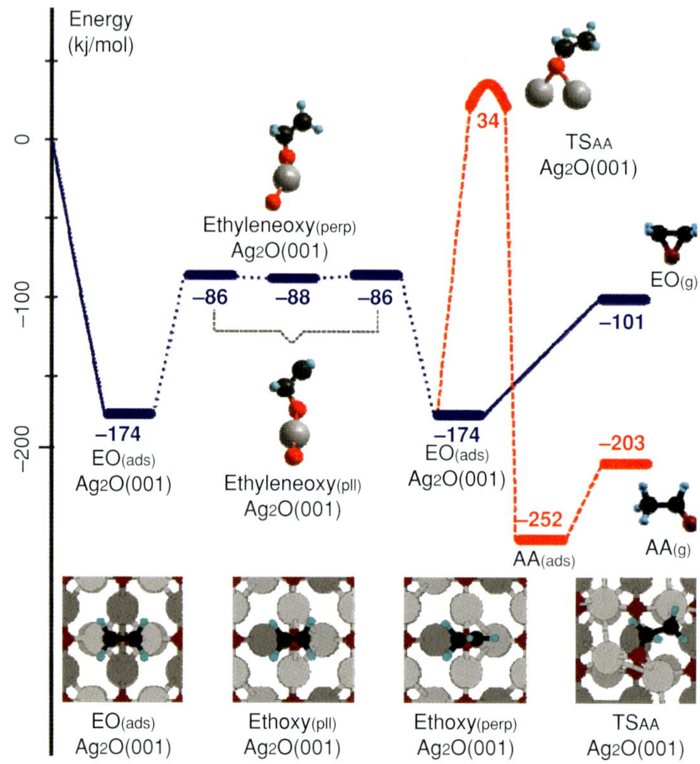

Figure 7.26 *DFT-computed reaction energy diagram of ethylene epoxidation over Ag_2O (001) surface. Reprinted with permission from M.O. Ozbek, I. Önal, R.A. van Santen., Journal of Catalysis, 284(2): p. 230–235 © 2011 Elsevier.*

EO. As long as the oxygenated surface has metallic sites, ethylene can access the Ag atoms. This activates the ethylene carbon−hydrogen bond to produce AA. On the oxide surface this activation does not occur because direct activation with Ag cations is hampered. Selective ethylene epoxidation will only occur on highly oxidized surfaces that will not allow for OMC formation.

Nevertheless, as soon as EO is formed an oxygen vacancy is generated on the surface. This leads to the decrease of the electrophilicity of the oxygen adatoms and therefore their lower affinity toward ethylene as well as to the formation of open Ag sites, which are able to promote the formation of OMC intermediates. Their decomposition will ultimately lead to the nonselective conversion routes. To suppress these undesired pathways, the high electrophilicity of the surface oxygen species and low accessibility of silver sites have to be maintained.

Figure 7.27 shows how coadsorption of Cl on $Ag_2O(100)$ surface directs the EO formation. When adsorbed on the oxygen vacancies, on such a modified catalyst surface, ethylene is not able to access the silver cations and therefore the OMC intermediate cannot be formed. Now, the mechanism of ethylene oxidation again involves a direct reaction between the surface oxygen atoms and the π-electron

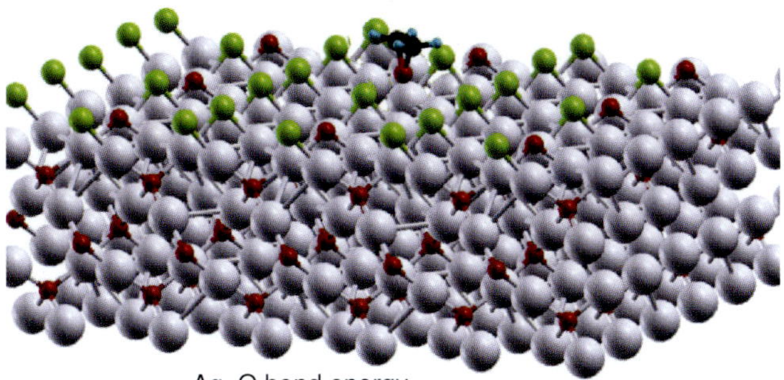

Figure 7.27 *Model of ethylene epoxidation on chlorinated Ag_2O surface.*

system of the alkene. Ethylene conversion over such a surface composition leads solely to the EO formation.

The presence of Cl atoms on the catalyst surface is crucial. Early experimental studies have demonstrated the promoting effect of this anion on the selectivity of ethylene epoxidation [77]. In order to obtain high selectivity of the reaction, chlorinated hydrocarbons are added at ppm levels to the reaction mixture. This leads to the adsorption of Cl on the surface upon the combustion of the hydrocarbon part.

The important selectivity principle discovered here is that a single-point activation of ethylene directly leads to a highly selective EO formation. On the other hand, the two-point activation as in the OMC complex, in which Ag activates the CH bond and O is inserted into the π-systems, is a reaction path that leads to lower selectivity because it opens the possibility for the alternative conversion routes of the reaction intermediates. We will see in Section 7.5.2 that a similar selectivity principle defines the preference for single-site catalysis.

7.5.2 Benzene Oxidation by Iron-Modified Zeolite

Selective oxidation of benzene to phenol by N_2O with iron-modified zeolitic catalysts is called the Panov reaction [73, 78, 79]. The reaction is catalyzed by Fe-containing extraframework species stabilized in the zeolite micropores preferentially of MFI topology (Fe/ZSM-5). The main reason for low reactivity is rapid deactivation of the catalyst through the formation of oligomerization products of phenol. Because of the largely heterogeneous nature of the iron species present in the zeolitic catalyst, the direct determination of the catalytically active iron species and the mechanism of the catalytic reaction by experimental methods were not possible. Nevertheless, many studies have concluded that the specific benzene oxidation activity of Fe/ZSM-5 catalysts is associated with the presence of highly dispersed Fe^{2+} extraframework cations [74, 80–82]. The questions of the nuclearity of these species as well as the role of the predominant Fe^{3+}-containing

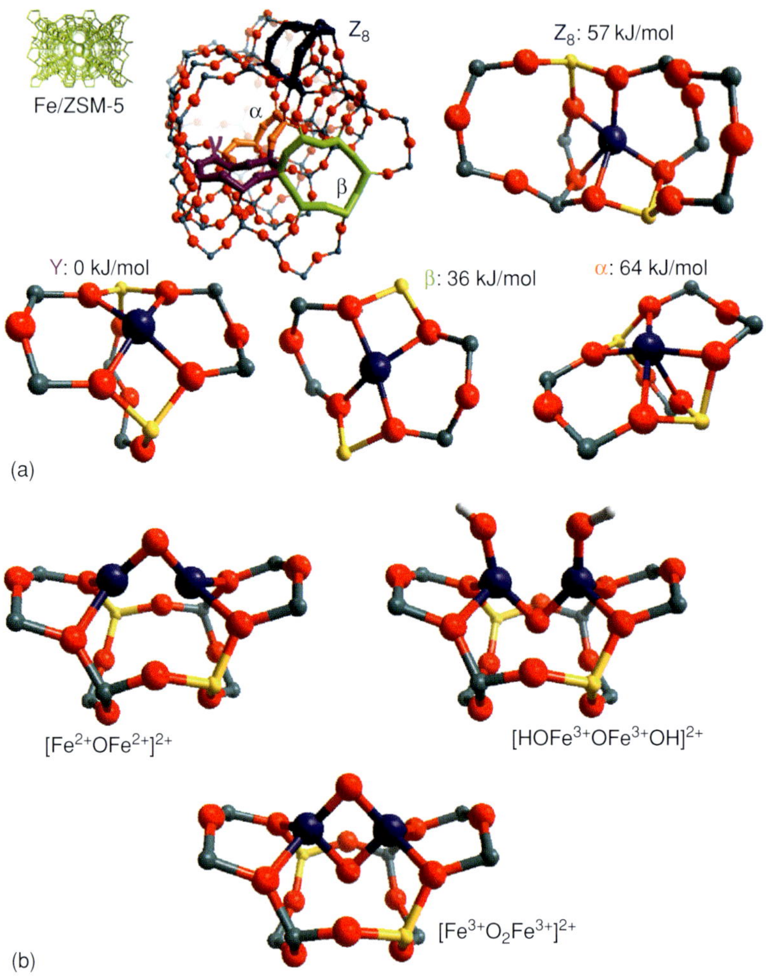

Figure 7.28 *(a) The most stable Fe/ZSM-5 configurations containing Fe^{2+}-isolated extraframework cations as the charge-compensating species. Their relative stabilities as predicted by periodic DFT calculations and the representative local optimized geometries are shown. (b) A selection of binuclear Fe(II) and Fe(III) oxygenated and hydroxylated complexes stabilized at the Z_8 site of ZSM-5 zeolite considered as the active sites for benzene oxidation [83].*

extraframework complexes were still open. By performing a comprehensive periodic DFT study, these questions were to a large extent addressed [75, 83]. A wide range of different extraframework Fe^{2+} and Fe^{3+} species were considered as the catalytic sites for benzene oxidation (Figure 7.28).

A very important result obtained from these calculations is that the elementary processes involved in the catalytic benzene oxidation, namely, the initial N_2O decomposition, benzene activation, and the phenol formation, are influenced only

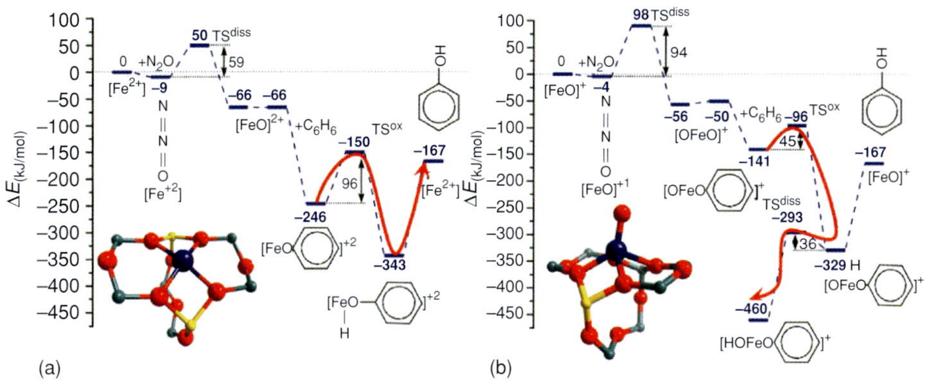

Figure 7.29 *A comparison of DFT-computed van der Waals corrected reaction energy diagrams for benzene oxidation over mononuclear (a) Fe(II) and (b) Fe(III) extraframework sites in ZSM-5 zeolite [83].*

slightly by the nature and the structural properties of the catalytic site. This can be seen from the two reaction energy diagrams calculated for the mononuclear Fe^{2+} and Fe^{3+} sites (Figure 7.29). The reactivity of binuclear sites is very similar to that of the mononuclear iron complexes.

The striking difference between the isolated Fe(II) species and all the alternative oxygen-containing Fe(II) and Fe(III) sites arises only at the stage when phenol is actually formed and it has to be removed from the reaction site (Figure 7.29). Similar to the Ga-modified zeolites considered above, the extraframework oxygen ligand is highly basic and therefore can play a role of a proton-accepting site. It therefore readily reacts with the acidic proton of phenol, resulting in the formation of stable grafted phenolate species and a hydroxyl group. The resulting iron complex is inactive toward further benzene oxidation. Whereas the negatively charged oxygen atoms of the zeolite framework are also basic, transfer of the proton to the zeolite framework is highly endothermic. The protonation of extraframework oxygen species within different iron complexes is always favored thermodynamically and exhibits a very low activation barrier (Figure 7.29b). This process is always preferred in the case of oxygen-containing iron sites over the desorption of phenol and the regeneration of the initial active site Once these grafted phenol species are formed, they will undergo further oligomerization with other phenol molecules resulting in a long-term deactivation of the catalyst.

7.6 Conclusion

In this chapter, we have illustrated recent advances in computational catalysis by highlighting recent theoretical insights into the mechanism of heterogeneous catalytic systems chosen for their interest to energy conversion processes or reduction of raw material use. In the past 10 years, the drastically increased power of computational hardware along with the improvements in methodologies based on DFT allowed a realistic theoretical molecular level description of complex reaction sequences on surface of heterogeneous catalysts. The state-of-the-art computational methodologies allow a detailed investigation of highly sophisticated catalytic processes on atomistic models of the catalyst reaction centers that are structurally very close to postulated reality.

In the first part of the chapter, we highlighted some of the concepts arising from the results obtained by means of computational catalysis, which are now the cornerstones of the physical chemical basis of heterogeneous catalysis. In the second part of the chapter, we discussed chemical reactivity principles for some important heterogeneous catalytic systems.

The detailed mechanistic information on particular catalytic systems and especially insights into relationships between the structural properties and the reactivity of specific catalytic sites obtained so far through the still increasing body of computational data have made it possible to create a modern theory of heterogeneous catalysis. The new fundamental concepts underlying the reactivity and selectivity principles of catalysis by solids allow the qualitative interpretation of the experimental data and help discovering new and improved catalytic systems. The increasing accuracy of computed results and possibilities to accurately model the catalytic reactive sites are making the computational approach of increasing importance to the holy grail of rational design of catalytic systems.

References

1. Davis, B.H. (2008) *Handbook of Heterogeneous Catalysis* (eds G. Ertl, H. Knözinger, F. Schüth and J. Weitkamp), Wiley-VCH Verlag GmbH, Weinheim, p. 16.
2. Somorjai, G.A. (1994) *Introduction to Surface Chemistry and Catalysis*, John Wiley & Sons, Ltd, Chichester.
3. Bluhm, H., Hävecker, M., Kleimenov, E. *et al.* (2003) In situ surface analysis in selective oxidation catalysis: n-butane conversion over VPP, *Top. Catal.* **23**(1), 99–107.
4. van Santen, R.A. and Neurock, M. (2006) *Molecular Heterogeneous Catalysis*, Wiley-VCH Verlag GmbH, Weinheim.
5. Reuter, K. and Scheffler, M. (2003) First-principles atomistic thermodynamics for oxidation catalysis: surface phase diagrams and catalytically interesting regions, *Phys. Rev. Lett.* **90**(4), 046103.
6. Teschner, D., Révay, Z., Borsodi, J. *et al.* (2008) Understanding palladium hydrogenation catalysts: when the nature of the reactive molecule controls the nature of the catalyst active phase, *Angew. Chem. Int. Edit.* **47**(48), 9274–9278.
7. van Santen, R.A. and Sautet, P. (2009) *Computational Methods in Catalysis and Materials Science*, Wiley-VCH Verlag GmbH, Weinheim.

8. Kramers, H.A. (1940) Brownian motion in a field of force and the diffusion model of chemical reactions. *Physica* **7**(4), 284–304.
9. Frenkel, D. and Smit, B. (2002) *Understanding Molecular Simulation*, 2nd edn, Academic Press, San Diego, CA.
10. Dumesic, J.A., Rudd, D.F., Aparicio, L.M. *et al.* (1992) *The Microkinetics of Heterogeneous Catalysis*, American Chemical Society.
11. Jansen, A.P.J. (2009) Kinetic Monte Carlo, in *Computational Methods in Catalysis and Materials Science*, Wiley-VCH Verlag GmbH, Weinheim, pp. 183–197.
12. van Duin, A. (2009) Reactive force fields: concepts of ReaxFF, in *Computational Methods in Catalysis and Materials Science*, Wiley-VCH Verlag GmbH, Weinheim, pp. 167–181.
13. Sabatier, P. (1913) *La Catalyse en Chimie Organique*, Librairie Polytechnique, Paris.
14. Bligaard, T. and Nørskov, J.K. (2008) Heterogeneous catalysis, in *Chemical Bonding at Surfaces and Interfaces*, Elsevier, Amsterdam, pp. 255–321.
15. van Santen, R.A., Neurock, M. and Shetty, S.G. (2009) Reactivity theory of transition-metal surfaces: a Brønsted–Evans–Polanyi linear activation energy–free-energy analysis, *Chem. Rev.* **110**(4), 2005–2048.
16. Henry, C.R. (1997) Size effects in heterogeneous catalysis: a surface science approach, in *Chemisorption and Reactivity on Supported Clusters and Thin Films* (eds R.M. Lambert and G. Pacchioni), Kluwer Academic Publishers, Dordrecht, The Netherlands.
17. van Santen, R.A. (2008) Complementary structure sensitive and insensitive catalytic relationships, *Acc. Chem. Res.* **42**(1), 57–66.
18. van Grootel, P.W., Hensen, E.J.M. and van Santen, R.A. (2010) The CO formation reaction pathway in steam methane reforming by Rhodium, *Langmuir* **26**(21), 16339–16348.
19. van Santen, R.A., Ghouri, M.M., Shetty, S. and Hensen, E.J.M. (2011) Structure sensitivity of the Fischer-Tropsch reaction: molecular kinetics simulations, *Catal. Sci. Technol.* **1**(6), 891–911.
20. van Santen, R.A., Ciobîcă, I.M., van Steen, E. and Ghouri, M.M. (2011) Mechanistic issues in Fischer-Tropsch catalysis, *Adv. Catal.* **54**, 127.
21. van Santen, R.A. and Kuipers, H.P.C.E. (1987) The mechanism of ethylene epoxidation, in *Advances in Catalysis* (eds H.P.D.D. Eley and B.W. Paul), Academic Press, San Diego, CA, pp. 265–321.
22. Ozbek, M.O., Önal, I. and van Santen, R.A. (2011) Why silver is the unique catalyst for ethylene epoxidation, *J. Catal.* **284**(2), 230–235.
23. Özbek, M.O., Önal, I. and van Santen, R.A. (2011) Ethylene epoxidation catalyzed by silver oxide, *ChemCatChem* **3**(1), 150–153.
24. van Santen, R.A. and Kramer, G.J. (1995) Reactivity theory of zeolitic broensted acidic sites, *Chem. Rev.* **95**(3), 637–660.
25. Catlow, C.R.A., French, S.A., Sokol, A.A. and Thomas, J.M. (2005) Computational approaches to the determination of active site structures and reaction mechanisms in heterogeneous catalysts, *Phil. Trans. R. Soc. A* **363**(1829), 913–936.
26. Thomas, J.M., Catlow, C.R.A. and Sankar, G. (2002) Determining the structure of active sites, transition states and intermediates in heterogeneously catalysed reactions, *Chem. Commun.* 24, 2921–2925.

27. Dugal, M., Sankar, G., Raja, R. and Thomas, J.M. (2000) Designing a heterogeneous catalyst for the production of adipic acid by aerial oxidation of cyclohexane, *Angew. Chem. Int. Edit.* **39**(13), 2310–2313.
28. Gómez-Hortigüela, L., Corà, F., Sankar, G. *et al.* (2010) Catalytic reaction mechanism of Mn-doped nanoporous aluminophosphates for the aerobic oxidation of hydrocarbons, *Chemistry* **16**(46), 13638–13645.
29. Shustorovich, E. (1986) Chemisorption phenomena: analytic modeling based on perturbation theory and bond-order conservation, *Surf. Sci. Rep.* **6**(1), 1–63.
30. Evgeny, S. (1986) Bond making and breaking on transition-metal surfaces: Theoretical projections based on bond-order conservation, *Surf. Sci. Lett.* **176**(3), L863–L872.
31. Thomas, J.M. and Thomas, W.J. (1997) *Principles and Practice of Heterogeneous Catalysis*, Wiley-VCH Verlag GmbH, Weinheim.
32. Jones, G., Jakobsen, J.G., Shim, S.S. *et al.* (2008) First principles calculations and experimental insight into methane steam reforming over transition metal catalysts, *J. Catal.* **259**(1), 147–160.
33. Ciobîca, I.M., Frechard, F., van Santen, R.A. *et al.* (2000) A DFT study of transition states for C−H activation on the Ru(0001) surface†, *J. Phys. Chem. B* **104**(14), 3364–3369.
34. Ciobîca, I.M. and van Santen, R.A. (2002) A DFT study of CHx chemisorption and transition states for C−H activation on the Ru(1120) surface, *J. Phys. Chem. B* **106**(24), 6200–6205.
35. Liu, Z.-P. and Hu, P. (2003) General rules for predicting where a catalytic reaction should occur on metal surfaces: a density functional theory study of C−H and C−O bond breaking/making on flat, stepped, and kinked metal surfaces, *J. Am. Chem. Soc.* **125**(7), 1958–1967.
36. Diefenbach, A., de Jong, G.T. and Bickelhaupt, F.M. (2005) Activation of H−H, C−H, C−C and C−Cl bonds by Pd and $PdCl^-$. Understanding anion assistance in C−X bond activation, *J. Chem. Theory Comput.* **1**(2), 286–298.
37. Kratzer, P., Hammer, B. and Norskov, J.K. (1996) A theoretical study of CH_4 dissociation on pure and gold-alloyed Ni(111) surfaces, *J. Chem. Phys.* **105**(13), 5595–5604.
38. Michaelides, A. and Hu, P. (2001) Catalytic water formation on platinum: a first-principles study, *J. Am. Chem. Soc.* **123**(18), 4235–4242.
39. Michaelides, A., Liu, Z.P., Zhang, C.J. *et al.* (2003) Identification of general linear relationships between activation energies and enthalpy changes for dissociation reactions at surfaces, *J. Am. Chem. Soc.* **125**(13), 3704–3705.
40. Henkelman, G. and Jónsson, H. (2001) Theoretical calculations of dissociative adsorption of CH_4 on an Ir(111) surface, *Phys. Rev. Lett.* **86**(4), 664–667.
41. van Grootel, P.W., Hensen, E.J.M. and van Santen, R.A. (2009) DFT study on H_2O activation by stepped and planar Rh surfaces, *Surf. Sci.* **603**(22), 3275–3281.
42. Rane, N., Overweg, A.R., Kazansky, V.B. *et al.* (2006) Characterization and reactivity of Ga^+ and GaO^+ cations in zeolite ZSM-5, *J. Catal.* **239**(2), 478–485.
43. Hensen, E.J.M., Pidko, E.A., Rane, N. and van Santen, R.A. (2007) Water-promoted hydrocarbon activation catalyzed by binuclear gallium sites in ZSM-5 zeolite, *Angew. Chem. Int. Edit.* **46**(38), 7273–7276.
44. Almutairi, S.M.T., Mezari, B., Magusin, P.C.M.M. *et al* (2011) Structure and reactivity of Zn-modified ZSM-5 zeolites. The importance of clustered cationic Zn complexes, *ACS Catal.* **2**(71), 83.

45. Rozanska, X., van Santen, R.A., Demuth, T. *et al.* (2003) A periodic DFT study of isobutene chemisorption in proton-exchanged zeolites: dependence of reactivity on the zeolite framework structure, *J. Phys. Chem. B* **107**(6), 1309–1315.
46. De Moor, B.A., Reyniers, M.-F., Gobin, O.C. *et al.* (2010) Adsorption of C2–C8 n-Alkanes in Zeolites, *J. Phys. Chem. C* **115**(4), 1204–1219.
47. Eder, F., Stockenhuber, M. and Lercher, J.A. (1997) Brønsted acid site and pore controlled siting of alkane sorption in acidic molecular sieves, *J. Phys. Chem. B* **101**(27), 5414–5419.
48. Bates, S.P., van Well, W.J.M., van Santen, R.A. and Smit, B. (1996) Energetics of n-alkanes in zeolites: a configurational-bias Monte Carlo investigation into pore size dependence, *J. Am. Chem. Soc.* **118**(28), 6753–6759.
49. Smit, B. and Maesen, T.L.M. (2008) Towards a molecular understanding of shape selectivity, *Nature* **451**(7179), 671–678.
50. Tuma, C. and Sauer, J. (2004) A hybrid MP2/planewave-DFT scheme for large chemical systems: proton jumps in zeolites, *Chem. Phys. Lett.* **387**(4–6), 388–394.
51. Tuma, C. and Sauer, J. (2006) Treating dispersion effects in extended systems by hybrid MP2:DFT calculations-protonation of isobutene in zeolite ferrierite, *Phys. Chem. Chem. Phys.* **8**(34), 3955–3965.
52. Svelle, S., Tuma, C., Rozanska, X. *et al.* (2009) Quantum chemical modeling of zeolite-catalyzed methylation reactions: toward chemical accuracy for barriers, *J. Am. Chem. Soc.* **131**(2), 816–825.
53. Grimme, S. (2004) Accurate description of van der Waals complexes by density functional theory including empirical corrections, *J. Comput. Chem.* **25**(12), 1463–1473.
54. Grimme, S. (2006) Semiempirical GGA-type density functional constructed with a long-range dispersion correction, *J. Comput. Chem.* **27**(15), 1787–1799.
55. Pidko, E.A., van Santen, R.A. and Hensen, E.J.M. (2009) Multinuclear gallium-oxide cations in high-silica zeolites, *Phys. Chem. Chem. Phys.* **11**(16), 2893–2902.
56. Pidko, E.A., Hensen, E.J.M., Zhidomirov, G.M. and van Santen, R.A. (2008) Non-localized charge compensation in zeolites: A periodic DFT study of cationic gallium-oxide clusters in mordenite, *J. Catal.* **255**(2), 139–143.
57. Pidko, E.A., Kazansky, V.B., Hensen, E.J.M. and van Santen, R.A. (2006) A comprehensive density functional theory study of ethane dehydrogenation over reduced extra-framework gallium species in ZSM-5 zeolite, *J. Catal.* **240**(1), 73–84.
58. Pidko, E.A. and van Santen, R.A. (2009) Structure–reactivity relationship for catalytic activity of gallium oxide and sulfide clusters in zeolite, *J. Phys. Chem. C* **113**(11), 4246–4249.
59. Pidko, E.A. and van Santen, R.A. (2007) Activation of light alkanes over zinc species stabilized in ZSM-5 zeolite: a comprehensive DFT study, *J. Phys. Chem. C* **111**(6), 2643–2655.
60. Petersen, M.A., van den Berg, J.-A. and van Rensburg, W.J. (2010) Role of step sites and surface vacancies in the adsorption and activation of CO on χ-Fe5C2 surfaces, *J. Phys. Chem. C* **114**(17), 7863–7879.
61. Shetty, S., Jansen, A.P.J. and van Santen, R.A. (2009) Direct versus hydrogen-assisted CO dissociation, *J. Am. Chem. Soc.* **131**(36), 12874–12875.
62. Shetty, S.G., Ciobica, I.M., Hensen, E.J.M. and van Santen, R.A. (2011) Site regeneration in the Fischer-Tropsch synthesis reaction: a synchronized CO dissociation and C-C coupling pathway, *Chem. Commun.* **47**(35), 9822–9824.

63. Chang, C.D. and Silvestri, A.J. (1977) The conversion of methanol and other O-compounds to hydrocarbons over zeolite catalysts, *J. Catal.* **47**(2), 249–259.
64. van den Berg, J.P., Wolthuizen, J.P. and Hooff, J.M.C.v. (1980) Proceedings of the 5th International Conference on Zeolites. Naples, Heyden, London.
65. Olah, G.A., Doggweiler, H., Felberg, J.D. *et al.* (1984) Onium Ylide chemistry. 1. Bifunctional acid-base-catalyzed conversion of heterosubstituted methanes into ethylene and derived hydrocarbons. The onium ylide mechanism of the C1 -> C2 conversion, *J. Am. Chem. Soc.* **106**(7), 2143–2149.
66. Blaszkowski, S.R. and van Santen, R.A. (1997) Theoretical study of C–C bond formation in the methanol-to-gasoline process, *J. Am. Chem. Soc.* **119**(21), 5020–5027.
67. Lesthaeghe, D., Van Speybroeck, V., Marin, G.B. and Waroquier, M. (2006) Understanding the failure of direct C–C coupling in the zeolite-catalyzed methanol-to-olefin process, *Angew. Chem. Int. Edit.* **45**(11), 1714–1719.
68. Lesthaeghe, D., De Sterck, B., Van Speybroeck, V. *et al.* (2007) Zeolite shape-selectivity in the gem-methylation of aromatic hydrocarbons, *Angew. Chem. Int. Edit.* **46**(8), 1311–1314.
69. Kvisle, S., Fuglerud, T., Kolboe, S. *et al.* (2008) Methanol to hydrocarbons, in *Handbook of Heterogeneous Catalysis* (eds G. Ertl, H. Knözinger, F. Schüth and J. Weitkamp), Wiley-VCH Verlag GmbH, Weinheim, pp. 2950–2957.
70. Pidko, E.A. and van Santen, R.A. (2010) Synthesis, reactions and applications, in *Zeolites and Catalysis* (eds J. Cejka, A. Corma and S.I. Zones), Wiley-VCH Verlag GmbH, Weinheim, pp. 301–333.
71. Hemelsoet, K., Nollet, A., Vandichel, M. *et al.* (2009) The effect of confined space on the growth of naphthalenic species in a chabazite-type catalyst: a molecular modeling study, *ChemCatChem* **1**(3), 373–378.
72. Linic, S. and Barteau, M.A. (2003) Construction of a reaction coordinate and a microkinetic model for ethylene epoxidation on silver from DFT calculations and surface science experiments, *J. Catal.* **214**(2), 200–212.
73. Linic, S. and Barteau, M.A. (2003) Control of ethylene epoxidation selectivity by surface oxametallacycles, *J. Am. Chem. Soc.* **125**(14), 4034–4035.
74. Torres, D., Illas, F. and Lambert, R.M. (2008) Towards an understanding of promoter action in heterogeneously catalyzed ethene epoxidation: why chlorine is the best halogen, *J. Catal.* **260**(2), 380–383.
75. Bocquet, M.-L. and Loffreda, D. (2005) Ethene epoxidation selectivity inhibited by twisted oxametallacycle: a DFT study on Ag surface-oxide, *J. Am. Chem. Soc.* **127**(49), 17207–17215.
76. Bocquet, M.-L., Michaelides, A., Loffreda, D. *et al.* (2003) New insights into ethene epoxidation on two oxidized Ag{111} surfaces, *J. Am. Chem. Soc.* **125**(19), 5620–5621.
77. Campbell, C.T. (1985) Atomic and molecular oxygen adsorption on Ag(111), *Surf. Sci.* **157**(1), 43–60.
78. Panov, G.I., Sobolev, V.I. and Kharitonov, A.S. (1990) The role of iron in N_2O decomposition on ZSM-5 zeolite and reactivity of the surface oxygen formed, *J. Mol. Catal.* **61**(1), 85–97.
79. Panov, G.I., Sheveleva, G.A., Kharitonov, A.S. *et al.* (1992) Oxidation of benzene to phenol by nitrous oxide over Fe-ZSM-5 zeolites, *Appl. Catal. A Gen.* **82**(1), 31–36.

80. Sun, K., Xia, H., Feng, Z. *et al.* (2008) Active sites in Fe/ZSM-5 for nitrous oxide decomposition and benzene hydroxylation with nitrous oxide, *J. Catal.* **254**(2), 383–396.
81. Chernyavsky, V.S., Pirutko, L.V., Uriarte, A.K. *et al.* (2007) On the involvement of radical oxygen species O^- in catalytic oxidation of benzene to phenol by nitrous oxide, *J. Catal.* **245**(2), 466–469.
82. Pérez-Ramírez, J., Santhosh Kumar, M. and Brückner, A. (2004) Reduction of N_2O with CO over FeMFI zeolites: influence of the preparation method on the iron species and catalytic behavior, *J. Catal.* **223**(1), 13–27.
83. Li, G., Pidko, E.A., van Santen, R.A. *et al.* (2011) Stability and reactivity of active sites for direct benzene oxidation to phenol in Fe/ZSM-5: A comprehensive periodic DFT study, *J. Catal.* **284**(2), 194–206.

8

Energy Conversion: Solid-State Lighting

E. Kioupakis[1,2], *P. Rinke*[1,3], *A. Janotti*[1], *Q. Yan*[1], *and C.G. Van de Walle*[1]
[1]*Materials Department, University of California, Santa, Barbara, USA*
[2]*Department of Materials Science and Engineering, University of Michigan, Ann Arbor, USA*
[3]*Fritz-Haber-Institut der Max-Planck-Gesellschaft, Berlin, Germany*

8.1 Introduction to Solid-State Lighting

Artificial illumination is a catalyst for development and improves the quality of life, but it also consumes a large fraction of the world's energy resources [1]. Lighting has been one of the earliest applications of electric power and today utilizes approximately 20% of the world's grid-based electricity production. The inefficiency of existing light sources that waste most of the power they consume is the reason for this large share. The worldwide demand for lighting is increasing rapidly, in particular in the developing world, and is projected to compound its strain on available energy sources. Therefore, there is a strong motivation for improving the efficiency of light sources and mitigating their effect on worldwide energy consumption and global climate change.

Historically, electric light bulbs have been of the incandescent type. Although this technology was developed more than 100 years ago, it is still in use today. Incandescent light bulbs operate by heating a tungsten wire in a vacuum bulb to a temperature of 3500 K. As a result of resistive heating, the wire glows white hot and emits a broad black-body radiation spectrum that covers the entire visible

Computational Approaches to Energy Materials, First Edition. Edited by Aron Walsh, Alexey A. Sokol and C. Richard A. Catlow.

range. A variation is the halogen lamp, which contains a mix of halogen and inert gases in the bulb. The halogen redeposits the evaporated tungsten back on to the wire, which allows the bulb to operate at a higher temperature. However, the efficiency of incandescent light bulbs is remarkably low, of the order of 5%, and their lifetimes are typically on the order of a thousand hours. This is the main reason why governments around the world are implementing legislation to phase out incandescent light bulbs for general lighting in the near future.

Fluorescent light bulbs are a more efficient and long-lasting alternative, but they also have their limitations. The light bulb is made of a fluorescent tube that contains an inert gas and small amounts of mercury. As the electric current passes through the gas, it excites the mercury atoms, causing them to emit ultraviolet radiation, which gets converted to visible light by the tube's fluorescent coating. Fluorescent light bulbs have efficiencies of the order of 20% and typical lifetimes of 10 000 hours. This is much better than for incandescent bulbs, but still leaves considerable room for improvement. In addition, the mercury contained in the light bulb poses a severe health hazard and needs to be safely recycled or disposed.

Solid-state lighting based on light-emitting diodes (LEDs) is a promising recent development and has the potential to reshape the lighting industry [2]. LEDs can reach energy-conversion efficiencies exceeding 50%, significantly reducing the energy cost of lighting. They can also achieve lifetimes as long as 50 000 hours, therefore reducing maintenance costs and the amount of electronic waste associated with lighting, and do not contain toxic elements. LEDs have long found applications as indicator lights, for example, in watches and calculators. Currently they are used for display backlights in portable computers and cellphones, and are also available for bicycle and car headlights. The total cost of ownership of LED bulbs is currently comparable with that of fluorescent ones and their market share for general lighting applications is increasing rapidly.

Semiconductors are the materials of choice for solid-state lighting applications [3]. Light emission from a semiconductor was first observed in SiC in 1907. The rapid developments in groups III–V compound semiconductor research in the 1960s and 1970s led to the fabrication of efficient LEDs in the long-wavelength part of the visible spectrum. However, white-light emission for lighting applications requires LEDs emitting in the blue or violet part of the spectrum. Research efforts initially focused on ZnSe, but defects in the material deteriorated the device performance and limited their lifetimes. A breakthrough happened in the early 1990s, when Shuji Nakamura developed the first viable blue LEDs based on group III nitrides [4], which are the dominant short-wavelength LED materials to this day.

The major group III nitrides used for LEDs, namely AlN, GaN, and InN, have been the subject of extensive research [5]. All three crystallize in the hexagonal wurtzite structure with four atoms in the primitive unit cell. Their electronic band gaps, which determine the light-emission wavelength, range from 6.2 eV for AlN to 3.4 eV for GaN and 0.7 eV for InN. Therefore, alloying GaN with InN yields

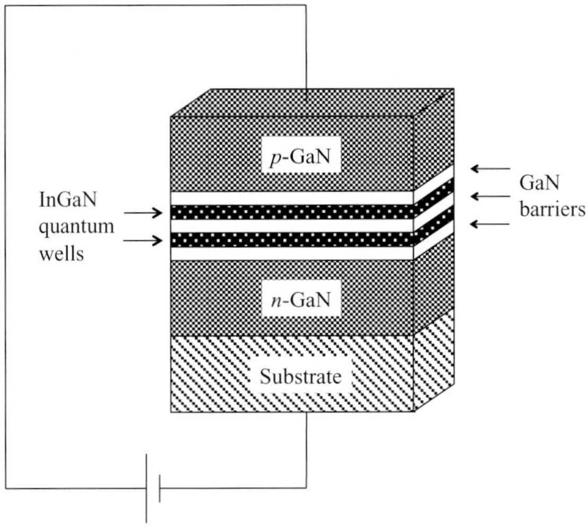

Figure 8.1 *Schematic illustration of a typical nitride light-emitting diode. The active region, consisting of InGaN quantum wells and GaN barriers, is sandwiched between the n- and p-type GaN layers. The device is usually grown on top of a foreign substrate, typically sapphire or SiC.*

a material with a direct band gap that is tunable in the entire visible range—a remarkable property that cannot be achieved with any other materials system at the moment. Doping GaN with Si and Mg yields *n*- and *p*-type-doped materials, respectively, that can be used for the fabrication of diodes.

A typical LED structure is shown in Figure 8.1. Bulk nitrides are difficult to fabricate, so devices are usually grown on a foreign substrate, typically sapphire (Al_2O_3) or SiC. The *n*-type Si:GaN layer is deposited first, followed by the growth of InGaN quantum wells and GaN barriers, and finally the *p*-type Mg:GaN layer. Not shown in Figure 8.1 are the transparent oxide top contact and the AlGaN electron-blocking layer that is commonly employed to enhance electron capture in the quantum wells. Connecting the device to a power source injects electrons and holes from the adjacent doping layers into the quantum wells, where they recombine and produce light at the wavelength corresponding to the band gap of the alloy material.

There are several approaches to achieve white-light emission appropriate for indoor lighting applications with nitride LEDs [2]. The simplest approach is to combine a blue LED with a yellow phosphor coating, which downconverts the high-energy blue photons to other visible wavelengths and produces a broad emission spectrum. However, the color quality that can be achieved with this approach is limited. Combining an ultraviolet LED with a mix of phosphors produces white light more suitable for indoor use but is less efficient. Another approach is to combine three or more LEDs of different colors, which is the best option for efficiency and color quality but comes at a higher cost.

The growth of nitride LEDs along nonpolar or semipolar crystallographic directions is a recent development that can improve the efficiency of devices and is a subject of active research [6]. Nitride LEDs are typically grown along the (0001) crystallographic direction of the hexagonal wurtzite structure, also known as the c-axis orientation. The nitride materials, however, are polar along this axis and, as a result of strained growth as well as spontaneous polarization, strong internal electric fields develop inside the InGaN quantum wells. These polarization fields keep the injected electrons and holes spatially separated and impede their efficient recombination, which affects the device efficiency. Different crystallographic growth orientations, however, can produce devices with reduced or even zero internal polarization fields, a fact that can substantially enhance their performance.

The study of nitride materials for light emission is a highly active research area with strong growth prospects. Current research focuses on improving the efficiency and reducing the cost of LEDs in order to deliver affordable light bulbs that can replace the existing solutions for general lighting. Growth of semipolar and nonpolar devices offers additional control over the device properties and is a promising area for new developments. Moreover, novel light-emitting devices have been fabricated based on even smaller nitride nanostructures, such as nanowires [7] or quantum dots [8]. These new developments guarantee that research interest in the field of nitride materials will remain strong.

First-principles computational approaches based on density-functional and many-body perturbation theory, introduced in Chapter 1, are a powerful tool set that has complemented and guided the experimental research efforts on nitride materials for LEDs. These methods yield equilibrium crystal structures of materials accurate to within 1% as well as accurate electronic properties, such as band gaps and carrier effective masses. In addition, these first-principles methods can be employed to understand and predict the properties of defects that are crucial for optoelectronic performance and device longevity. Finally, the radiative efficiency and the origin of the nonradiative losses in LEDs can be investigated and understood based on the results of theoretical calculations. In the following, we will present the corresponding computational methodology, along with specific results for group III nitrides used for solid-state lighting. These examples highlight the importance and predictive power of first-principles computational techniques for the study of technologically important materials and the performance of modern optoelectronic devices.

8.2 Structure and Electronic Properties of Nitride Materials

8.2.1 Density Functional Theory and Ground-State Properties

The most widely used computational method for electronic structure studies of materials properties is density functional theory (DFT, Section 1.3.3 and [9]). The central quantity of the theory is the total energy $E_{\mathrm{tot}}[n]$, which is a functional of the electron density $n(r)$ and attains its minimum at the exact ground-state density. Using the Born–Oppenheimer approximation, the functional for the electronic energy can be divided into different contributions:

$$E_{\mathrm{tot}}[n] = T_{\mathrm{s}}[n] + E_{\mathrm{ext}}[n] + E_{\mathrm{H}}[n] + E_{\mathrm{xc}}[n], \tag{8.1}$$

where $T_{\mathrm{s}}[n]$ is the noninteracting kinetic energy, $E_{\mathrm{ext}}[n]$ is the potential energy due to interaction with the ions, $E_{\mathrm{H}}[n]$ is the classical electrostatic interaction (Hartree) term, and $E_{\mathrm{xc}}[n]$ is the exchange–correlation energy.

Since the exact form of the exchange–correlation functional, which contains all the complexity of the quantum many-body interactions, is unknown, suitable approximations have to be found for practical calculations. Common approximations are the local density approximation (LDA), which is based on the correlation energy of the electron gas calculated with quantum Monte Carlo [10], or the generalized gradient approximation (GGA), which includes higher order gradient corrections [11]. Recent developments include the incorporation of exact exchange in the so-called hybrid functionals, which mix LDA or GGA with exact Hartree–Fock exchange. More advanced functionals combine exact exchange with correlation in the random phase approximation (RPA) [12, 13]. The exact-exchange energy cancels the spurious self-interaction error present in the Hartree term exactly, while the RPA correlation energy is fully nonlocal and accurately includes long-range van der Waals interactions.

Practical DFT calculations typically proceed in the context of Kohn–Sham theory, which maps the real system of interacting particles onto a fictitious system of noninteracting particles that have the same ground-state electron density and total energy as the real one. The Kohn–Sham equation that describes the motion of these noninteracting particles in the effective Kohn–Sham potential $v_{\mathrm{KS}}[n](r)$ is

$$\left[-\frac{\hbar^2\nabla^2}{2m} + v_{\mathrm{KS}}[n](r)\right]\varphi_i(r) = \varepsilon_i\varphi_i(r), \tag{8.2}$$

where ε_i and $\varphi_i(r)$ are, respectively, the energy eigenvalues and wavefunctions of the Kohn–Sham orbitals. The Kohn–Sham potential is a local potential that depends on the density and is given by

$$v_{\mathrm{KS}}[n](r) = v_{\mathrm{ext}}(r) + v_{\mathrm{H}}[n](r) + v_{\mathrm{xc}}[n](r),$$

where $v_{\mathrm{ext}}(r)$ is the applied external potential (e.g., the potential of the ions), $v_{\mathrm{H}}[n](r)$ is the Hartree term, and $v_{\mathrm{xc}}[n](r)$ is the exchange–correlation potential.

The electron density of the Kohn–Sham system, $n(r) = \sum_i^{\mathrm{occ}} |\varphi_i(r)|^2$, which comes from the solution of (8.2) is then inserted in (8.1) to yield the total energy of the electron system. The energy in (8.1) depends parametrically on the position of the nuclei. These can then subsequently be optimized to minimize the total energy. This yields equilibrium lattice parameters and atomic geometries, while the curvature

Table 8.1 *Equilibrium lattice parameters (a, c, and u) and bulk modulus (B) of the wurtzite-structured nitride materials used for solid-state lighting devices, as calculated with various exchange–correlation functionals within the DFT and measured by experiment ([14, 15] and references therein)*

Material	Method	a/Å	c/Å	c/a	u	B/Mbar
AlN	Theory, LDA	3.057	4.943	1.617	0.3802	2.09
	Theory, GGA	3.113	5.041	1.619	0.3798	1.92
	Theory, HSE	3.102	4.971	1.606	0.3819	–
	Experiment	3.111	4.978	1.601	0.385	1.85–2.12
GaN	Theory, LDA	3.193	5.218	1.634	0.376	2.02
	Theory, GGA	3.245	5.296	1.632	0.3762	1.72
	Theory, HSE	3.182	5.173	1.626	0.3772	–
	Experiment	3.180	5.166	1.624	0.375	1.88–2.45
InN	Theory, LDA	3.544	5.762	1.626	0.377	1.40
	Theory, GGA	3.614	5.884	1.628	0.377	1.161
	Theory, HSE	3.548	5.751	1.621	0.3796	–
	Experiment	3.533	5.693	1.611	0.375	1.25

of the energy-versus-volume curve provides the bulk modulus of the material. In general, structural parameters calculated within DFT are quite accurate and usually lie within a small percentage of their corresponding experimental values.

DFT within the LDA or GGA for the exchange–correlation functional provides accurate structural properties for the nitride materials [14]. The lattice parameters of the wurtzite crystal structure for AlN, GaN, and InN are listed in Table 8.1. The calculated results are in good agreement with experiment and the deviations are on the order of the typical 1% error of these functionals. The structural parameters obtained from more advanced hybrid functionals, such as Heyd, Scuseria, and Ernzehof (HSE) discussed later in Section 8.2.3, are in even better agreement with experiment than LDA or GGA.

8.2.2 Electronic Excitations: *GW* and Exact Exchange

The computational description of electronic excitations was introduced in Chapter 2; however, we will approach the problem from a complementary point of view. Although DFT performs exceptionally well in predicting the ground-state properties of materials, it fails to accurately describe properties that involve electronic excitations. In contrast to the electron density and total energy, which represent the exact properties of the interacting system, the Kohn–Sham eigenvalues in (8.2) are primarily mathematical constructs that do not formally correspond to excitations of the real system. In practice, however, they are a good zeroth-order approximation [16] to the quasi-particle energies as measured in direct and inverse photoemission and are therefore frequently used to describe the band structure of materials. The most prominent deficiency of Kohn–Sham band structures is the significant underestimation (in LDA and GGA typically by 50%) of band gaps. This error is a feature of the Kohn–Sham approach [17, 18] and arises even if the exact exchange–correlation potential is used.

As introduced in Chapters 1 and 2, many-body perturbation theory within the *GW* approximation for the self-energy operator can formally account for electronic excitations and is the method of choice for calculating quasi-particle band structures in solids [19–26]. Within many-body perturbation theory, the band structure of a material is given by the solution of the quasi-particle equation:

$$\left[-\frac{\hbar^2\nabla^2}{2m} + v_{\text{ext}}(r) + v_{\text{H}}[n](r)\right]\psi_i(r) + \int \Sigma(r, r'; \varepsilon_i^{\text{QP}})\psi_i(r')\text{d}r' = \varepsilon_i^{\text{QP}}\psi_i(r),$$

where $\Sigma(r, r'; \varepsilon)$ is the self-energy operator. The self-energy is a non-Hermitian, non-local, and energy-dependent operator that contains all the complexity of exchange and correlation of the many-electron system. The *GW* approximation consists of expanding the self-energy operator in terms of Feynman diagrams with respect to the screened Coulomb interaction and keeping only the first-order term:

$$\Sigma \approx \Sigma_{GW} = iGW,$$

where G is the Green's function and W the screened Coulomb interaction. The dielectric function $\varepsilon(r, r'; \omega)$ needed to describe the screened interaction between charge carriers at r and r' is then given within the RPA. In order to properly describe screening in semiconductors, it is important to account for local field effects, that is, the explicit dependence on r and r', as well as the frequency dependence of the dielectric function. The calculation of the dielectric function makes the *GW* approach computationally very demanding, and several computational techniques and approximations have been developed to make these calculations tractable [26].

Band structure studies within the *GW* method usually start with a DFT calculation for the ground-state properties and proceed by applying the self-energy operator as a first-order perturbation to the Kohn–Sham system. The Green's function and the screened interaction are calculated from the Kohn–Sham orbitals and eigenvalues. The quasi-particle energies $\varepsilon_i^{\text{QP}}$ are given by the following perturbation theory expression in terms of matrix elements of the self-energy operator and exchange–correlation potential with respect to the Kohn–Sham orbitals:

$$\varepsilon_i^{\text{QP}} = \varepsilon_i + \left\langle \varphi_i \left| \Sigma_{GW}(\varepsilon_i^{\text{QP}}) - v_{xc} \right| \varphi_i \right\rangle.$$

The quasi-particle wavefunctions ψ_i can be calculated in terms of the Kohn–Sham states by diagonalizing the full self-energy matrix, including off-diagonal elements. However, in *sp*-bonded solids, the Kohn–Sham orbitals typically differ by less than 0.1% from the corresponding quasi-particle wavefunctions and can therefore be taken as a very good approximation for the latter [27]. The quasi-particle energies and wavefunctions could in principle also be used to update the Green's function and the screening until self-consistency is achieved. A more detailed discussion of quasi-particle wavefunctions and the advantages and disadvantages of self-consistency would go beyond the scope of this chapter and we refer the interested reader to the literature [21, 22, 24, 26]. *GW* calculations for bulk materials are therefore usually performed as a one-shot correction, a procedure denoted G_0W_0.

Although LDA and GGA are most commonly used as the starting point for G_0W_0 in solids [20–22], orbital-dependent functionals [16] may provide a better description for certain materials. For example, ground-state functionals that address the self-interaction error of DFT have proven to be particularly advantageous for materials with *d* and *f* electrons [24, 28–30]. For the nitrides this becomes most apparent for the band gap of InN, for which LDA and GGA result in an overlap between the conduction and the valence bands and thus an effectively metallic state [31, 32]. A parametrization derived from such a band structure would not appropriately reflect the properties of bulk InN.

Orbital-dependent functionals have been used as the starting point of G_0W_0 calculations for group III nitrides and yield good agreement with experiment. In this context, the optimized effective potential approach applied to exact-exchange (OEPx) produces the variationally best local potential to the nonlocal Fock operator [33]. In OEPx(cLDA) standard LDA correlation is added to OEPx. Unlike in LDA and GGA, the self-interaction error in OEPx(cLDA) is greatly reduced (OEPx is fully self-interaction free) and correctly predicts InN to be semiconducting with the right band ordering in the wurtzite phase [34, 35]. Combining OEPx(cLDA) with G_0W_0, G_0W_0@OEPx(cLDA), yields band gaps for groups II and VI compounds, Ge, ScN, and group III nitrides that agree to better than 0.3 eV with experiment (Table 8.2) [24, 25, 32, 35, 36]. This is currently the best accuracy one can expect from a state-of-the-art first-principles theory that does not rely on fitting parameters.

The shallow cation semicore *d* levels of nitride materials strongly affect the calculated electronic structure [37]. First-principles approaches typically make use of pseudopotentials to describe the interaction between electrons and ions. The pseudopotential approximation significantly reduces the computational cost of plane-wave calculations because it freezes the inner shell electrons in the core and reduces the number of Fourier components needed to describe the resulting smooth valence wavefunctions. However, core-polarization effects are important

Table 8.2 *Band gaps of group III nitride materials as calculated with DFT and GW [15, 24] and measured by experiment*

Material	LDA	HSE	OEPx(cLDA)	G_0W_0@OEPx(cLDA)	Experiment
AlN	4.29	5.64	5.73	6.47	6.0–6.3
GaN	1.78	3.24	3.15	3.24	3.5
InN	–	0.68	1.00	0.69	0.65

All values are in eV.

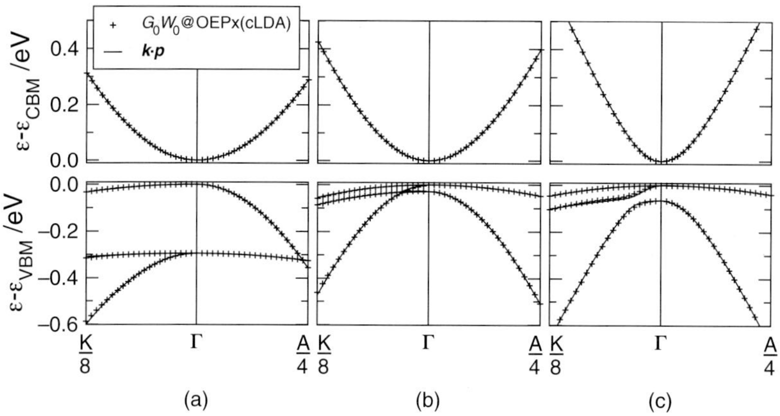

Figure 8.2 *Electronic band structure of (a) AlN, (b) GaN, and (c) InN as calculated with* G_0W_0*@OEPx(cLDA) (symbols) and parameterized with a* $k \cdot p$ *Hamiltonian (lines).*

for materials with shallow *d* levels, such as group III nitrides. *GW* calculations starting from LDA for systems with semicore *d* electrons in general show poor agreement with experiment, unless the *s* and *p* orbitals of the semicore *d* shell are also treated explicitly in the valence [37, 38]. This accounts for the polarization of the ionic cores, but the strong spatial localization of these semicore orbitals substantially increases the cost of plane-wave calculations. On the other hand, pseudopotentials calculated with the OEPx(cLDA) method properly account for the exchange interaction between the localized semicore orbitals and allow the semicore *s* and *p* states to be frozen in the pseudopotential core [24, 25].

The electronic structure near the band edges determines the carrier effective masses and hence the performance of optoelectronic devices. *GW* calculations can also determine the band parameters of nitrides that can be used in device simulations. The G_0W_0@OEPx(cLDA) band structures of group III nitrides near the band edges are shown in Figure 8.2 [32]. A single state forms the lowest conduction band, but the top of the valence band involves three states—two degenerate heavy- and light-hole bands, and the crystal-field-split level. While the bands are parabolic near the Γ point, nonparabolicities develop away from Γ in particular for InN. The calculated crystal-field splittings and carrier effective masses are also in good agreement with available experimental data [15, 32].

Materials parameters can be derived from first-principles electronic structure methods for bulk phases, but the size and complexity of structures required for device simulations currently exceeds the capabilities of first-principles electronic structure tools. To bridge this gap, first-principles calculations can be used to parameterize simplified methods such as the $k \cdot p$ approach [39]. This methodology can then easily be implemented in device simulation software to determine the band structure as a function of position and operating conditions for the complex layered heterostructures used in modern optoelectronic devices. Figure 8.2 displays the band structures of group III nitrides in the vicinity of the Γ point as determined by a 4×4 $k \cdot p$ model. The band dispersion parameters of the model have been fitted to the G_0W_0@OEPx(cLDA) bands for the corresponding material [32], and capture both the nonparabolicity of the conduction band and the complexity of the valence band structure. This consistent set of band parameters can in turn be used to model the electronic properties of nitride alloys, for example, by interpolating from the pure phases.

8.2.3 Electronic Excitations: Hybrid Functionals

Another method that can simultaneously yield reliable structural and electronic properties of nitride semiconductors is hybrid density functional theory. Hybrid functionals mix the usual local exchange potentials, such as LDA or GGA, with the exact nonlocal Hartree–Fock exchange potential, while correlation is described within the LDA or GGA. This approach originated in quantum chemistry, and in particular the B3LYP functional, which combines Becke's exchange functional (B3) with Lee, Yang, and Parr (LYP) correlation, has been quite successful in describing structural properties and energetics of molecules [40]. Range-separated hybrid functionals can be used to study bulk semiconductors by introducing a screening length, as proposed by Heyd, Scuseria, and Ernzehof (HSE) [41,42]. The HSE functional divides the exchange energy into a short- and a long-range part. The short-range part mixes the GGA exchange of Perdew, Burke, and Ernzerhof (PBE) [11] with Hartree–Fock exchange in a 3:1 ratio. The long-range exchange-correlation potential is described by the PBE functional. The additional computational cost involved with HSE calculations, compared to the more usual functionals, stems from the calculation of four-center integrals for the Hartree–Fock potential, and typical HSE runs require at least an order of magnitude, more computational resources than standard DFT calculations.

Hybrid functionals produce accurate structural parameters while significantly reducing the band-gap error of local and semilocal DFT functionals. The equilibrium lattice parameters for group III nitrides as obtained with HSE are in better agreement with experiment than the corresponding LDA or GGA values, as shown in Table 8.1 [15]. In addition, the electronic band gaps calculated with HSE are in much better agreement with experiment than LDA or GGA and of comparable accuracy to G_0W_0@OEPx(cLDA) (Table 8.2). Hybrid functional calculations are thus a powerful technique that can simultaneously yield accurate structural and electronic properties for group III nitrides.

8.2.4 Band-gap Bowing and Band Alignments

Band structure calculations using the HSE hybrid functional have been performed to investigate the band-gap bowing of InGaN alloys [43]. When GaN with a band gap of 3.24 eV is alloyed with InN with a band gap of 0.68 eV, an alloy forms that can have a gap value anywhere between these two extremes. The equation that determines the alloy band gap is

$$E_g(\mathrm{In}_x\mathrm{Ga}_{1-x}\mathrm{N}) = (1-x)E_g^{\mathrm{GaN}} + xE_g^{\mathrm{InN}} - bx(1-x),$$

where the first two terms represent the linear average between the two extrema (Vegard's law), while the nonlinear term expresses the band-gap bowing. The nonlinear behavior originates from composition fluctuations due to the statistical randomness in cation-site occupation either by In or Ga. The randomness leads to fluctuations of the local electrostatic potential arising from the different atomic potentials as well as to variations of the local microscopic structure (bond lengths and angles). The bowing parameter b quantifies the impact of these two effects on the band gap. HSE calculations have been employed to determine the band gaps of random InGaN alloy supercells of varying composition. The calculations reveal a strong band-gap bowing, which depends on In composition. The value of b is 1.10 eV if a single bowing parameter is used for the entire alloy range. There is, however, stronger bowing at low In content, that is, the regime usually employed for optoelectronic devices, and the bowing parameter is $b = 2.29$ eV at $x = 6.125\%$ and $b = 1.79$ eV at $x = 12.5\%$.

The band alignment between two different nitride materials, which is important for the operation of optoelectronic devices, can also be determined with hybrid functionals [43]. The natural band alignments between different nitrides are referenced to the vacuum level and reflect the relative band arrangement of the unstrained materials. These can be derived from separate bulk and surface calculations. The positions of the valence and conduction bands can be calculated with respect to the average electrostatic potential in the material from bulk calculations, while the alignment of the electrostatic potential relative to the vacuum level can be determined from a surface calculation in a nonpolar slab geometry. The calculations reveal that the valence band offset between GaN and InN is 0.62 eV, in good agreement with experiment, and varies linearly with alloy composition, while most of the band-gap bowing is manifested in the conduction band. This is expected since the valence band is composed of N orbitals, while the conduction band consists of randomly ordered cation orbitals. The natural band alignments account for bulk effects but do not consider the effect of interfaces between the two materials, which can be sizeable and must be studied explicitly for each specific interface geometry.

8.2.5 Strain and Deformation Potentials

One effect of the interface between dissimilar nitride materials, such as the InGaN quantum wells grown between GaN layers in heterostructures, is the development of strain. Vegard's law predicts that the equilibrium lattice constant of the alloy will be at a linearly interpolated value between the two extreme phases and will in general be different from either of them. As a result, nitride alloy layers that are pseudomorphically grown on GaN will be strained, which will affect their electronic properties. Strain changes the absolute position of the valence and conduction band extrema and hence modifies the quantum well depth and confinement of electrons and holes in the active region [44]. It also induces piezoelectric polarization, which greatly lowers the electron–hole overlap and hence the radiative recombination rate in quantum wells [45]. In addition, biaxial strain modifies the details of the valence band structure of nitride alloys in c-plane InGaN/GaN devices and changes the transition energies and effective masses of carriers [46]. Finally, biaxial strain in nonpolar and semipolar quantum wells drastically modifies the subband structure and wavefunction characteristics and can induce polarized light emission [47].

The effects of strain on the band structure of group III nitrides can be described by deformation potentials, defined as the linear coefficient in the expansion of the band energies with respect to strain. The concept of deformation potentials originates from $k \cdot p$ perturbation theory [48]. In this approach, the lowest conduction and topmost three valence bands are the eigenstates of an 8×8 Hamiltonian. The effects of a strain perturbation are described by two conduction band (a_{ct} and a_{cz}) and six valence band deformation potentials (D_1 to D_6). The deformation potentials a_{cz}, D_1, and D_3 are related to strain in the c-plane ($\varepsilon_{\perp} = \varepsilon_{xx} + \varepsilon_{yy}$), while a_{ct}, D_2, and D_4 are related to the out-of-plane component ($\varepsilon_{\parallel} = \varepsilon_{zz}$). The deformation potential D_5 determines the effect of anisotropic in-plane strain $\left|\varepsilon_{xx} - \varepsilon_{yy}\right|$ on the band structure, especially the splitting of the heavy- and light-hole valence bands, and the D_6 component accounts for shear strain (ε_{xy}, ε_{xz}, and ε_{yz}) in semipolar nitride materials.

The experimental determination of deformation potentials is difficult. The measured data for GaN are scattered over a wide range of values, while for InN only one set of measured data have been reported so far [49]. For AlN no data are available yet. Most of the experimental data so far have been obtained by measuring the change of optical transition energies under biaxial stress induced by the c-plane substrate. However, uniaxial and biaxial stress cannot be applied separately, and hence a combination of several deformation potentials results. The quasi-cubic approximation is then applied to extract the deformation potentials from the combined data. Another experimental approach is to carry out optical measurements under uniaxial stress. The advantage of this approach is that various strain conditions can be applied without having to resort to the quasi-cubic approximation. This technique has produced a complete set of deformation potentials for GaN [50].

A consistent set of deformation potentials for group III nitrides can be obtained from first-principles calculations [51, 52]. The band structure of each material is calculated within LDA, GGA, or HSE for the corresponding relaxed crystal structure under various applied strain conditions. In particular, the internal structural parameter u sensitively affects the magnitude of the crystal field splitting and must be fully relaxed. The deformation potentials obtained by fitting the analytical $k \cdot p$ eigenenergies to the calculated band structures at the Γ point are listed in Table 8.3 and compared to experiment. The band-gap-related deformation potentials

Table 8.3 *Deformation potentials for group III nitrides as calculated with various first-principles methods and measured by experiment*

	Method	$a_{cz} - D_1$	$a_{ct} - D_2$	D_3	D_4	D_5	D_6
AlN	LDA	−3.44	−11.39	8.97	−3.95	−3.36	–
	GGA	−3.39	−11.38	9.12	−4.01	−3.37	–
	HSE	−4.21	−12.07	9.22	−3.74	−3.30	−4.49
GaN	LDA	−4.56	−8.03	5.61	−3.03	−2.94	–
	GGA	−4.46	−8.08	5.83	−2.98	−3.13	–
	HSE	−6.02	−8.98	5.45	−2.97	−2.87	−3.95
	GW	−5.49	−8.84	5.80	−3.10	–	–
	Experiment	−9.6 to −3.1	−11.8 to −8.1	1.4 to 8.2	−4.1 to −0.7	−4.7 to −2.4	–
InN	HSE	−3.64	−4.58	2.68	−1.78	−2.07	−3.02

All values are in eV.

($a_{cz} - D_1$ and $a_{ct} - D_2$) for GaN and AlN calculated with LDA and GGA agree with each other but appreciably deviate from the HSE results. This discrepancy has been attributed to the band-gap problem of these functionals. For InN, the negative band gap predicted by the same functionals inhibits the determination of deformation potentials. In contrast, first-principles approaches that account for corrections to the band gap, such as HSE and G_0W_0@OEPx(cLDA), give accurate deformation potentials for group III nitrides. Cross-checks show that the deformation potentials of GaN obtained with HSE are in good agreement with G_0W_0@OEPx(cLDA) calculations (Table 8.3), which validates the reliability of the HSE approach in studying strain effects in nitride materials. It has been found that the transition energies depend nonlinearly on the lattice parameters, hence the equilibrium lattice parameters around which the linear expansion is constructed significantly affect the deformation potentials. The set of deformation potentials presented in Table 8.3 was therefore obtained by constraining the strain range to realistic conditions in the linear regime around the experimental lattice parameters.

A complete set of deformation potentials constitutes essential input for the modeling of nitride optoelectronic devices. It allows the prediction of subband structures under realistic strain conditions especially for light-emitting devices grown along nonpolar or semipolar crystallographic directions, for which the strain is more complicated compared to c-plane devices and crucial for the device performance.

8.3 Defects in Nitride Materials

Point defects and impurity atoms can strongly affect the properties of materials and the performance of optoelectronic devices. The intentional addition of donor and acceptor atoms to semiconductors yields n- and p-type doped materials, respectively, that are necessary for the fabrication of diodes. However, the free carrier density that can be achieved by doping is limited by compensating defects that spontaneously form at high doping concentrations. Moreover, certain materials exhibit unintentional doping due to either intrinsic defects or unintentional incorporation of foreign impurity atoms. In addition, defects mediate nonradiative recombination and loss in optoelectronic devices and cause device degradation. These examples highlight the need to study defects in materials and understand their effect on device performance.

8.3.1 Methodology

First-principles methods based on DFT have already been successful in the study of defects in semiconductors and insulators [53], but the band-gap error limits the predictive power of the commonly used functionals. The severe underestimation of band gaps by LDA and GGA may introduce large errors in calculated defect formation energies (and, thus, defect concentrations), transition levels, and optical and emission energies associated with defects or impurities. A predictive theory to describe defects must therefore incorporate an accurate treatment of excited-state materials properties.

Recent developments in hybrid functionals have led to great improvements over the traditional DFT approaches for the study of defects [54]. As discussed in Section 8.2.2, hybrid functionals such as HSE [41, 42] significantly improve the description of band gaps, and therefore, aid in the investigation of basic electronic properties of host semiconductors and the electronic and structural properties of defects and impurities. The main advantage of hybrid functionals over other excited-state methods, such as the *GW* approach, is that hybrid functionals also provide accurate total energies and atomic forces. These quantities are essential for calculating formation energies and lattice relaxations associated with defects, impurities, interfaces, and surfaces.

The HSE functional describes the band gaps with great accuracy for many semiconductor materials [55], but it may still lead to band gaps noticeably smaller than experimental values for some wide band-gap semiconductors and insulators if used with the standard mixing parameter of 25% (GGA exchange mixed with Hartree–Fock in a 3 : 1 ratio). In these cases, band-gap values can be brought into closer agreement with experiment by increasing the mixing parameter beyond the standard 25% value [56–58]. It has been shown that the mixing parameter that reproduces the experimental band-gap value also predicts defect levels that quantitatively agree with experimental observations [58, 59].

Isolated defects and impurities are typically calculated using the supercell approach, in which the defect is created in a section of the material that is periodically repeated in three spatial directions. Supercell calculations containing up to thousands of atoms are nowadays achievable within LDA and GGA, but present computational resources limit HSE calculations to a few hundred atoms. These methods can compute total energies and forces for a solid containing an isolated defect or impurity, and can fully take into account atomic relaxations around the lattice imperfection.

The formation energy is the key quantity that determines the stability and concentration of defects in the dilute limit, that is, when defect–defect interactions are negligible. The formation energy of a defect D in charge state q is defined as a total energy difference and is given by [53]

$$E^f(D^q) = E_{\text{tot}}(D^q) - E_{\text{tot}}(\text{host}) + \sum n_X \mu_X + q\varepsilon_F,$$

where $E_{\text{tot}}(D^q)$ and $E_{\text{tot}}(\text{host})$ are total energies calculated for supercells containing the defect in charge state q and the perfect host crystal, respectively, n_X is the number of atoms of species X that were removed from the crystal to form the defect, and μ_X is the chemical potential of the species X reservoir. In the case of a nitrogen vacancy in GaN, for example, X is nitrogen and $n_X = +1$. The chemical potential for species X is conventionally referenced to the energy per atom of the standard phase of the element. It is bound by the stability condition of the host and the requirement of avoiding the formation of secondary phases. In the case of GaN, μ_N is referenced to the total energy of an isolated N_2 molecule, and must satisfy the following condition:

$$\mu_{\text{N}} + \mu_{\text{Ga}} = \Delta H^f(\text{GaN}),$$

where $\Delta H^f(\text{GaN})$ is the formation enthalpy of GaN [53]. Charged defect calculations, for which $q \neq 0$, involve the exchange of one or more electrons with an electron reservoir. The chemical potential of the reservoir is the Fermi level ε_F, which is conventionally referenced to the valence band maximum of the host. In practical calculations, the valence band maximum is referenced to the averaged electrostatic potential, rendering it necessary to align the average potentials in the host and the defect-containing supercell. Additional corrections are necessary to correct for finite-size effects in charged defect supercells. Calculations that employ periodic boundary conditions represent the charged defect as an infinite array of charges in a uniform neutralizing background. The interactions between defects in neighboring supercells can lead to errors on the order of 1 eV in the formation energy of defects, especially for highly charged defects ($|q| > 2$). Several approaches have been developed to correct for interactions between defects in neighboring supercells [60, 61]. The most straightforward, but computationally expensive, is to perform calculations for progressively increasing supercell sizes and extrapolate the results to infinite limit. A more rigorous approach for performing charge-state corrections, based on an exact treatment of the electrostatic problem, has recently been proposed [62], circumventing the need to perform calculations for various supercell sizes.

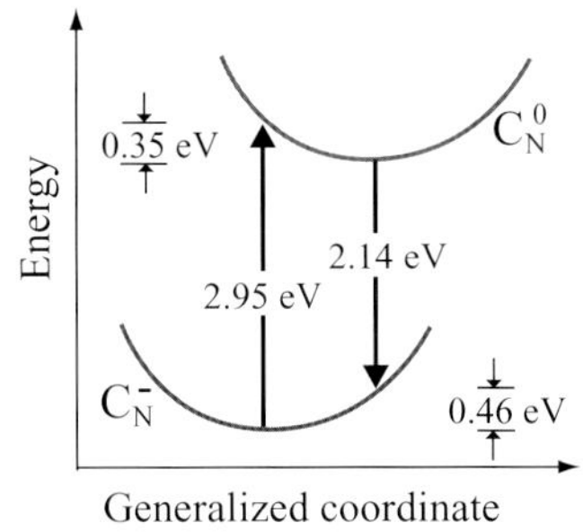

Figure 8.3 *Calculated configuration–coordinate diagram for the C_N impurity in GaN using the HSE hybrid functional [59]. The optical absorption by the negatively charged defect has a zero phonon line at 2.60 eV and a peak at 2.95 eV. The emission process ($C_N^0 + e \rightarrow C_N^- + \hbar\omega$) has a zero-phonon line at 2.60 eV and a peak at 2.14 eV, in the yellow part of the visible spectrum.*

The electrical activity of defects, which determines for instance whether a defect is a deep or shallow donor or acceptor, is defined by the position of transition levels relative to the band edges. The thermodynamic transition levels of a defect can be derived from formation energies [53]:

$$\varepsilon(q/q') = \frac{E^f(D^q) - E^f(D^{q'})}{q' - q}.$$

For Fermi-level values below the transition level $\varepsilon(q/q')$ the defect is stable in charge state q, while for values above the transition level it is stable in charge state q'. The difference between thermodynamic and optical transition levels of defects is that the former involve local lattice relaxations while the latter do not. The optical transition levels are associated with optical absorption and luminescence. Since optical transitions occur instantaneously compared to the time scale associated with lattice relaxations, they can be represented with vertical lines in a configuration–coordinate diagram (Figure 8.3). The optical transition energy between two charge states is the difference in formation energies of the corresponding charged defects calculated at the initial-state lattice geometry. It is also necessary to specify whether during the optical transition the defect exchanges electrons with the valence or with the conduction band. If the transition involves placing an electron in the conduction band, the Fermi level ε_F is taken at the conduction band minimum, while if a hole is lifted from the valence band to the defect level, then the Fermi level is placed at the top of the valence band.

8.3.2 Example: C in GaN

The carbon impurity in GaN presents an example of application of first-principles techniques to study defects and impurities in the nitride materials. The role of

C in GaN has been put under scrutiny since the early explorations of GaN for LEDs. Ion implantation with various elements demonstrated that C was associated with yellow luminescence in GaN [63]. However, the microscopic structure of the C impurity—whether it was substituting for a Ga or a N atom or sitting on an interstitial site, or possibly associated with a point defect—remained inconclusive [64, 65]. Based on atomic size and valence considerations, it was thought that C substituting on the N site (C_N) would act as a shallow acceptor in GaN [66, 67], yet no experimental evidence of *p*-type GaN using C as an acceptor exists. Intentional C doping actually leads to semi-insulating GaN layers, which are used in AlGaN/GaN heterojunction field-effect transistors [68]. The semi-insulating behavior was tentatively explained by the incorporation of the C_N acceptor and the C_{Ga} donor in equal concentrations [68, 69].

First-principles calculations employing hybrid functionals provide an accurate description of C impurities in GaN and can explain all the experimental observations, including the source of the yellow luminescence. Previous DFT-LDA/GGA calculations found C_N to be a shallow acceptor with a transition level in the range of 200–300 meV [69–72]. The band-gap error in LDA/GGA calculations introduces large errors in the calculated transition levels, precluding definite conclusions on the exact position of the C acceptor level with respect to the valence band. In addition, hole localization tends to be underestimated in LDA or GGA, a problem that is also alleviated by the use of a hybrid functional [73]. Calculations employing the HSE hybrid functional [59] show that C is likely to incorporate on the N site and gives rise to an acceptor $\varepsilon(0/-)$ level located at 0.90 eV above the valence band maximum (Figure 8.4). C substituting on the Ga site (C_{Ga}) acts as a shallow donor, in accord with previous results [72, 74]. These results indicate that C_N itself can explain the semi-insulating behavior of C-doped GaN without the necessity of invoking equal incorporation of C_N acceptors and C_{Ga} donors.

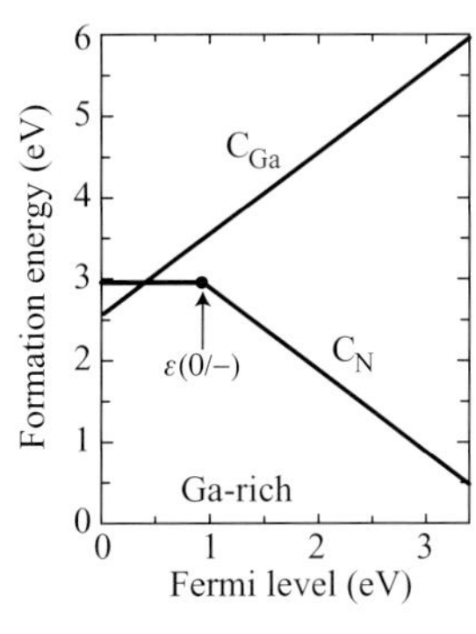

Figure 8.4 *Calculated formation energy versus Fermi level for the C impurity in GaN under Ga-rich conditions using the HSE hybrid functional [59]. The transition level $\varepsilon(0/-)$ for the C_N acceptor is indicated.*

The HSE-calculated optical transition levels of the C_N acceptor in GaN are represented in the configuration–coordinate diagram in Figure 8.3. An absorption peak at 2.95 eV is predicted for the optical transition from the negatively charged acceptor to the conduction band ($C_N^- + \hbar\omega \rightarrow C_N^0 + e$). The inverse process ($C_N^0 + e \rightarrow C_N^- + \hbar\omega$) exhibits an emission peak at 2.14 eV, which is in the yellow part of the visible spectrum. The difference between absorption and emission energies originates from the large differences in local lattice relaxations around C_N in the negative and neutral charge states. These results indicate that C_N can explain both the semi-insulating behavior of C-doped GaN and the yellow luminescence.

8.4 Auger Recombination and Efficiency Droop Problem of Nitride LEDs

8.4.1 Efficiency Droop

The peak efficiencies of nitride LEDs far exceed those of incandescent and fluorescent light bulbs. However, the remarkable efficiency of LEDs decreases dramatically when they are operating at the high power that general lighting requires (Figure 8.5). This phenomenon is called *efficiency droop* and is a universal feature of nitride LEDs [75]. Moreover, droop becomes progressively worse for devices operating at longer wavelengths, an effect known as the *green gap problem*. This efficiency loss limits the light output power and increases the operating cost of LEDs. To compensate for the light output reduction, more individual LEDs per light bulb need to be employed, which increases the total cost. The efficiency droop problem prevents LED light bulbs from taking over the lighting market. Hence, there is a strong incentive to understand its origin and devise engineering solutions to mitigate its impact.

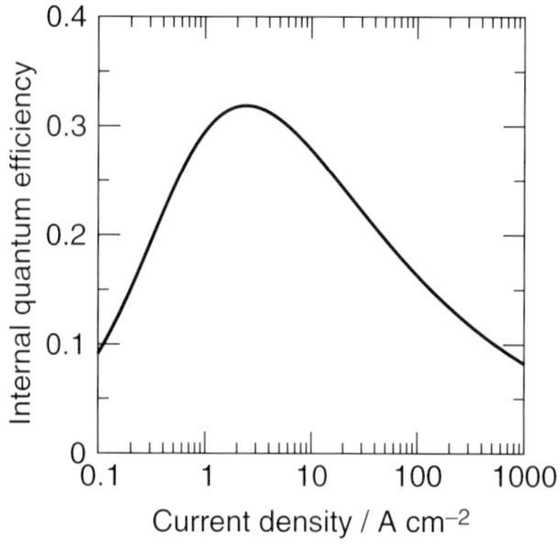

Figure 8.5 *Typical internal quantum efficiency versus injected current density for a nitride LED. The curve is generated from representative experimental data [76]. The efficiency exhibits a peak at low current but decreases dramatically at the high power needed for lighting.*

Several loss mechanisms have been proposed as the origin of the droop, but the topic is still a subject of active research and intense debate. The most prominent hypotheses for the loss include electron leakage from the quantum wells [77], defect-related mechanisms [78, 79], or Auger recombination [80–83]. It is difficult to distinguish the contribution by each source of loss experimentally and to identify the precise microscopic mechanism responsible for the droop. Therefore, insight gained from first-principles calculations can provide valuable input on the importance of the various loss mechanisms and their effect on the efficiency of nitride LEDs.

8.4.2 Auger Recombination

Auger recombination is one of the more plausible theories that have been put forward as the cause of the efficiency droop. In the Auger process, an electron and a hole recombine in a nonradiative manner, while the energy released by the recombination is transferred via electron–electron scattering to a third carrier that gets excited to a higher conduction band or lower valence band state (Figure 8.6). The third carrier involved in the transition can either be an electron (electron–electron–hole process) or a hole (hole–hole–electron). Since three carriers are involved in each Auger recombination event, the overall recombination rate scales with the third power of the free carrier density, $R_{\text{Auger}} = Cn^3$, and becomes dominant at high injected current densities. On the other hand, the radiative recombination rate increases with the square of the free carrier density, $R_{\text{rad}} = Bn^2$, while Shockley–Read–Hall nonradiative recombination at defects depends linearly on

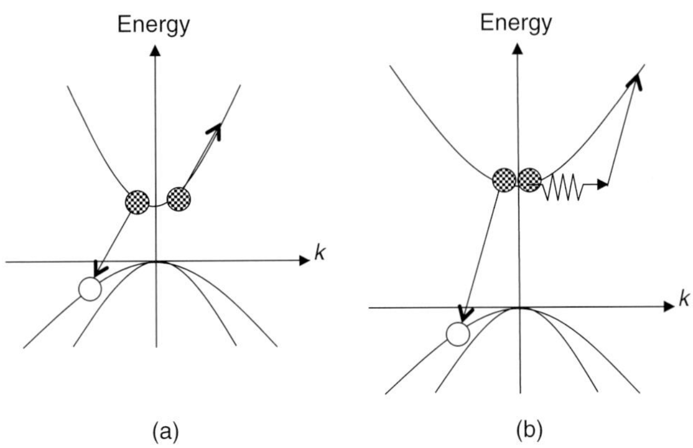

Figure 8.6 Schematic illustration of Auger recombination of free carriers in semiconductors. (a) During the direct Auger process, an electron and a hole recombine nonradiatively while transferring the excess energy to a third carrier. (b) Indirect Auger recombination is a higher order process that is assisted by a scattering mechanism, which provides additional momentum and facilitates the energy transfer. Indirect Auger recombination is the dominant nonradiative recombination mechanism in nitride LEDs at high power.

the carrier concentration, $R_{\text{SRH}} = An$. As a result, the internal quantum efficiency η_{IQE}, defined as the ratio of emitted photons over total recombined electron–hole pairs,

$$\eta_{\text{IQE}} = \frac{R_{\text{rad}}}{R_{\text{SRH}} + R_{\text{rad}} + R_{\text{Auger}}} = \frac{Bn^2}{An + Bn^2 + Cn^3}, \tag{8.3}$$

decreases at high injected carrier densities and explains the efficiency reduction in Figure 8.5. The Auger mechanism was first proposed after photoluminescence experiments showed that efficiency droop behavior described by equation (8.3) occurs even for optically excited samples [80], excluding any carrier leakage effects. The efficiency droop was subsequently observed for bulk GaN samples [84], indicating that the mechanism for the droop is intrinsic to nitride materials and not a consequence of the heterostructure. Experimentally measured values for the Auger coefficient of InGaN are on the order of 10^{-31}–10^{-30} cm^6 s^{-1} and are large enough to explain the observed efficiency droop in LEDs [75].

However, the significance of Auger recombination has been put into question due to an apparent discrepancy between experiment and calculations. Theoretical investigations that focused on the direct Auger recombination process (Figure 8.6a) found an Auger coefficient of the order of 10^{-34} cm^6 s^{-1} [85, 86]. This value is three to four orders of magnitude smaller than the experimental results and cannot account for the measurements. This is expected because the Auger coefficient of a direct band-gap semiconductor decays exponentially with increasing band gap and will therefore be small for wide band-gap materials such as the nitrides. Moreover, the direct Auger recombination rate is very sensitive to temperature, a fact that is at variance with experimental measurements [81]. These observations rule out direct Auger recombination as the cause of the droop.

Higher order indirect Auger processes are also possible, however, and can potentially be of importance in nitride materials. Indirect Auger recombination requires the assistance of a microscopic carrier-scattering mechanism to provide additional momentum and facilitate the Auger transition (Figure 8.6b). As a result, indirect processes can become the dominant Auger recombination mechanism in wide band-gap materials and can be strong enough to explain the efficiency droop. It is therefore important to investigate these indirect Auger processes theoretically and to evaluate their effect on LED performance.

Several carrier-scattering mechanisms can potentially be at work in nitride materials. One mechanism is the interaction of electrons and holes with lattice vibrations. Electron–phonon coupling is a universal carrier-scattering mechanism in every material and responsible for several well-known electronic phenomena such as resistivity, superconductivity, and indirect optical absorption. This coupling is particularly strong in the nitrides because of the ionic nature of the material as well as the nature of the nitrogen valence orbitals. Since the $2p$ orbitals are the innermost p orbitals of an atom, they feel the core potential almost unscreened and are thus very sensitive to atomic vibrations [87]. Another scattering mechanism is alloy scattering, due to the disordered distribution of In and Ga atoms in the InGaN quantum wells where carriers recombine in devices. Finally, the long-range Coulomb potential of charged defects can scatter charge carriers and enable Auger recombination.

8.4.3 Computational Methodology

First-principles calculations are a powerful set of computational tools that can elucidate the impact of indirect Auger recombination on the performance of nitride LEDs. These computational techniques allow explicit investigations of microscopic phenomena such as electron–phonon coupling and alloy scattering and their effect on the Auger recombination rate. Moreover, Auger recombination in these wide band-gap materials generates hot carriers at excited states that are very high in energy and cannot be described with near-band-edge models such as the $k \cdot p$ theory. First-principles methods, in contrast, can accurately provide the full band structure throughout the Brillouin zone, and can readily describe these high-energy states of interest. First-principles approaches are therefore a suitable and predictive computational formalism for the description of indirect Auger recombination in the nitride materials.

The transition probability rate for Auger recombination is calculated from Fermi's golden rule [88]:

$$R_{\text{Auger}} = 2\frac{2\pi}{\hbar}\sum_{1234} P\,|M_{1234}|^2\,\delta(\varepsilon_1 + \varepsilon_2 - \varepsilon_3 - \varepsilon_4),$$

where we used composite band and wave-vector indices $i = (n_i, \mathrm{k}_i)$, and P is a statistics factor that ensures transitions are only allowed from occupied to empty states,

$$P = f_1 f_2 (1 - f_3)(1 - f_4),$$

where f_1 is the Fermi–Dirac occupation number of state i. The perturbation Hamiltonian is the screened Coulomb interaction between charge carriers, and the matrix elements involve a direct M^d and an exchange M^x term (Figures 8.7a and b) to

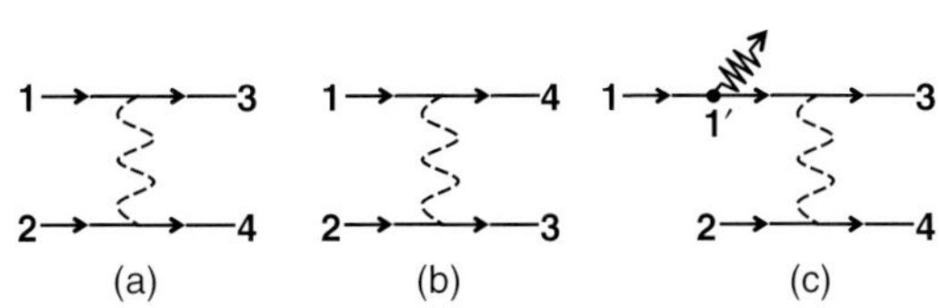

Figure 8.7 Electron–electron scattering diagrams for the (a) direct and (b) exchange terms of direct Auger recombination. (c) One of the 16 diagrams involved in the phonon-assisted Auger process.

account for fermion exchange symmetry:

$$|M_{1234}|^2 = \left|M^d_{1234} - M^x_{1234}\right|^2 + \left|M^d_{1234}\right|^2 + \left|M^x_{1234}\right|^2,$$

where $M^d_{1234} = M^x_{1234} = \iint \psi_1^*(r_1)\psi_2^*(r_2)W(r_1, r_2)\psi_3(r_1)\psi_4(r_2)\mathrm{d}r_1\mathrm{d}r_2$ are matrix elements of the screened Coulomb interaction W. The band energies ε_i and wavefunctions $\psi_i(r)$ are readily calculated with DFT. The dielectric function and its wave-vector dependence needed for W can be described with a model [89] or, alternatively, can be calculated entirely from first principles within the RPA [20].

The transition rate equations for indirect Auger recombination are slightly modified to account for the additional phonon emission or absorption process. In this case the probability rate is given by second-order Fermi's golden rule:

$$R^{\text{Indirect}}_{\text{Auger}} = 2\frac{2\pi}{\hbar}\sum_{1234;\mathrm{vq}}\left(n_{\mathrm{vq}} + \frac{1}{2} \pm \frac{1}{2}\right)P\left|\tilde{M}_{1234;\mathrm{vq}}\right|^2 \delta(\varepsilon_1 + \varepsilon_2 - \varepsilon_3 - \varepsilon_4 \mp \hbar\omega_{\mathrm{vq}}),$$

where n_{vq} are the phonon Bose occupation numbers, $\hbar\omega_{\mathrm{vq}}$ are the phonon frequencies, and the upper or lower sign corresponds to phonon emission or absorption, respectively. The perturbation matrix elements $\tilde{M}_{1234;\mathrm{vq}}$ are also slightly modified to account for the electron–phonon coupling interaction. The overall rate involves 16 terms to account for all possible orderings of electron–electron scattering and electron–phonon coupling diagrams. For example, the matrix element that corresponds to the diagram of Figure 8.7c is given by

$$\tilde{M}_{1234;\upsilon\mathrm{q}} = \sum_{1'} \frac{g_{11';\upsilon}M^d_{1'234}}{\varepsilon_{1'} - \varepsilon_1 \pm \hbar\omega_{\upsilon\mathrm{q}}},$$

where $g_{11';\mathrm{v}}$ is the electron–phonon coupling matrix element from state 1 to state $1'$ by phonon mode v.

8.4.4 Results

First-principles calculations have been applied to determine the alloy-scattering-assisted Auger recombination rate in InGaN alloys [90]. A common method to study alloys is to employ the virtual crystal approximation, which models the alloy potential as an interpolated average between the two pure phases. This approximation, however, ignores the effect of atomistic disorder that gives rise to alloy scattering and enables Auger recombination. In order to capture these effects, calculations need to explicitly model the composition fluctuations and randomness of the alloy on the atomic scale. This can be accomplished with quasi-random alloy crystal structures, which are periodic alloy supercells of the desired composition with an atomic distribution that has been optimized to reproduce the short-range correlation function of the fully random alloy [91]. The calculations were done for the 32-atom structure of 25% composition and interpolated for other alloy configurations by considering the dependence of the alloy-scattering potential on

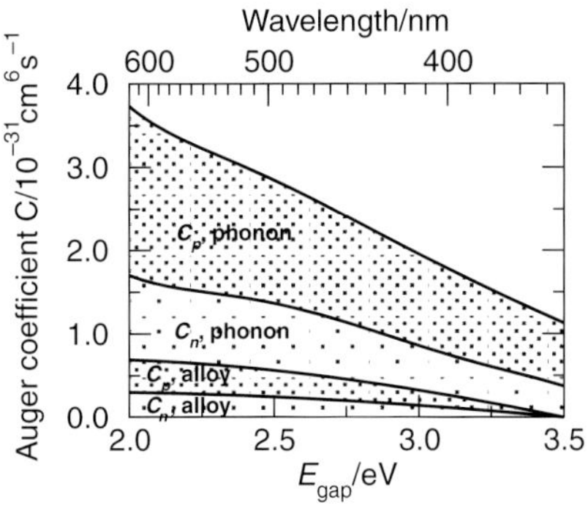

Figure 8.8 *Indirect Auger recombination coefficients for the electron–electron–hole (C_n) and hole–hole–electron (C_p) processes due to electron–phonon coupling and alloy scattering as a function of the InGaN alloy band gap.*

composition, $V^2 \propto x(1 - x)$ [88]. Alloy scattering was found to have a significant effect on Auger recombination. The calculated Auger coefficients shown in Figure 8.8 are significantly enhanced compared to the direct Auger process and increase for increasing In mole fraction.

Phonon-assisted Auger recombination has also been studied with first-principles methods and can account for the efficiency droop of nitride LEDs [90]. The vibrational parameters, such as the phonon frequencies and electron–phonon coupling matrix elements, can be calculated with density functional perturbation theory for arbitrary phonon wave vectors in the first Brillouin zone [92]. The electron wavefunctions and energies were calculated for GaN with DFT and the LDA functional. The LDA energy eigenvalues were corrected with a scissors shift that was adjusted to account for varying In mole fractions and to model various alloy compositions. The calculated phonon-assisted Auger coefficients are shown in Figure 8.8. The values for the phonon-assisted case are also strongly enhanced compared to direct Auger recombination and can explain the experimentally measured values from droop experiments. Moreover, the corresponding Auger coefficients increase for higher In content and for increasing temperature, since more phonons are present at elevated temperatures. Phonon-assisted Auger is also possible even at absolute zero temperature because of the phonon-emission processes.

It is informative to examine which vibrational modes are the ones that primarily contribute to phonon-assisted Auger recombination. Group III nitrides are ionic materials, for which the longitudinal-optical (LO) modes are known to be strongly coupled to charge carriers [88]. However, an analysis of the contribution of the various modes to the Auger coefficient reveals that they are not the dominant ones. Significant contributions arise from the acoustic as well as the other optical phonon modes. This is to be expected because phonon-assisted Auger recombination for these wide band-gap materials excites charge carriers to higher electronic states, which happen to be near the edges of the Brillouin zone. These transitions involve scattering by short-range phonons. Although the LO phonons are indeed strongly coupled to electrons, they are so only for long-range scattering, while for short-range scattering all the phonon modes are important. Therefore, a first-principles analysis that accounts for short-range phonon effects is needed to properly describe phonon-assisted Auger for the nitrides.

Scattering by charged defects can also enable indirect Auger recombination, but the resulting values (assuming realistic defect concentrations) are much weaker than those due to electron–phonon coupling or alloy scattering [90]. The interaction of carriers with charged defects can be modeled with a screened Coulomb potential that accounts for dielectric screening by the bulk material as well as metallic screening by the free carriers [88]. However, the calculated charged defect-assisted Auger coefficient values are weak and become significant only for high defect concentrations, of the order of 10^{20} cm^{-3}, which are not encountered in actual devices. This kind of indirect Auger process can therefore be ruled out as a significant source of energy loss in LEDs.

The cumulative Auger recombination coefficient is plotted in Figure 8.8. The overall value ranges from 1.5×10^{-31} cm^6 s^{-1} for violet devices to 3.0×10^{-31} cm^6 s^{-1} for green and is consistent with the experimentally measured range of values. This first-principles result provides evidence that indirect Auger recombination assisted by alloy disorder and electron–phonon scattering is the primary LED efficiency loss mechanism at high power. This finding also suggests strategies to mitigate the impact of Auger recombination on devices. Since Auger is an intrinsic loss mechanism, it is hard to eliminate entirely. However, the device efficiency can improve if the carrier density in the quantum wells is kept as low as possible. This can be achieved by either growing thicker quantum wells or using a larger number of them and spreading the charge carriers over a larger volume. Another approach is to employ nonpolar or semipolar growth directions and eliminate or reduce the polarization fields that keep electrons and holes apart inside the quantum wells. This will enhance the overall recombination rate and reduce the steady-state carrier density in the quantum wells.

8.5 Summary

We discussed recent developments in first-principles computational methods for the study of nitride materials employed for solid-state lighting. We also presented examples that showed the wide range of applications of first-principles calculations in this field, ranging from the basic structural and electronic properties of the nitride materials to the effects of strain, defects, and nonradiative recombination on the optoelectronic device performance. First-principles methods are, therefore, a powerful explanatory and predictive computational tool that can assist and guide the experimental development of efficient solid-state optoelectronic devices and can help reduce the impact of general lighting on the world's energy resources.

Acknowledgments

We acknowledge fruitful discussions with K. Delaney, M. Scheffler, P.G. Moses, J. Speck, and C. Weisbuch. EK was supported as part of the Center for Energy Efficient Materials, an Energy Frontier Research Center funded by the U.S. DOE, BES under Award No. DE-SC0001009. PR acknowledges the support of the Deutsche Forschungsgemeinschaft, the UCSB-MPG Exchange Program and the NSF-IMI Program (Grant No. DMR04-09848). Additional support was provided by the NSF (Award No. DMR-0906805) and by the UCSB Solid State Lighting and Energy Center. Computational resources were provided by the CNSI Computing Facility under NSF Grant No. CHE-0321368, the DOE NERSC facility, and Teragrid.

References

1. Humphreys, C.J. (2008) Solid-state lighting, *MRS Bull.* **33**, 459–470.
2. Pimputkar, S., Speck, J.S., DenBaars, S.P. and Nakamura, S. (2009) Prospects for LED lighting, *Nat. Photonics* **3**, 180–182.
3. Schubert, E.F. (2006) *Light-Emitting Diodes*, Cambridge University Press, New York.
4. Nakamura, S., Senoh, M. and Mukai, T. (1993) P-GaN/n-InGaN/n-GaN double-heterostructure blue-light-emitting diodes, *Japan. J. Appl. Phys.* **32**, L8–L11.
5. (2007) *Nitride Semiconductor Devices: Principles and Simulation* (ed. J. Piprek), Wiley-VCH Verlag GmbH, Berlin.
6. Speck, J.S. and Chichibu, S.F. (2009) Nonpolar and semipolar group III materials, *MRS Bull.* **34**, 304–309.
7. Guo, W., Banerjee, A., Bhattacharya, P. and Ooi, B.S. (2011) InGaN/GaN disk-in-nanowire white light emitting diodes on (001) silicon, *Appl. Phys. Lett.* **98**, 193102.
8. Zhang, M., Bhattacharya, P. and Guo, W. (2010) InGaN/GaN self-organized quantum dot green light emitting diodes with reduced efficiency droop, *Appl. Phys. Lett.* **97**, 011103.
9. Martin, R.M. (2004) *Electronic Structure, Basic Theory and Practical Methods*, Cambridge University Press, Cambridge.
10. Ceperley, D.M. and Alder, B. (1980) Ground state of the electron gas by a stochastic method, *Phys. Rev. Lett.* **45**, 566–569.
11. Perdew, J., Burke, K. and Ernzerhof, M. (1996) Generalized gradient approximation made simple, *Phys. Rev. Lett.* **77**, 3865–3868.
12. Harl, J. and Kresse, G. (2009) Accurate bulk properties from approximate many-body techniques, *Phys. Rev. Lett.* **103**, 056401.
13. Ren, X., Tkatchenko, A., Rinke, P. and Scheffler, M. (2011) Beyond the random-phase approximation for the electron correlation energy: the importance of single excitations, *Phys. Rev. Lett.* **106**, 153003.
14. Stampfl, C. and Van de Walle, C.G. (1999) Density-functional calculations for III-V nitrides using the local-density approximation and generalized gradient approximation, *Phys. Rev. B* **59**, 5521–5535.

15. Yan, Q., Rinke, P., Winkelnkemper, M. *et al.* (2011) Band parameters and strain effects in ZnO and group-III nitrides, *Semicond. Sci. Technol.* **26**, 014037.
16. Kümmel, S. and Kronik, L. (2008) Orbital-dependent density functionals: theory and applications, *Rev. Mod. Phys.* **80**, 3–60.
17. Perdew, J. and Levy, M. (1983) Physical content of the exact Kohn-Sham orbital energies: band gaps and derivative discontinuities, *Phys. Rev. Lett.* **51**, 1884–1887.
18. Sham, L. and Schluüter, M. (1983) Density-functional theory of the energy gap, *Phys. Rev. Lett.* **51**, 1888–1891.
19. Hedin, L. and Lundqvist, S. (1970) Effects of electron-electron and electron-phonon interactions on the one-electron states of solids, *Solid State Phys.* **23**, 1–181.
20. Hybertsen, M.S. and Louie, S.G. (1986) Electron correlation in semiconductors and insulators: Band gaps and quasiparticle energies, *Phys. Rev. B* **34**, 5390–5413.
21. Aulbur, W.G., Jönsson, L. and Wilkins, J.W. (2000) Quasiparticle calculations in solids, *Solid State Phys.* **54**, 1–218.
22. Onida, G., Reining, L. and Rubio, A. (2002) Electronic excitations: density-functional versus many-body Green's-function approaches, *Rev. Mod. Phys.* **74**, 601–659.
23. Louie, S.G. (2006) Predicting materials and properties: theory of the ground and excited state, in *Conceptual Foundations of Materials: A Standard Model for Ground- and Excited-State Properties* (eds S.G. Louie and M.L. Cohen), Elsevier, Amsterdam.
24. Rinke, P., Qteish, A., Neugebauer, J. and Scheffler, M. (2008) Exciting prospects for solids: Exact-exchange based functionals meet quasiparticle energy calculations, *Phys. Status Solidi B* **245**, 929–945.
25. Rinke, P., Qteish, A., Neugebauer, J. *et al.* (2005) Combining GW calculations with exact-exchange density-functional theory: an analysis of valence-band photoemission for compound semiconductors, *New J. Phys.* **7**, 126.
26. Giantomassi, M., Stankovski, M., Shaltaf, R. *et al.* (2011) Electronic properties of interfaces and defects from many-body perturbation theory: recent developments and applications, *Phys. Status Solidi B* **248**, 275–289.
27. Bruneval, F., Vast, N. and Reining, L. (2006) Effect of self-consistency on quasiparticles in solids, *Phys. Rev. B* **74**, 045102.
28. Fuchs, F., Furthmüller, J., Bechstedt, F. *et al.* (2007) Quasiparticle band structure based on a generalized Kohn-Sham scheme, *Phys. Rev. B* **76**, 115109.
29. Jiang, H., Gomez-Abal, R., Rinke, P. and Scheffler, M. (2009) Localized and itinerant states in lanthanide oxides united by GW@LDA + U, *Phys. Rev. Lett.* **102**, 126403.
30. Bechstedt, F., Fuchs, F. and Kresse, G. (2009) Ab-initio theory of semiconductor band structures: new developments and progress, *Phys. Status Solidi B* **246**, 1877–1892.
31. Janotti, A. and Van de Walle, C.G. (2007) Absolute deformation potentials and band alignment of wurtzite ZnO, MgO, and CdO, *Phys. Rev. B* **75**, 121201.
32. Rinke, P., Winkelnkemper, M., Qteish, A. *et al.* (2008) Consistent set of band parameters for the group-III nitrides AlN, GaN, and InN, *Phys. Rev. B* **77**, 075202.
33. Städele, M., Majewski, J.A., Vogl, P. and Görling, A. (1997) Exact Kohn-Sham exchange potential in semiconductors, *Phys. Rev. Lett.* **79**, 2089.
34. Qteish, A., Al Sharif, A.I., Fuchs, M. *et al.* (2005) Role of semicore states in the electronic structure of group-III nitrides: An exact-exchange study, *Phys. Rev. B* **72**, 155317.

35. Rinke, P., Qteish, A., Winkelnkemper, M. *et al.* (2006) Band gap and band parameters of InN and GaN from quasiparticle energy calculations based on exact-exchange density-functional theory, *Appl. Phys. Lett.* **89**, 161919.
36. Qteish, A., Rinke, P., Neugebauer, J. and Scheffler, M. (2006) Exact-exchange-based quasiparticle energy calculations for the band gap, effective masses, and deformation potentials of ScN, *Phys. Rev. B* **74**, 245208.
37. Tiago, M.L., Ismail-Beigi, S. and Louie, S.G. (2004) Effect of semicore orbitals on the electronic band gaps of Si, Ge, and GaAs within the GW approximation, *Phys. Rev. B* **69**, 125212.
38. Rohlfing, M., Krüger, P. and Pollmann, J. (1995) Quasiparticle band structure of CdS, *Phys. Rev. Lett.* **75**, 3489–3492.
39. Kane, E.O. (1982) *Band Theory and Transport Properties, Handbook on Semiconductors*, vol. 1 (ed. W. Paul), North-Holland, Amsterdam.
40. Becke, A.D. (1993) A new mixing of Hartree-Fock and local density-functional theories, *J. Chem. Phys.* **98**, 1372–1377.
41. Heyd, J., Scuseria, G.E. and Ernzerhof, M. (2003) Hybrid functionals based on a screened Coulomb potential, *J. Chem. Phys.* **118**, 8207–8215.
42. Heyd, J., Scuseria, G.E. and Ernzerhof, M. (2006) Erratum: 'Hybrid functionals based on a screened Coulomb potential' [*J. Chem. Phys.* **118**, 8207 (2003)], *J. Chem. Phys.* **124**, 219906.
43. Moses, P.G. and Van de Walle, C.G. (2010) Band bowing and band alignment in InGaN alloys, *Appl. Phys. Lett.* **96**, 021908.
44. Van de Walle, C.G. and Martin, R.M. (1989) 'Absolute' deformation potentials: Formulation and ab initio calculations for semiconductors, *Phys. Rev. Lett.* **62**, 2028.
45. Romanov, A.E., Baker, T.J., Nakamura, S. and Speck, J.S. (2006) Strain-induced polarization in wurtzite III-nitride semipolar layers, *J. Appl. Phys.* **100**, 023522.
46. Shan, W., Hauenstein, R.J., Fischer, A.J. *et al.* (1996) Strain effects on excitonic transitions in GaN: deformation potentials, *Phys. Rev. B* **54**, 13460.
47. Ueda, M., Funato, M., Kojima, K. *et al.* (2008) Polarization switching phenomena in semipolar $In_xGa_{1-x}N$/GaN quantum well active layers, *Phys. Rev. B* **78**, 233303.
48. Bir, G.L. and Pikus, G.E. (1974) *Symmetry and Strain-Induced Effects in Semiconductors*, John Wiley & Sons, Inc., New York.
49. Gil, B., Moret, M., Briot, O. *et al.* (2009) InN excitonic deformation potentials determined experimentally, *J. Cryst. Growth* **311**, 2798.
50. Ishii, R., Kaneta, A., Funato, M. *et al.* (2010) All deformation potentials in GaN determined by reflectance spectroscopy under uniaxial stress: Definite breakdown of the quasicubic approximation, *Phys. Rev. B* **81**, 155202.
51. Yan, Q., Rinke, P., Scheffler, M. and Van de Walle, C.G. (2009) Strain effects in group-III nitrides: Deformation potentials for AlN, GaN, and InN, *Appl. Phys. Lett.* **95**, 121111.
52. Yan, Q., Rinke, P., Scheffler, M. and Van de Walle, C.G. (2010) Role of strain in polarization switching in semipolar InGaN/GaN quantum wells, *Appl. Phys. Lett.* **97**, 181102.
53. Van de Walle, C.G. and Neugebauer, J. (2004) First-principles calculations for defects and impurities: Applications to III-nitrides, *J. Appl. Phys.* **95**, 3851.

54. Van de Walle, C.G. and Janotti, A. (2011) Advances in electronic structure methods for defects and impurities in solids, *Phys. Status Solidi B* **248**, 19–27.
55. Marsman, M., Paier, J., Stroppa, A. and Kresse, G. (2008) Hybrid functionals applied to extended systems, *J. Phys. Condens. Matter.* **20**, 064201.
56. Oba, F., Togo, A., Tanaka, I. *et al.* (2008) Defect energetics in ZnO: a hybrid Hartree-Fock density functional study, *Phys. Rev. B* **77**, 245202.
57. Clark, S.J., Robertson, J., Lany, S. and Zunger, A. (2010) Intrinsic defects in ZnO calculated by screened exchange and hybrid density functionals, *Phys. Rev. B* **81**, 115311.
58. Lyons, J.L., Janotti, A. and Van de Walle, C.G. (2009) Why nitrogen cannot lead to *p*-type conductivity in ZnO, *Appl. Phys. Lett.* **95**, 252105.
59. Lyons, J.L., Janotti, A. and Van de Walle, C.G. (2010) Carbon impurities and the yellow luminescence in GaN, *Appl. Phys. Lett.* **97**, 152108.
60. Makov, G. and Payne, M.C. (1995) Periodic boundary conditions in ab initio calculations, *Phys. Rev. B* **51**, 4014.
61. Shim, J., Lee, E.-K., Lee, Y.J. and Nieminen, R.M. (2005) Density-functional calculations of defect formation energies using supercell methods: defects in diamond, *Phys. Rev. B* **71**, 035206.
62. Freysoldt, C., Neugebauer, J. and Van de Walle, C.G. (2009) Fully ab initio finite-size corrections for charged-defect supercell calculations, *Phys. Rev. Lett.* **102**, 016402.
63. Ogino, T. and Aoki, M. (1980) Mechanism of yellow luminescence in GaN, *Japan. J. Appl. Phys.* **19**, 2395.
64. Pankove, J.I. and Moustakas, T.D. (eds) (1998) *Gallium Nitride (GaN) I, Semiconductors and Semimetals*, vol. 50, Academic Press, San Diego, CA.
65. Pankove, J.I. and Moustakas, T.D. (eds) (1999) *Gallium Nitride (GaN) II, Semiconductors and Semimetals*, vol. 57, Academic Press, San Diego, CA.
66. Fischer, S., Wetzel, C., Haller, E.E. and Meyer, B.K. (1995) On p-type doping in GaN-acceptor binding energies, *Appl. Phys. Lett.* **67**, 1298.
67. Reshchikov, M.A. and Morkoç, H. (2005) Luminescence properties of defects in GaN, *J. Appl. Phys.* **97**, 061301.
68. Armstrong, A., Arehart, A.R., Moran, B. *et al.* (2004) Impact of carbon on trap states in n-type GaN grown by metalorganic chemical vapor deposition, *Appl. Phys. Lett.* **84**, 374.
69. Seager, C.H., Wright, A.F., Yu, J. and Götz, W. (2002) Role of carbon in GaN, *J. Appl. Phys.* **92**, 6553.
70. Neugebauer, J. and Van de Walle, C.G. (1996) Gallium vacancies and the yellow luminescence in GaN, *Appl. Phys. Lett.* **69**, 503.
71. Bogusławski, P. and Bernholc, J. (1997) Doping properties of C, Si, and Ge impurities in GaN and AlN, *Phys. Rev. B* **56**, 9496.
72. Ramos, L.E., Furthmüller, J., Leite, J.R. *et al.* (2002) Carbon-based defects in GaN: doping behaviour, *Phys. Status Solidi B* **234**, 864.
73. Lany, S. and Zunger, A. (2009) Polaronic hole localization and multiple hole binding of acceptors in oxide wide-gap semiconductors, *Phys. Rev. B* **80**, 085202.
74. Wright, A.F. (2002) Substitutional and interstitial carbon in wurtzite GaN, *J. Appl. Phys.* **92**, 2575.
75. Piprek, J. (2010) Efficiency droop in nitride-based light-emitting diodes, *Phys. Status Solidi A* **207**, 2217–2225.

76. Laubsch, A., Sabathil, M., Baur, J. *et al.* (2010) High-power and high-efficiency InGaN-based light emitters, *IEEE Trans. Electron Dev.* **57**, 79–97.
77. Kim, M.-H., Schubert, M.F., Dai, Q. *et al.* (2007) Origin of efficiency droop in GaN-based light-emitting diodes, *Appl. Phys. Lett.* **91**, 183507.
78. Monemar, B. and Sernelius, B.E. (2007) Defect related issues in the 'current roll-off' in InGaN based light emitting diodes, *Appl. Phys. Lett.* **91**, 181103.
79. Hader, J., Moloney, J.V. and Koch, S.W. (2010) Density-activated defect recombination as a possible explanation for the efficiency droop in GaN-based diodes, *Appl. Phys. Lett.* **96**, 221106.
80. Shen, Y.C., Mueller, G.O., Watanabe, S. *et al.* (2007) Auger recombination in InGaN measured by photoluminescence. *Appl. Phys. Lett.* **91**, 141101.
81. Laubsch, A., Sabathil, M., Bergbauer, W. *et al.* (2009) On the origin of IQE-'droop' in InGaN LEDs, *Phys. Status Solidi C* **6**, S913–S916.
82. Zhang, M., Bhattacharya, P., Singh, J. and Hinckley, J. (2009) Direct measurement of auger recombination in $In_{0.1}Ga_{0.9}N$/GaN quantum wells and its impact on the efficiency of $In_{0.1}Ga_{0.9}N$/GaN multiple quantum well light emitting diodes, *Appl. Phys. Lett.* **95**, 201108.
83. David, A. and Grundmann, M.J. (2010) Droop in InGaN light-emitting diodes: A differential carrier lifetime analysis, *Appl. Phys. Lett.* **96**, 103504.
84. David, A. and Gardner, N.F. (2010) Droop in III-nitrides: comparison of bulk and injection contributions, *Appl. Phys. Lett.* **97**, 193508.
85. Hader, J., Moloney, J.V., Pasenow, B. *et al.* (2008) On the importance of radiative and Auger losses in GaN-based quantum wells, *Appl. Phys. Lett.* **92**, 261103.
86. Bertazzi, F., Goano, M. and Bellotti, E. (2010) A numerical study of Auger recombination in bulk InGaN, *Appl. Phys. Lett.* **97**, 231118.
87. Cardona, M. (2005) Electron-phonon interaction in tetrahedral semiconductors, *Solid State Commun.* **133**, 3–18.
88. Ridley, B.K. (1999) *Quantum Processes in Semiconductors*, Oxford University Press, Oxford.
89. Cappellini, G., Del Sole, R., Reining, L. and Bechstedt, F. (1993) Model dielectric function for semiconductors, *Phys. Rev. B* **47**, 9892–9895.
90. Kioupakis, E., Rinke, P., Delaney, K.T. and Van de Walle, C.G. (2011) Indirect Auger recombination as a cause of efficiency droop in nitride light-emitting diodes, *Appl. Phys. Lett.* **98**, 161107.
91. Shin, D., Arróyave, R., Liu, Z.-K. and Van de Walle, A. (2006) Thermodynamic properties of binary hcp solution phases from special quasirandom structures, *Phys. Rev. B* **74**, 1–13.
92. Baroni, S., de Gironcoli, S., Dal Corso, A. and Giannozzi, P. (2001) Phonons and related crystal properties from density-functional perturbation theory, *Rev. Mod. Phys.* **73**, 515–562.

9

Toward the Nanoscale

Phuti E. Ngoepe[1], Rapela R. Maphanga[1], and Dean C. Sayle[2]
[1]Materials Modelling Centre, University of Limpopo, Sovenga, South Africa
[2]Defence College of Management and Technology, Cranfield University, Shrivenham, UK

9.1 Introduction

One of the great challenges in the twenty-first century is to respond to the needs of modern society and emerging ecological concerns, by finding new, low-cost and environmentally friendly energy conversion and storage systems. The performance of such devices depends on, and can be tuned by, the properties of the underlying materials. When the dimension of a material decreases from the bulk limit to the nanoscale, the fundamental physical, chemical and mechanical properties can be significantly modified. These modifications can be particularly profound for small clusters, quantum dots, larger clusters, nanorods, wires, belts, sheets, porous, tubes and so on and can, in part, be attributed to changes in their electronic and atomistic structure compared to their bulk counterparts. For example, surface relaxation and surfaces exposed can change as a function of curvature. In particular, the curvature increases as the size of the nanomaterial reduces promising size-tuneable properties, enabling environmentally benign solutions to energy conversion and storage compared to traditional routes such as doping. Much progress has been made in the study of such nanoarchitectures for energy applications [1,2]; however, a major challenge is that nanomaterials are not easily amenable to direct experimental investigation. Here, computational modeling can aid experiment in unraveling the structure and properties of nanomaterials.

Computational Approaches to Energy Materials, First Edition. Edited by Aron Walsh, Alexey A. Sokol and C. Richard A. Catlow.

The scope of this work is broad and could include nanoalloys [3], carbon nanotubes [4], quantum dots and so on [5], which cannot all be covered in this chapter. Accordingly, to demonstrate the important role of computational modeling in nanoscaling of energy materials, we will focus on four inorganic compounds, namely ZnO, ZnS, MnO_2 and TiO_2, with emphasis on structural generation and characterization. In particular, the initial stages of nucleation and growth—need to be better understood together with how cluster size and morphology—could be related to the energies and intensities of photo-absorption and luminescence. Lastly, we will also explore how the nanoarchitecture can impact significantly upon their electrochemical properties and hence their exploitation in storage materials.

ZnO is known to have many useful optical applications; the cluster size adds an additional degree of freedom that can be used to tune its optical and other properties. As an example, recent developments report enhanced optical absorption of photovoltaic dye solar cells [6–9]. These were achieved by coupling TiO_2 nanostructured films in the cell with ZnO nanocrystallite aggregates, yielding an impressive increase in the overall conversion efficiency. Consequently, studies have been conducted on the synthesis, characterization and surface modification of ZnO aggregates [10]. Accordingly, ZnO nanoparticle prediction is an important area of computational science.

ZnS is one of the most important materials in optoelectronic applications due to its good photoluminescent properties [11]; it is also used as a photocatalyst [12]. Unlike ZnO, it can adopt two crystal structures: sphalerite (cubic) and wurtzite (hexagonal) [13]. The relative stability of the two phases is indeed influenced by the size and morphology of the material. For example, thermodynamic analysis, which makes use of surface energy data, shows that smaller wurtzite nanoparticles are more thermodynamically stable than sphalerite-structured particles [14].

New applications, ranging from mobile units to electric vehicles, have accelerated developments of alkaline, lithium ion, lithium air batteries and supercapacitors where MnO_2 plays a significant role in electrode technology. In addition to being a low-cost, high-capacity compound, its structural variety, ranging from α- to δ-MnO_2, provides the possibility for controlling its thermal and electrochemical properties [15]. Furthermore, nanoarchitectures of MnO_2, such as nanowires, nanorods, nanobelts and nanoflowers [16, 17], have recently been synthesized, providing an opportunity for size-tuneable device performance spanning various energy applications.

TiO_2 has been studied extensively owing to its photocatalytic properties [18, 19] enjoying numerous applications [20, 21, 22] and demonstrating a wide variety of polymorphs. Currently, nanoparticulate TiO_2 is used in dye solar cells [6]. In order to attain high-rate capabilities of the TiO_2 anode in Li–ion batteries, different TiO_2 nanostructures, such as nanoparticles [23], nanotubes [24], nanowires [25] and mesoporous structures [26], are studied. As the work on nanostructured TiO_2 gains importance [27], computational studies on TiO_2 clusters could provide valuable insights into its structure and properties.

The role of computational modeling in generating, predicting and characterizing structures is well established. However, prediction of clusters and certain complex structures including nanoarchitectures such as nanoporous structures together with complex microstructures, as exhibited by γ-MnO_2, comprising polymorphic intergrowths, grain boundaries, dislocations and point defects is demanding (see [28]) and are not easily amenable to conventional computational modeling methods. Here, novel evolutionary methods, such as global minimization [29] and simulated amorphization and recrystallization [30] can play an important role in capturing such complex structures within an atomistic model.

9.2 Review of Simulation Methods

In general, four main classes of technique have been employed in the current literature on nanoscaled energy materials: atomistic (static lattice), quantum mechanical (QM), global minimization (GM) and simulated amorphization and recrystallization (A + R) methods; descriptions of these techniques will be brief as comprehensive reviews are available elsewhere [31–34].

9.2.1 Established Computational Methods

The computational methods based on interatomic potentials (IPs) minimization, Monte Carlo (MC) and molecular dynamics (MD) have all played a significant role in modeling materials; and there have been very extensive efforts over the last 30 years in developing and refining potential models particularly for oxide materials used in energy technologies. The last 15 years has seen a very rapid expansion in the use of electronic structure methods, based on both density functional theory (DFT) and Hartree–Fock (HF) techniques, with the former being particularly prominent. More detailed discussions of these methods are available in Chapter 1 and elsewhere [29, 35–38]. All of these techniques will continue to be needed in investigating the complex and varied problems posed by nanostructured energy materials.

9.2.2 Evolutionary Methods

While established computational simulation and modeling methods have provided valuable insights for materials systems with well-defined crystal structures, evolutionary techniques are necessary for prediction of structure without prior knowledge. A brief overview of two such methods, GM and simulated A + R, is given.

9.2.2.1 GM Methods

Different search algorithms can be implemented in GM methods; namely simulated annealing, Monte Carlo, basin hopping, genetic algorithm and evolutionary algorithm, and their description is provided elsewhere [29]. The multistage procedure,

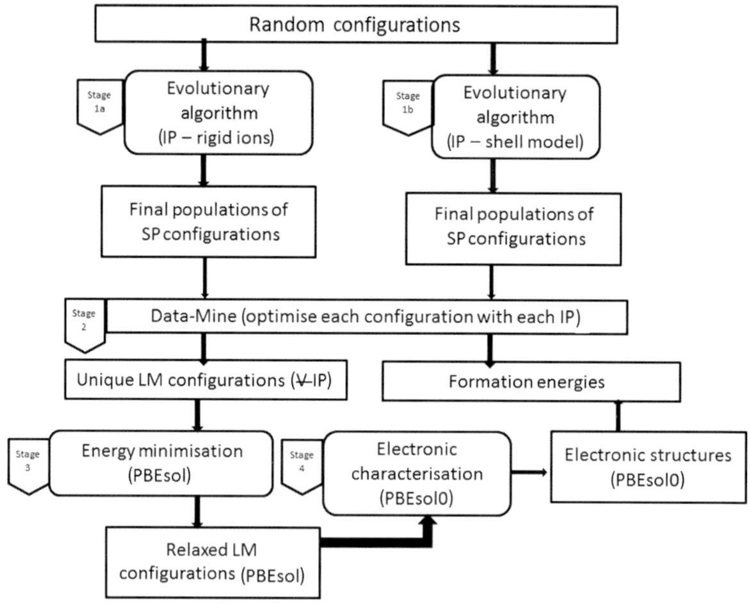

Figure 9.1 *Schematic representation of the multistage procedure used for nanostructure prediction based on a combination of interatomic potentials (IP), semi-local (PBEsol) and nonlocal (PBEsol0) density functionals. Courtesy of S. M. Woodley.*

shown in Figure 9.1, has been developed to predict low-energy configurations of nanostructures.

An evolutionary algorithm method for GM in stage 1 is employed to search for lower local minima (LM) on the energy hypersurface to predict stable and low-energy metastable atomic configurations. Different relaxed configurations, initially random stationary points on the energy hypersurface, form a population in which competition to survive and procreate is simulated. The probability of success on any current configuration is based on its energy relative to the other configurations in the population. Each LM corresponds to a configuration of atoms of the cluster, referred to as plausible structure. As a second stage, each set of plausible structures found for the IP is refined using standard local optimization techniques. In the final stage, the global minima structures and a number of metastable LM configurations, as defined by the IPs, were refined using a DFT method. Global optimization studies [29, 39–41] have predicted the lowest energy structures for cluster size of different compounds with remarkable success.

9.2.2.2 *Amorphization and Recrystallization*

Properties of a nanomaterial are governed by its hierarchical structural complexity including (polymorphic) crystal structure, morphology (surfaces exposed), microstructure (dislocations, grain boundaries and point defects) and nanoarchitecture. Accordingly, if one has to generate an atomistic model that can reliably predict properties for experiment directly then one must capture such structural complexity in a single model. Such models will enable direct comparison with experiment. For example, grain boundaries and surface relaxation in a particle, with nanometer dimensions, can result in lattice strain and facilitate change in the reactivity and ionic transport; such phenomenon is spawning the field of nanoionics. Similarly, the reactivity is influenced by the surfaces exposed: CeO_2 (100) is catalytically more reactive compared to (111). Moreover, as the size of the nano(particle) reduces, the curvature increases necessitating an increase in, for example, the concentration of steps to facilitate such curvature. Experimentally, the microstructure evolves during "crystallization" and therefore perhaps the easiest way to capture such microstructure is to simulate the processes that lead to its evolution.

Crystallization is a phase change from the liquid to solid states and involves the atoms rearranging periodically with long-range order. However, crucially during crystallization, microstructural features evolve within the crystallizing material including dislocations, grain boundaries, defects and surfaces exposed (morphology). Accordingly, if one could simulate crystallization, then potentially one might also capture important microstructural features within the model—together with their synergy of interaction—without recourse to introducing the structural features "artificially" using, for example symmetry operators that have traditionally been used to capture the microstructure. Specifically, if one were able to simulate a process, which occurs during synthesis to facilitate the complex microstructure, then arguably the microstructural features should be an accurate reflection (structurally) of the real material because the models were generated by simulating their experimental mode of manufacture.

An evolutionary simulation technique that exploits this philosophy is simulated A + R [30]. This procedure [42], attempts to avoid the inherent limitations of MD when applied to crystalline solids, which, owing to the barrier heights for ionic mobility, cannot explore configurational space for the system within the time scales accessible. The A + R is currently based on IPs, and the procedure starts by generating an amorphous configuration, which is then allowed to nucleate and recrystallize. During this process, the ions are able to move into low-energy configurations within the time scales (typically nanoseconds) accessible to the simulation. The material, as it recrystallizes, will evolve microstructural features observed experimentally, including, for example, grain boundaries, dislocations, point defects, surface structures, faceting and morphology [43]. A + R is, therefore, an appropriate simulation tool to explore systems with complex microstructures, such as MnO_2 and TiO_2, because it does not require one to propose and generate a realistic structure that includes all the microstructural features, prior to simulating with dynamical or static methods.

A general and systematic simulation strategy for creating models (atom coordinates) of nanomaterials has been developed [44]. In particular, amorphized nanoparticles are assembled into periodic arrays, using crystallographic rules, and

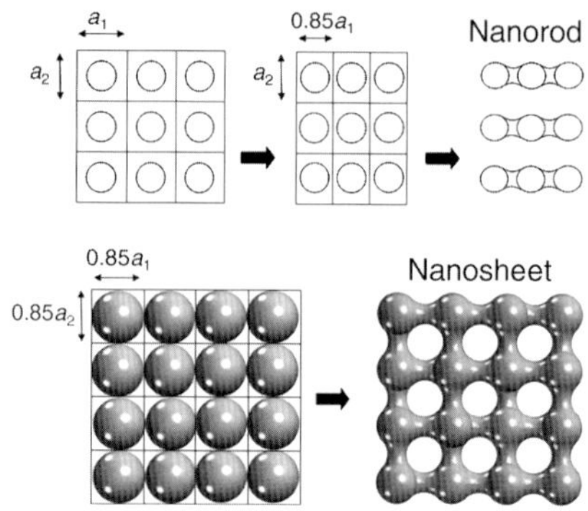

Figure 9.2 Schematic diagram illustrating the strategy used to construct extended nanostructures. A periodic array of nanoparticles is constructed in which the nanoparticles are introduced into simulation cells of size $a_1 \times a_2$ (top). The simulation cell sizes are then reduced to $0.85a_1 \times a_2$ (top middle). The close proximity between nanoparticles along a_1 causes the nanoparticles, under MD, to start to agglomerate (top right), facilitating the evolution of a one-dimensional nanorod. On the other hand, if the sizes of the simulation cells are reduced to $0.85a_1$ and $0.85a_2$ (bottom left), the nanoparticles can agglomerate in two directions facilitating the evolution of a nanosheet (bottom right). If the nanoparticles agglomerate in all three directions, a porous architecture results. Clearly, if distance between neighboring nanoparticles is high, then the nanoparticles are not able to agglomerate. Reprinted with permission from T.X.T. Sayle, R.R. Maphanga, P.E. Ngoepe and D.C. Sayle., J. American Chem. Soc., vol. 131, no. 17, pp. 6161–6173 © 2009 American Chemical Society.

facilitate their aggregation and crystallization into framework nanoarchitectures, as illustrated in Figure 9.2.

9.3 Applications

9.3.1 Nanoclusters

9.3.1.1 ZnO

The lowest energy structures for each cluster size of different 1 : 1 stoichiometry semiconductors, $(MX)_n$, have been predicted using GM [39, 40] and remarkable similarities were found. In ZnO different structural motifs can be found for the global minimum in four different size ranges. As shown in Figure 9.3, the smallest clusters are typically planar, for $n = 1$–5, whereas in the next range of cluster size, the dominant structural motif takes the form of spherical bubbles, for $n =$ 6 and greater, where all atoms are threefold coordinated. Then stable clusters are found where some atoms are fourfold coordinated, thus forming onion-like structures, before the structural motif starts to resemble cuts from bulk phases, for $n > 75$. One particularly stable cluster is found for $n = 12$ (Figure 9.3), a spheroid configuration, which is often called a sodalite cage as it is the basic tile of the sodalite structure (SOD). The relative stability of different $(MX)_n$ clusters can be

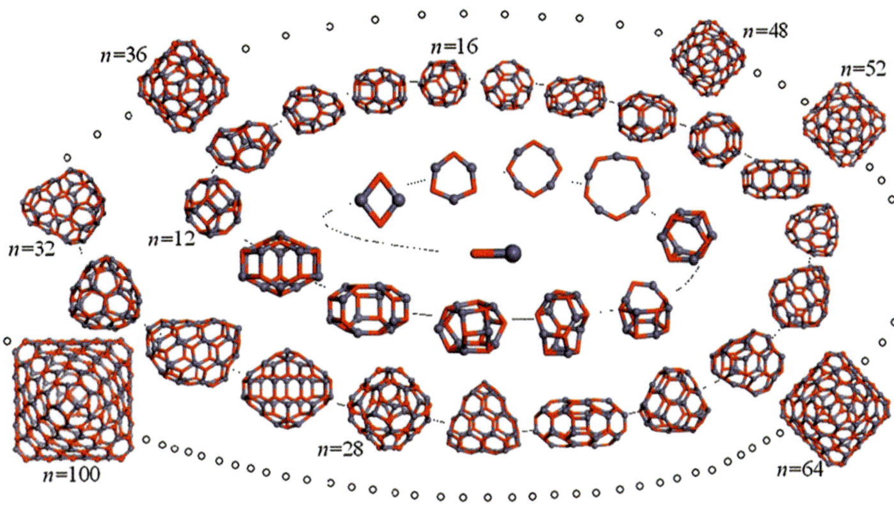

Figure 9.3 Ground-state clusters, or global minima, for $(ZnO)_n$ for n = 1–32, extended to larger sized octahedral bubbles. Grey represents zinc atoms and red oxygen atoms. Reproduced from C.R.A. Catlow, S.T. Bromley, S. Hamad, M. Mora-Fonz, A.A. Sokol and S.M. Woodley., Phys. Chem. Chem. Phys., vol. 12, pp. 786–811, 2010 with permission of the PCCP Owner Societies.

characterized by the energy of formation per formula unit with respect to one unit (for a dimer, M–N). Considering the most stable structure for each n in the lower range, an LM at $n = 12$ for ZnO, and the next LM was previously found, although much shallower, at $n = 16$ [45]. The sodalite cage at $n = 12$ is the first appearance of a spherical (also referred to as fullerene) cage with high symmetry (T_h): for ZnO, the stability of small clusters is often associated with high symmetry. On addition or removal of more units, similar sized, $n = 11$ and 13, stable spherical cages are not possible. However, larger sized stable cages that are similar to those of the SOD, that is with the T_h point symmetry group and where the number of surface hexagons is maximized, are possible [46]. The $n = 16$ cluster is also a spherical cage, but has a different point symmetry group, T_d. Similarly, there is a series of ever-larger cages that resemble the $n = 16$ cluster and have T_d, rather than T_h, point symmetry [46]. Both sets of cages can be viewed as being constructed from eight sheets, in which the ions are arranged in a honeycomb, or hexagonal, pattern (Figure 9.4). The sheets are bowed to create octahedron-based cages where, as in the sodalite cage, the strain induced by the curvature of the cages is reduced by the inclusion of just six tetragons. Finally, there is the special case of the $n =$ 24 cluster, which possesses the O point symmetry group. The energy of formation is much lower for stoichiometric cuts that are perfect cuboids. Interestingly, the sizes, n cuboid, where perfect cuboids are possible, match the sizes, n magic, which are more abundant in a plume of small particles created by laser ablation of the surface [47].

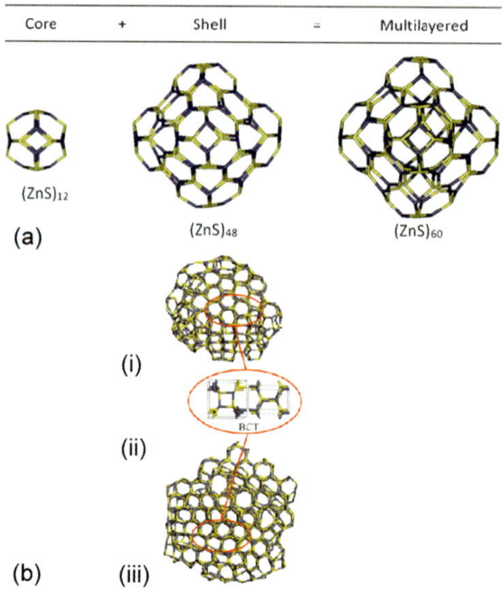

Figure 9.4 *(a) Structure of the core-shell cluster, the $(ZnS)_{60}$ double bubble composed of the two T_h GM bubbles. The complete structure is shown in the right panel, while the innermost and outermost clusters are shown in the left and central panels, respectively. (b) Structure of (i) $(ZnS)_{256}$ and (iii) $(ZnS)_{512}$ clusters obtained with simulated annealing, and the (ii) BCT unit cell that matches the structure obtained within these clusters. Yellow sticks represent sulphur and grey sticks zinc. Reproduced from C.R.A. Catlow, S.T. Bromley, S. Hamad, M. Mora-Fonz, A.A. Sokol and S.M. Woodley., Phys. Chem. Chem. Phys., vol. 12, pp. 786–811, 2010 with permission of the PCCP Owner Societies.*

9.3.1.2 ZnS

As with ZnO, there are a large number of theoretical studies suggesting that small ZnS clusters do not have bulk-like structures. Among the first DFT studies, using the B3LYP exchange–correlation functional [48], were those of Matxain *et al.* [49] for $(ZnS)_n$ clusters ($n = 1$–9). Again, the GM structures of the smaller clusters ($n = 2$–5) are planar, ring structures, whereas for $n = 6$–9 the most stable clusters are "bubbles." In a later DFT study (B3LYP/6-311 + G$*$) of a large number of $(ZnS)_n$ and $(ZnS)_n{}^+$ clusters, $n = 1$–16, Burnin *et al.* [50] have found that, for $n < 6$, neutral and positively charged clusters show planar structures, while large neutral clusters are bubbles. Hamad and Catlow [51] have carried out a theoretical study of the stabilities of $(ZnS)_n$ clusters ($n = 18$–560) using IP and DFT techniques, both of which predict similar results. The small clusters ($n = 18$–47) are bubble clusters: hollow structures in which all atoms are threefold coordinated. For $n = 50$–80, the most stable clusters are double bubbles, onion-like structures where one bubble cluster is inserted inside another one and these tend to correspond to magic numbers. Furthermore, the formation of such double bubbles, which are either the GM or have an energy similar to the GM, is not possible for all sizes because there needs to be alignment of outer and inner shells as in $n = 60$, which is composed of 12 and 48 single-shell bubbles (Figure 9.4a).

The next clusters obtained by simulated annealing $(ZnS)_{256}$ and $(ZnS)_{512}$ had a crystalline body-centered tetragonal (BCT) structure, different from the two bulk phases, wurtzite and sphalerite, of ZnS. Several large clusters generated from the three bulk structures were optimized. The results show that clusters produced from the BCT structure are generally more stable than from the others. For ZnS nanoparticles, however, the presence of four and six rings in the BCT structure (Figure 9.4b) may help to reduce the strain in the structure, since it permits a smooth transition between the crystalline BCT interior and the bubble-like outer shell. The next set of clusters in terms of stability is formed by those generated from the wurtzite structure. Their energies are very close to those of the BCT structure clusters, and it is expected that their relative energies may be reversed by solvation effects. But despite the great computational evidence supporting the possibility of ZnS nanoparticles adopting the BCT structure, no experimental confirmation has yet been found.

A complete systematic search for larger clusters has not yet been reported. Different structural motifs have been proposed for the GM structure. Contrary to the trends described above, in a recent tight-binding DFT study [52] for $n = 10, 16, 37, 57$ and 68, Pal *et al.* have suggested that the most stable structures are ring- or band-like configurations, with large radii, and these tend to be quantum dots [52]. However, using a simulated annealing approach for clusters of size $n = 50, 60, 70$ and 80, Spano *et al.* [53] and Hamad *et al.* [54] predicted double-shell bubbles, that is, double bubbles, to be the GM.

There is also a growing interest in studying other ZnS structures, both experimentally [55–60] and using tight-binding density functional theory (TB-DFT) and DFT approaches [61–64], such as nanotubes or nanowires, which could show interesting optical properties for various applications. IP-based [65] and DFT the PBE functional [62, 64] calculations were also used to study ZnS nanotubes and nanowires. The simulations suggest that both nanotubes and nanowires have stabilities close to those of double bubbles, and therefore the type of structure formed will be largely determined by kinetic factors. Finally, nanoporous ZnS derived by the A + R method [66] has also shown the presence of four-membered rings observed from the GM method (discussed in Section 9.3.2.5).

9.3.1.3 MnO_2

The lowest energy LM configurations for $(MnO_2)_n$, clusters ($n = 1$–4), are presented in Figure 9.5 together with the order of stability as measured by IP and DFT (the PBEsol0 functional) (R.R. Maphanga, C.R.A. Catlow, P.E. Ngoepe, and S.M. Woodley, unpublished results). The analysis of MnO_2 clusters obtained by the IP method reveals that they are formed by three basic structural motifs: (i) a planar or quasi-planar MnO_2 rhombus, (ii) a planar or quasi-planar MnO_3 unit with a threefold-coordinated Mn atom and (iii) a distorted MnO_4 tetrahedron with a tetra-coordinated Mn atom. A linear (planar) structure predicted for the $n =$ 1–2 clusters using IP methods bends into a boomerang ($n1_a$) during relaxation when DFT approach is employed. The IP results are in agreement with previous

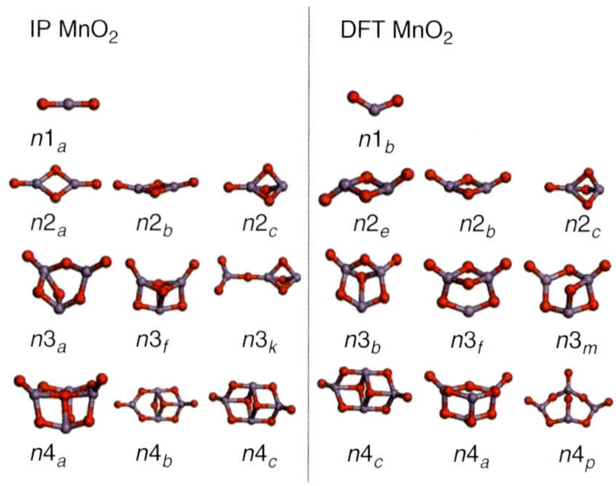

Figure 9.5 *Global minima for $(MnO_2)_n$, $n = 1$–4, from IP and DFT calculations. The grey represents manganese atoms and red oxygen atoms.*

IP TiO_2 studies [67] and DFT-relaxed MgF_2 clusters [68]. The smallest global minima configurations, with DFT, have a linear arrangement for germania, silica and stania, while titania, hafnia and zirconia have a boomerang arrangement. For $n = 2$, $n2_a$ is found to be the global minimum for the IP method. When the DFT method is employed to relax plausible structures, $n2_b$ configuration is obtained as the global minima for MnO_2 clusters. Both systems predict a configuration with two-terminal Mn–O bond as the global minima in agreement with previous studies on TiO_2 DFT nanocluster calculations [67]. The models predict the bond distance of the two Mn–O_1 pairs to be smaller than that of four Mn–O_2 pairs. On comparing the IP configurations for $(MnO_2)_2$ nanoclusters with those obtained using DFT, the atomic arrangement is similar with DFT configurations bending to form a quasi-like isomer for the first two LM, that is, $n2_e$ and $n2_b$, for the $n =$ 3 cluster. The lowest energy configuration, $n3_b$ for $(MnO_2)_3$ is composed of two terminal oxygen atoms and one threefold coordinated oxygen atom in agreement with previous DFT calculations. The DFT predicts that the GM configuration is composed of three tetragonal sides of a cuboid with two singly coordinated oxygen atoms. The quasi-linear arrangement was suggested to be the global minimum in the previous studies on TiO_2 isostructure. Small $(MnO_2)_n$ $n = 1$–4 clusters are isostructural to those predicted for titania, silica and hafnia and characterized by Mn ions concentrating in the core region of the cluster decorated by dangling, or singly coordinated, oxygen ions at the periphery. No GM studies have been carried out for $(MnO_2)n$ clusters for $n > 4$.

Simulation techniques have also been used to generate different defect models of electrolytic MnO_2 and relaxed using IP minimization [69]. The models that reproduced experimental X-ray diffraction (XRD) patterns were identified. Maphanga *et al.* [70] employed IP methods to predict the surface structures, stabilities and morphologies of pyrolusite and ramsdellite polymorphs of MnO_2. Recently,

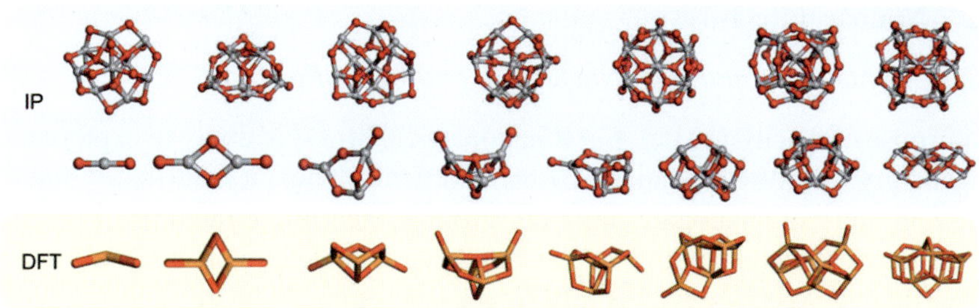

Figure 9.6 IP and DFT models of the lowest energy clusters found for TiO_2. For IP red indicates oxygen, grey titanium, and for DFT red indicates oxygen and orange titanium. Reproduced from C.R.A. Catlow, S.T. Bromley, S. Hamad, M. Mora-Fonz, A.A. Sokol and S.M. Woodley., Phys. Chem. Chem. Phys., vol. 12, pp. 786–811, 2010 with permission of the PCCP Owner Societies.

simulated A + R techniques were used to produce defect-rich microstructures of nano [71], bulk [28] and mesoporous nanoarchitectures of MnO_2 [16], whose structural properties compare reasonably with experimental results. MnO_2 seeds observed during nucleation and crystallization are discussed in Section 9.3.2.1. Owing to the complexity of the MnO_2, experimental investigations on the nucleation and small clusters have not been reported.

9.3.1.4 TiO_2

In an initial study, Hamad *et al.* [67] applied several GM techniques to generate plausible structures of $(TiO_2)_n$ clusters (n = 1–15). In particular, a combination of Monte Carlo basin hopping and simulated annealing MD, as well as different evolutionary algorithms, has been employed to search the energy landscapes defined using IPs. The plausible, or low energy, structures were subsequently refined using DFT (shown in Figure 9.6). The small clusters are isostructural to those predicted for silica [29] and MnO_2 in Section 9.3.1.3. They are characterized by Ti ions concentrating in the core region of the cluster decorated by dangling, or singly coordinated, oxygen ions at the periphery. However, when the energy of formation is calculated using a DFT approach, after relaxation, the dangling oxygen atoms for titania do not remain in the plane containing the nearest Ti_2O_2 tetragon. For the larger clusters, the structures are compact, with a central octahedral and a surrounding layer of fourfold- and fivefold-coordinated Ti atoms, although there seems to be some energy penalty for particles containing the fivefold-coordinated metal atoms with a square base pyramid structure and dangling oxygen atoms. Here, a conventional double-bond notation is sometimes used between the dangling oxygen and the titanium, but this species is more likely to be a charge-transfer complex—similar to that revealed recently for the silanone species in silica clusters [72]. Accurate predictions of the spectroscopic properties with respect to the cluster size have been made. The TiO_2 nanostructures produced by the A + R method are discussed in Section 9.3.2.4.

9.3.2 Nanoarchitectures

9.3.2.1 MnO_2 Nanoparticle (Nucleation and Crystallization)

The purpose of discussing the A + R nanoparticle MnO_2 is firstly to explore nucleation and crystal growth of an MnO_2 nanoparticle. Hence, it is surmised that other TiO_2, ZnS and ZnO nanoparticles would follow the same growth process, though variations could occur owing to the number of nucleating seeds. Furthermore, nanoparticulate MnO_2 serves as a basis for generation of other nanoarchitectures of MnO_2, and the same approach will be extended to TiO_2, ZnO and ZnS, as discussed in Section 9.2.2.2. Lastly, the A + R enables microstructural features to be introduced into the structural model, which governs a wide range of physical and chemical properties.

GM methods in Section 9.3.1.3 have shown minima configurations of MnO_2 for $n = 1$–4, and it may be suggested that nucleation in this compound could have its origins from similar clusters.

The production of nanoparticulate MnO_2 has been studied by Sayle *et al.* [71]. Figure 9.7 depicts variation of the configuration energy with time, for various nanostructures of MnO_2, and initially attention is focused on the red nanoparticle curve. The initial plateau, in the first 1000 ps, is related to the nucleation process. A corresponding snapshot of a movie at 500 ps (Figure 9.8a), derived from the trajectories of nanoparticle MnO_2, depicts an amorphous structure. A closer inspection reflects evolution of a small crystalline region with the pyrolusite structure at 750 ps (isostructural to rutile TiO_2 (Figure 9.8a). The structure of this seed was examined in closer detail to determine the smallest size that was crystalline. Prior to 750 ps, no crystalline "seed" could be identified. One might surmise that "crystalline" clusters did spontaneously evolve but were subsequently reamorphized and consumed back into the amorphous sea of ions. However, once a crystalline seed reaches a critical size, it nucleates the crystallization of the remaining (amorphous) nanoparticle.

The reduction in the configuration energy marked beyond 1000 ps (Figure 9.7) corresponds to crystallization of the nanoparticle, and crystalline patterns emerge in the related snapshot (Figure 9.8b) at 1800 ps. It is deduced that the nucleation seed initiates crystallization of the surrounding region, which is much larger (at 1750 ps), and octahedral patterns conform to the pyrolusite structure and grain boundaries and initial stages of microtwinning are also visible. A near constant plateau (Figure 9.7) is observed beyond 2 ns and fully crystallized pattern is noted in Figure 9.8c. The final, 0 K structure reveals a crystalline nanoparticle that predominantly accommodates domains that adopt the pyrolusite structure. A slice of the nanostructure was cut, and it is evident that the pyrolusite structure has evolved and comprises a wealth of twin boundaries rather than existing as a single crystal. Twin boundary regions are evidenced by observing where the parallel lines of Mn ions change direction.

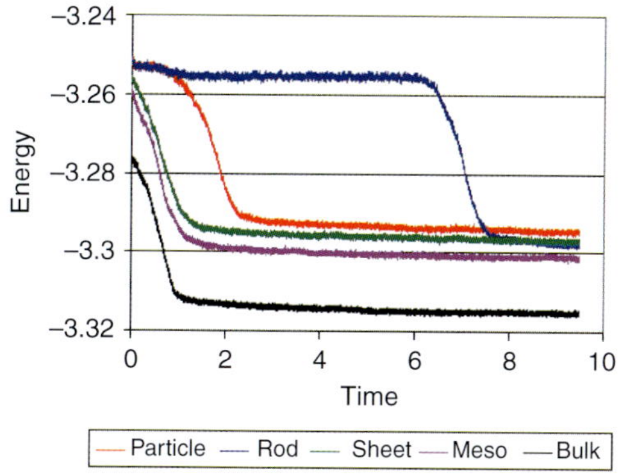

Structural notation	Nanosphere	Nanorod	Nanosheet	Nanoporous	Bulk
Box size (nm)	10.0 ×10.0 ×10.0	10.0 ×8.5 ×10.0	8.5 ×8.5 ×8.5	7.4 ×7.4 ×7.4	6.3 ×6.3 ×6.3
Dimension	0D 	1D 	2D 	3D 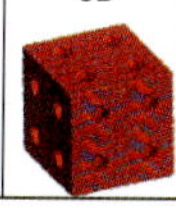	Bulk 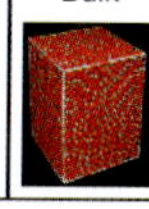

Figure 9.7 *Configuration energy (10^5 eV) calculated as a function of time (ns) for each of the nanostructures, that is, nanoparticle, nanorod, nanosheet, nanoporous/mesoporous and bulk. The abrupt change in energy reflects a phase change and is associated with the amorphous–crystalline transition; the difference in energy between the starting (amorphous) configuration and the final (crystalline state) loosely reflects the heat of crystallization. The box sizes and dimensions for each nanoarchitecture are shown below. Reprinted with permission from T.X.T. Sayle, R.R. Maphanga, P.E. Ngoepe and D.C. Sayle., J. American Chem. Soc., vol. 131, no. 17, pp. 6161–6173 © 2009 American Chemical Society.*

The whole crystal growing process is further illustrated in Figure 9.9a, which shows the radial distribution functions (RDFs) of the MnO_2 nanoparticle. The curve corresponding to 500 ps depicts the nucleation process, and the nanoparticle is predominantly amorphous. At 1500 ps, the structure begins to emerge as the crystallization occurs. After annealing the crystallized structure, at final 0 K, the RDF curve has well-defined sharp peaks.

Extended X-ray absorption fine structure (EXAFS) data on nanoparticles of MnO_2, about 10 nm in size (the model of MnO_2 nanoparticle is 8 nm in diameter), revealed nearest neighbor Mn–O and Mn–Mn interatomic distances of 1.91 and 2.90 Å, respectively [73], which are very close to calculated values of 1.88 and 2.88 Å, respectively (Figure 9.9b). The experimental studies therefore help validate the simulated nanoparticle models. Figure 9.7 also shows how other nanoarchitectures nucleate and crystallize.

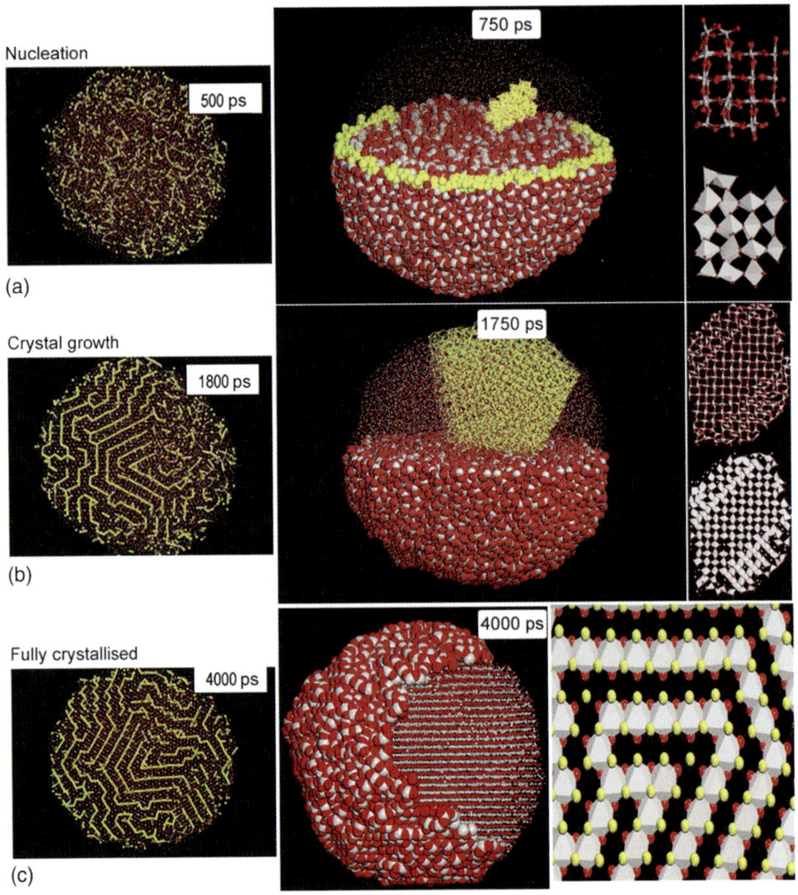

Figure 9.8 *An illustration of nanoparticle MnO_2 nucleation and crystal growth. (a) The snapshot of the atom positions, at 500 ps, followed by location of the MnO_2 nucleating "seed" (yellow) within the amorphous surroundings at 750 ps, and a ball-and-stick representation of the ion positions, to the right, showing pyrolusite structure. (b) The snapshot of the crystal growth for nano-MnO_2 at 1800 ps, and the crystalline region at 1750 ps; the corresponding atom representations are shown on the right, reflecting initial stages of microtwinning. (c) The final, 0 K, structure of the nanoparticle depicting the snapshot with full crystalline patterns at 4000 ps followed by a sphere model representation, and a related segment of a cut slice, with polyhedral rendering of the Mn ions with heavily twinned structure (manganese, grey; oxygen, red). Reprinted with permission from T.X.T. Sayle, R.R. Maphanga, P.E. Ngoepe and D.C. Sayle., J. American Chem. Soc., vol. 131, no. 17, pp. 6161–6173 © 2009 American Chemical Society.*

Lastly, as mentioned earlier, the nanoparticle plays a significant role in A + R methods since it is the basis for the production of other nanoarchitectures such as nanoporous rods and bulk, where simulations commence with amorphized nanoparticles, shaped into appropriate forms and crystallized, as discussed in Section 9.2.2.2. The generation and characterization of the bulk and nanoporous

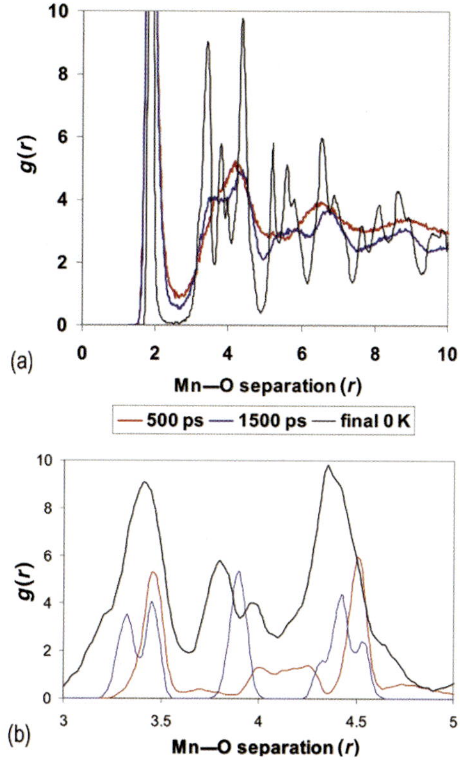

Figure 9.9 *(a) Mn–O RDF (separation in Å) of the MnO_2 nanoparticle calculated after 500 ps (amorphous structure), 1500 ps (partially recrystallized) and (fully crystallized) at the end of the simulation. (b) Mn–O RDF calculated at the end of the simulation (black trace) compared with the parent pyrolusite (blue trace) and ramsdellite (red) materials; the RDFs for the parent materials are not from crystallographic data taken from experiment—rather they were calculated based upon the (potential) models used to describe the pyrolusite and ramsdellite polymorphs. Note the Mn–O separation for (b) runs from 3 to 5 Å rather than 0–10 Å for (a). Reprinted with permission from T.X.T. Sayle, C.R.A. Catlow, R.R. Maphanga, P.E. Ngoepe and D.C. Sayle., J. Am. Chem. Soc., vol. 127, pp. 12828–12837 © 2005 American Chemical Society.*

structures will be explored together with their lithiation for Li–ion and Li–air battery applications.

9.3.2.2 MnO_2 Bulk

The bulk γ-MnO_2, known to accommodate a complex microstructure, including intergrowths of ramsdellite and pyrolusite, will now be discussed [28]. The corresponding nucleation and crystallization of the bulk structure is illustrated by the black line in Figure 9.7 of the configuration energy against time. The A + R generated simulation cell has dimensions 6.3 × 6.3 × 6.3 nm. The validation of the bulk MnO_2 is achieved by comparing XRD of one of the simulated models with the experimental results [69] (Figure 9.10). XRDs associated with the model are in

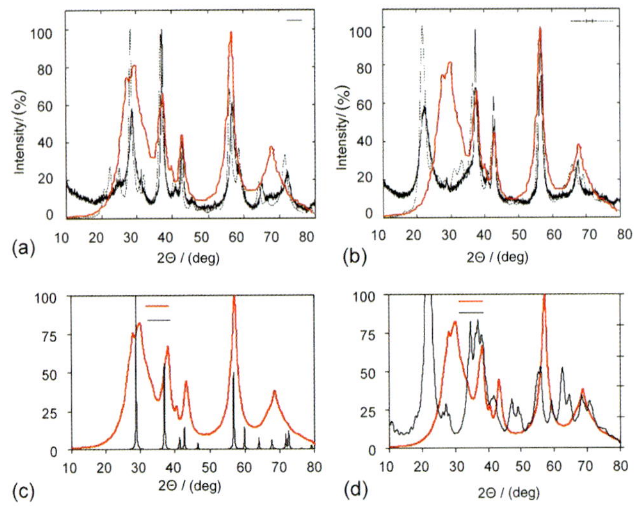

Figure 9.10 *Comparison between calculated and experimental MnO_2 XRD patterns for the model MnO_2 system: (a) calculated (red) compared with experimental XRD of CMD (bold black trace) and the theoretical model [11] (black dotted trace), (b) calculated and experimental XRD of EMD (bold black trace) and the theoretical model [11] (black dotted trace), (c) calculated (red) compared with the perfect ramsdellite parent (black) and (d) calculated compared with the perfect pyrolusite parent. The y-coordinate is intensity and abscissa, 2θ (degrees). Reprinted with permission from J.R. Hill, C.M. Freeman and M.H. Rossouw., J. Solid State Chem., vol. 177 , pp. 165–175 © 2004 Elsevier.*

better agreement with the γ-MnO_2, that is, chemical MnO_2 (CMD) and electrolytic MnO_2 (EMD), and very broad compared with the parent pyrolusite and ramsdellite structures, which confirm that the simulated structure accommodates a wealth of defects noted in a typical experimental microstructure. Furthermore, there is a better alignment of the model with the pyrolusite than the ramsdellite structure. The RDFs of the models correlate better with the pyrolusite structure than ramsdellite, and their energies lie between those of the two polymorphs [28]. Indeed, the simulated microstructures of the models predominantly exhibit the pyrolusite structure, as in Figure 9.11a. In addition, any domains conforming to the ramsdellite structure were scarce although a few superficial intergrowths of ramsdellite and pyrolusite, such as in Figure 9.11b, were observed. Inspection of the structures also revealed pyrolusite twin boundaries, heavily (fully) twinned regions, stacking faults and Mn^{4+} vacancies. It is noted that a fully twinned pyrolusite structure would be isostructural with α-PbO_2, and although this disordered structure has not yet been observed in MnO_2, it has been seen in a high-pressure polymorph of TiO_2. DFT method calculations [74] have shown that protons (Reutschi), which compensate for Mn^{4+} vacancies in MnO_2, tend to stabilize ramsdellite over pyrolusite, and their absence in the A + R simulated model could account for the prevalence of the latter structure.

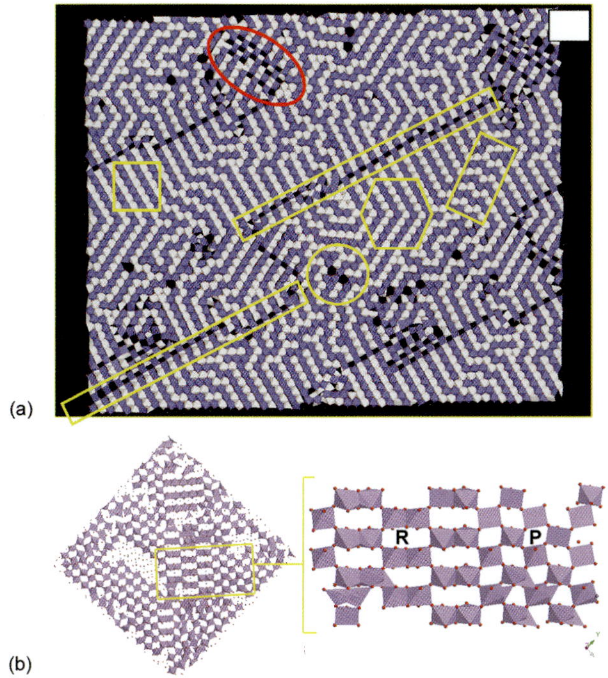

Figure 9.11 *(a) The microstructure is shown from a slice, cut perpendicular to [001], through the* MnO_2 *showing two (blue, upper; white, lower) planes of the Mn sublattice for a model system. The figure is presented using polyhedral rendering to illustrate, more clearly, the structure together with the wealth of twin boundaries. The figure is enlarged to help distinguish microstructural features including square, pyrolusite; hexagon, pyrolusite single twin; rectangle, pyrolusite heavily twinned; long rectangle, stacking fault; circle, two* Mn^{4+} *vacancies; red oval, this region appears to be isostructural with brookite* TiO_2. *(b) A slice of* MnO_2 *and its enlarged version show a shallow ramsdellite and pyrolusite intergrowth. Reprinted with permission from T.X.T. Sayle, C.R.A. Catlow, R.R. Maphanga, P.E. Ngoepe and D.C. Sayle., J. Crystal Growth, vol. 294, pp. 118–129 © 2006 Elsevier.*

The charging and discharging of a 3 V Li–ion battery with a bulk MnO_2 cathode is related to its lithiation and delithiation; hence, associated microstructural changes could provide valuable insights into degradation processes. Lithiation of the bulk MnO_2 has also been studied experimentally [75] and the simulated A + R method has been used to generate microstructural models of Li_xMnO_2 (with $x = 0.06$–0.33) [76]. Figure 9.12a gives a graphical representation of the final crystallized configuration for $Li_{0.105}MnO_2$ (880 Li ions) at 0 K, and the crystallized system is viewed from different orientations and representations in Figure 9.12b. As shown in Figure 9.12c, the A + R structural models and the simulated XRD they give rise to are in accord with literature data [77], which helps validate the models. The simulated results support the possibility of intrinsic defects facilitating the stabilization of ramsdellite domains within the system in accord with proposals

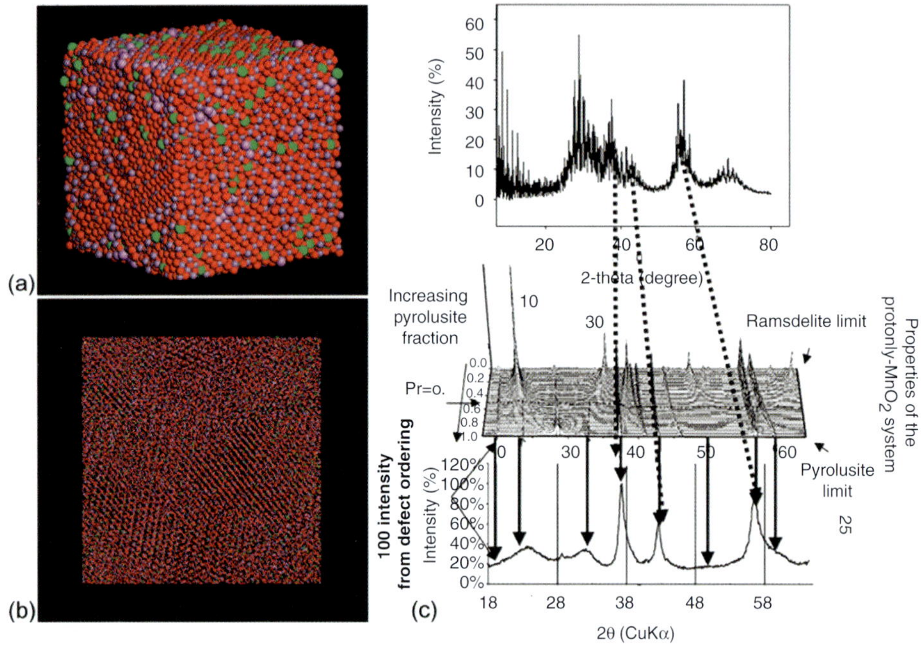

__Figure 9.12__ (a) The final configuration for $Li_{0.106}MnO_2$ (520 Li ions) Mn^{4+} is colored purple (small), Mn^{3+} is purple (big), oxygen is red and lithium green. (b) Configuration in (a) viewed from a different orientation and showing crystallinity. (c) Comparison between calculated and literature XRD patterns. The upper part depicts the A + R results and the lower part is taken from Whittingham et al. [77] obtained with simulated DIFFAX program. Dashed arrows show common peaks between the simulation and literature XRD patterns. Reproduced from R.R. Maphanga, D.C. Sayle, T.X.T. Sayle and P.E. Ngoepe., Phys. Chem. Chem. Phys., vol. 13, pp. 1307–1313, 2011 with permission of the PCCP Owner Societies.

based upon experiments. However, the simulated XRD patterns accommodate broad peaks as compared to experimental findings, which were attributed to the very high levels of microtwinning in the models.

Microstructural detail of a slice cut through the system perpendicular to [001] for $Li_{0.06}MnO_2$ is shown in Figure 9.13. Polyhedron representation is used to illustrate the structure and the wealth twin boundaries more clearly. The figure reveals microstructural features observed within the system. They include pyrolusite (square), pyrolusite single twin (triangle), pyrolusite heavily twinned (green rectangle), stacking fault, (thin rectangle), double chain formation (thick rectangle), grain boundary (yellow oval), defect cluster (green oval) and Mn^{4+} vacancies (circle). The red polyhedrons are the Mn^{3+} ions and the yellow are the lithium ions. The system tends to show a complex defect cluster—the green oval where the lithium ions are located. The cluster is comprised of several Mn^{4+} and oxygen vacancies.

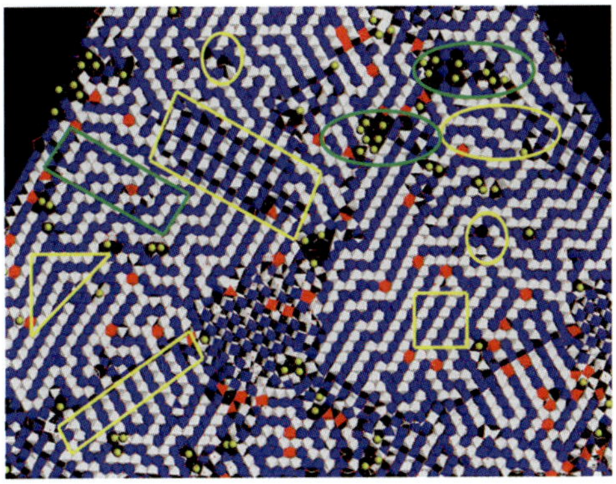

Figure 9.13 *Graphical representation of the final configuration for $Li_{0.06}MnO_2$ (520 Li ions) at 0 K. The blue and white octahedrons represent the upper and lower layers of Mn^{4+} ions. The red octahedrons represent the Mn^{3+} ions. The Li^+ ions are yellow. Reproduced from R.R. Maphanga, D.C. Sayle, T.X.T. Sayle and P.E. Ngoepe., Phys. Chem. Chem. Phys., vol. 13, pp. 1307–1313, 2011 with permission of the PCCP Owner Societies.*

9.3.2.3 MnO_2 Nanoporous

Nanoporous structures play an important role in facilitating high charge/discharge rates and for sustaining long life of Li–ion batteries. However, it is one of the most complex nanoarchitectures to model; and it has been simulated for MnO_2 [16]. The nanoporous structure of MnO_2 was also generated from amorphous nanoparticle, resulting in a 3D simulation cell box of dimensions 7.4 × 7.4 × 7.4 nm as shown in Figure 9.7. As with the bulk MnO_2, the validation of the simulated nanoporous structure was deduced from the simulated XRD [16], which is in full accord with experiment [78].

Further validation is evidenced in Figure 9.14, where the 0 K microstructure of the nanoporous MnO_2 model is shown; a slice cut through the nanoporous MnO_2 oriented with [111] normal to the plane of the page reveals a hexagonal array of channels. A perspective view, Figure 9.14b, more clearly reveals the nanoporous architecture of the model, which comprises cavities interconnected by channels. The MnO_2 model comprises crystalline walls (Figures 9.14c and d). Comparison between high-resolution transmission electron microscope micrographs (Figures 9.14e and f) and the model structure reveals that the model captures the basic architectural features of nanoporous MnO_2.

In its capacity as a component in a rechargeable battery, Li ions must be easily inserted, stored and deinserted into MnO_2, which acts as a host lattice. Such capability can be provided by model nanoporous MnO_2, consisting of internal pores and connecting channels and tunnels. To understand how the porous MnO_2 can facilitate Li cycling and maintain structural integrity, (internal) surfaces of the material are explored. To this end the MnO_2 model has been drawn at various length scales in Figure 9.15. In Figure 9.15a, an extended region of the nanoporous architecture is shown, and a small segment of this is enlarged in Figure 9.15b to

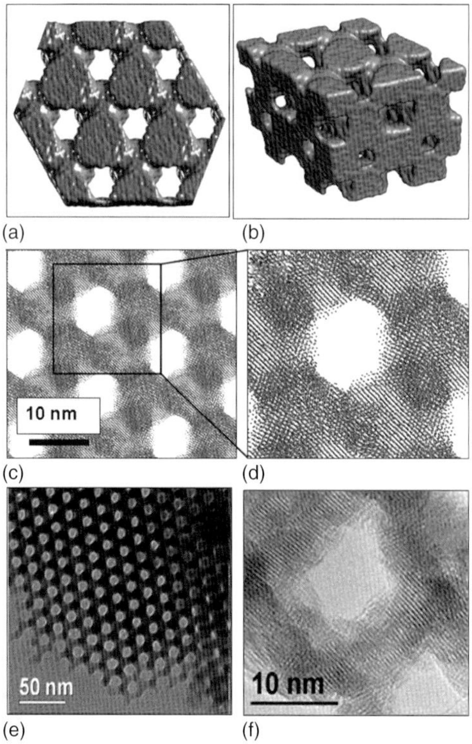

Figure 9.14 *Final low-temperature structures compared with experiment: (a) a surface-rendered model of a thin slice viewed along [111]; (b) perspective view of a surface-rendered model of the nanostructure; (c) sphere model representation of the atom positions revealing the hexagonal array of channels and the atomic planes of the MnO_2; (d) enlarged view of (c); (e) TEM; (f) high-resolution TEM of nanoporous MnO_2. Reprinted with permission from F. Jiao and P.G. Bruce, Advanced Materials, vol. 19, pp. 657–660 © 2007 Wiley-VCH Verlag GmbH & Co. KgaA. Reproduced with permission.*

reveal the repeat unit. If one then magnifies a region of the internal surface (Figure 9.15c), one can see the (1×1) tunnels that the Li must insert into; a schematic of the tunnel is shown in Figure 9.15d. Figure 9.16a shows a segment close to a channel, and considerable twinning is observed in the structural model, which enables Li to insert anywhere on the internal surface. Furthermore, Figure 9.16b shows that Li moves spatially in all three directions. Two Li ions (schematic), colored yellow, indicate a path that the Li might take once inside the MnO_2. Its change of direction at the microtwin boundary from x to y is noticed, and further change occurs at another twin boundary that is not in the plane of the page, y to z (as indicated in the magnified portion). Accordingly, by inspection of the structure, we can observe that the Li ion is free to insert at many positions on the internal surface and can move in all three spatial directions once inside the nanoporous material. This will facilitate an even expansion of the framework upon insertion/deinsertion and thus enable the material to expand elastically and protect against plastic deformation upon cycling. However, an even expansion necessitates that the porous material be periodic. For

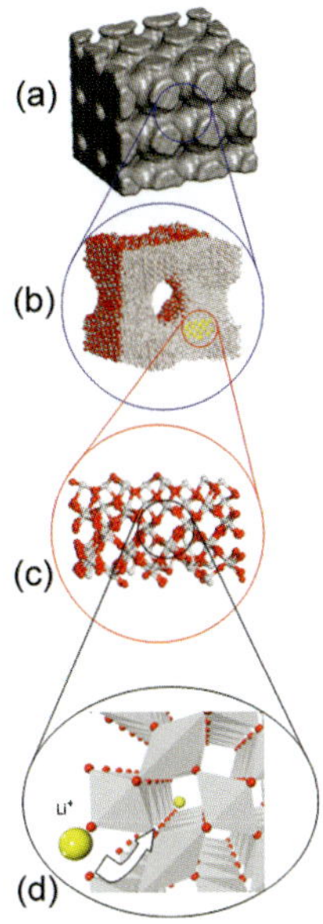

Figure 9.15 Hierarchical model structure of nanoporous MnO_2: (a) a surface-rendered model showing the porous structure; (b) a unit cell with the sphere model and polyhedral rendering of the atom positions; (c) atomistic structure of a small segment in (b); (d) schematic representation of the Mn polyhedra revealing the 1×1 tunnels along which (c direction) the Li ions move. Reprinted with permission from T.X.T. Sayle, R.R. Maphanga, P.E. Ngoepe and D.C. Sayle., J. American Chem. Soc., vol. 131, no. 17, pp. 6161–6173 © 2009 American Chemical Society.

example, if one region were larger than its neighbor, then it could accommodate a higher concentration of Li and thus expand more. A discontinuity in expansion would impose higher stress at a particular point and, if sufficiently high, would lead to plastic deformation. This argument is supported by experimental data from Lai and coworkers who found that the stress–strain curve for porous copper with a periodic (P) minimal surface structure revealed a higher elastic modulus and yield stress compared with its nonperiodic porous counterpart [80].

We now consider the simulation of lithium insertion in a nanoporous structure of MnO_2 [81]. The final low-temperature structure for the $Li_{0.16}MnO_2$ model, compared with experiment, is shown in Figure 9.17 and reveals Li ions located within the framework walls of the β-MnO_2 and also decorating its internal surfaces (Figure 9.17a). A symmetric distribution of the channels together with its magnified version is shown in Figures 9.17b and c respectively and a good correspondence with experiments is noted (Figures 9.17d and e). Simulations have further revealed, by applying hydrostatic and uniaxial strains, that for MnO_2 to be a viable host within rechargeable battery systems, the material must be symmetrically porous to enable it to expand linearly with Li intercalation [81]. Nonsymmetrically porous MnO_2 will suffer plastic deformation. An enlargement of a particular region of the β-MnO_2 surface (Figure 9.17f) reveals 1×1 tunnels, which terminate at the internal surfaces

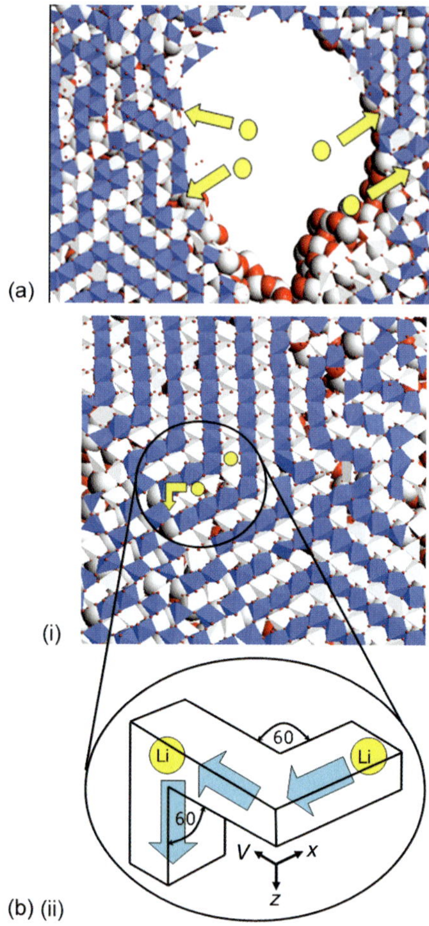

Figure 9.16 *A slice cut through the nanoporous MnO_2: (a) A channel and tunnels, revealing the microtwinning; depicting more clearly the positions (yellow arrows) that the Li ions can insert into the surface of the MnO_2 host via 1×1 tunnels. Mn planes above the page are colored blue, and polyhedra below the page are white. Oxygen is red, manganese is white and lithium is yellow. Li is shown as schematic. (b) An atomistic model and schematic illustrating the 3D mobility of Li in nanoporous MnO_2; Li atoms are schematic. (i) The atomistic model showing the 1×1 tunnels and (ii) schematic showing how the Li can move in all three spacial directions in the porous MnO_2. Reprinted with permission from T.X.T. Sayle, R.R. Maphanga, P.E. Ngoepe and D.C. Sayle., J. American Chem. Soc., vol. 131, no. 17, pp. 6161–6173 © 2009 American Chemical Society.*

and remain "open" enabling Li (charge carriers) to insert into the host; Figure 9.17g reveals several Li species that reside within the 1×1 tunnels of the host. A slice cut through the β-MnO_2 (Figure 9.18a) reveals complex (predominantly twin) grain boundary structures with curved, rather than planar, interfacial planes. The individual grains exhibit low-angle rotations with respect to one another, which further obscures identification of the boundaries. Such misorientation is attributed

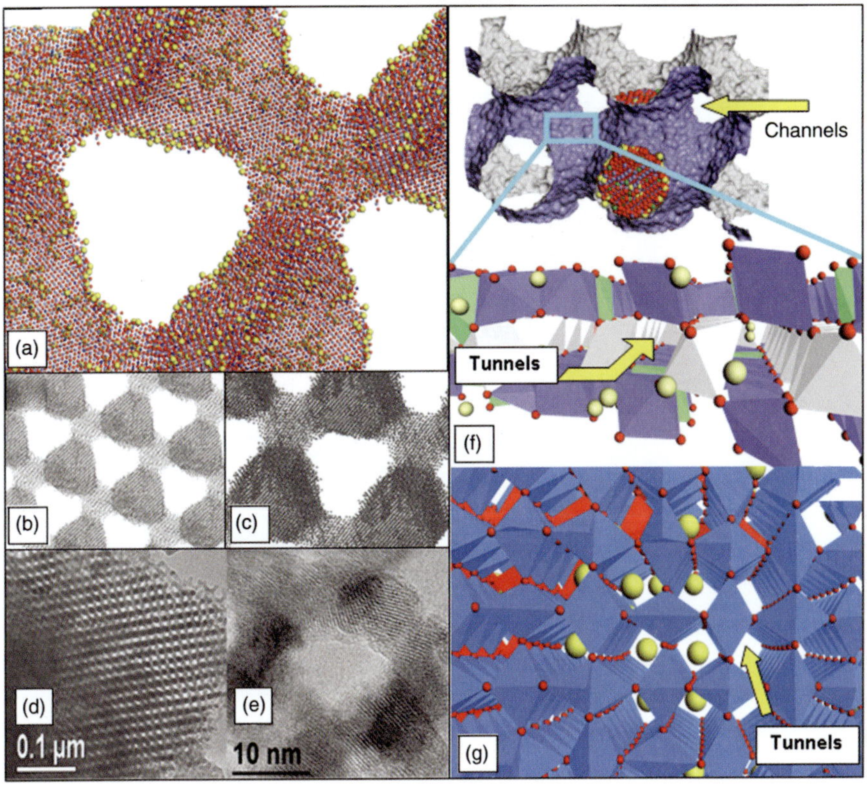

Figure 9.17 Atomistic models for nanoporous $Li_{0.16}MnO_2$: (a) slice cut through the atomistic model revealing Li ions, which locate within the host β-MnO_2 and decorate the (internal) surfaces of the β-MnO_2; (b) the atomistic model revealing the symmetrically porous architecture; (c) enlarged region of (b); (d and e) HRTEM of $Li_{0.92}MnO_2$ (to compare with atomistic models). Reprinted with permission from F. Jiao and P.G. Bruce, Advanced Materials, vol. 19, pp. 657–660 © 2007 Wiley-VCH Verlag GmbH & Co. KgaA. Reproduced with permission; (f) the surface-rendered model of the porous architecture revealing 1×1 tunnels exposed and open at the surface, which enable Li^+ ions to insert into the mesoporous β-MnO_2 framework; (g) perspective view looking along the 1×1 tunnels, some of which are occupied by Li^+, MnO_6 are shown as polyhedra. The images are not schematic; rather they represent atom positions corresponding to the atomistic model. Li is colored yellow; oxygen is red and manganese, blue. Reprinted with permission from T.X.T. Sayle, P.E. Ngoepe and D.C. Sayle, "Simulating mechanical deformation in nanomaterials with application for energy storage in nanoporous architectures," ACS Nano, vol. 3, pp. 3308–3314 © 2009 American Chemical Society.

to the convex and concave architecture of the framework. A slice cut through the $Li_{0.16}MnO_2$ framework, with polyhedral rendering of the individual MnO_6 octahedra to aid visualization of the 1×1 tunnels (Figures 9.18b and c) shows distribution of Mn^{3+} species within the lattice together with Li species located within the tunnels. A twin boundary is shown more clearly in Figure 9.18d. In Figure 9.18e, a region conforming to ramsdellite (2×2 tunnel structure) is visible.

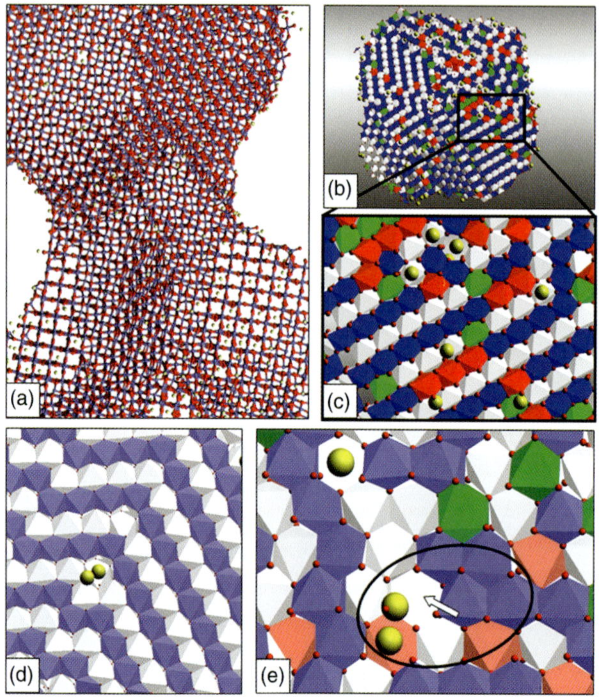

Figure 9.18 *Microstructural features: (a) slice cut through the nanoporous $Li_{0.16}MnO_2$ revealing the complex grain boundary structure; (b) polyhedral representation of the MnO_6 octahedra revealing the 1×1 tunnels and the Li ions that locate within the tunnels; (c) enlarged segment of (b); (d) region depicting a twin boundary; (e) region comprising a 2×2 tunnel (oval) indicative of a ramsdellite intergrowth; the arrow shows the position of a Mn vacancy. Li is colored yellow and Mn^{4+} is colored blue (upper plane) and white (lower plane); Mn^{3+} is green (upper plane) and red (lower plane). Mn^{3+} species in (a) and (d) are not distinguished from Mn^{4+}. Reprinted with permission from T.X.T. Sayle, P.E. Ngoepe and D.C. Sayle, "Simulating mechanical deformation in nanomaterials with application for energy storage in nanoporous architectures," ACS Nano, vol. 3, pp. 3308–3314 © 2009 American Chemical Society.*

Experimentally, ramsdellite intergrowths have been observed within pyrolusite (β-MnO_2) [82]. Specifically, it is proposed that the ramsdellite domains are stabilized with protons (Ruetschi defects). In previous simulations of β-MnO_2, no evidence of ramsdellite intergrowths was found using simulated $A + R$. It is surmised, therefore, that its appearance in the Li–MnO_2 model suggests that the Li species, which are located within the 1×1 tunnels, act as a "scaffold" stabilizing the 2×2 tunnel, similar to the proposal that protons stabilize ramsdellite intergrowths. Specifically, the 2×2 channel appears to be stabilized by two neighboring Li species, which are bound to the host MnO_2 lattice via Mn^{3+} species (colored red). Closer inspection of the ramsdellite intergrowth (Figure 9.18e) reveals the presence of a Mn vacancy (highlighted by the arrow). Similar to the twin boundaries, the evolution of vacancies within the structural model is achieved during the crystallization process. The stabilizing effect of Ruetschi defects on the ramsdellite structure was also reported by Balachandran *et al.* [74] on MnO_2.

9.3.2.4 TiO_2 Nanoporous

As discussed in Section 9.3.2.1, an amorphous nanoparticle of MnO_2 was used to produce various nanoforms based on the A + R method; nanoporous structures of TiO_2 can be generated using similar approaches. A TiO_2 nanoparticle was simulated by Sayle and Sayle [83], commencing from nucleation to a fully crystallized system. The final structure of the simulated nanocrystal comprised rutile- and α-PbO_2-structured TiO_2; the latter is known as a high-pressure phase of TiO_2. The nanocrystal expressed well-defined facets including (111), (010) and (001) and a small region corresponding to (1–10). Microstructural features include three grain boundaries conforming to (001)/(001) 30° twist, (001)/(111) twist and (010)/(111) and a grain junction between the three.

A simulated porous structure of TiO_2 has also been investigated [84]; the model was generated using A + R starting from amorphous nanoparticle precursors, as discussed in Section 9.3.2.1. A comparison with experimental XRD patterns suggests that the structure is likely to be brookite. In Figure 9.19a, a porous structure with well-defined crystalline patterns and grain boundaries is noted. Tunnels that are perpendicular and parallel to the surface together with the channel are observed. These concur well with the high-resolution transmission electron microscopy (TEM) micrographs [85] reflected in Figure 9.19b. Figure 9.19c depicts microstructural features with magnified tunnels that are connected to the channel, and the tunnels have zigzag patterns that were noted in the earlier A + R calculations of a TiO_2 nanosphere [83] and high-resolution TEM results in Figure 9.19d. As with the porous MnO_2, it may be deduced that the tunnels and channels provide accessible paths for Li ions, which will ultimately enhance the rate capability of related Li–ion batteries, and simultaneously enable many cycles to occur with the reduced level of degradation because the host lattice is able to expand and contract into the pores, thus minimizing the propensity for plastic deformation and fracture.

9.3.2.5 ZnS and ZnO Nanoporous

Model ZnS and ZnO nanoporous structures, generated using A + R, have been reported by Sayle *et al.* [44, 66]. Amorphous nanobuilding blocks of ZnS and ZnO were placed, with primitive filling, into a hexagonal supercell. The nanoparticles were then aggregated into a mesoporous framework, following the simulation strategy described in Section 9.2.2.2, and crystallized at 1700 K [66]. In Figure 9.20a, an array of channels in ZnS are observed, with the channel cross-section appearing more octagonal than hexagonal. The crystallization fronts, emanating from the individual (misoriented) seeds of ZnS, impinged upon each other, resulting in the evolution of a polycrystalline framework, comprising a plethora of

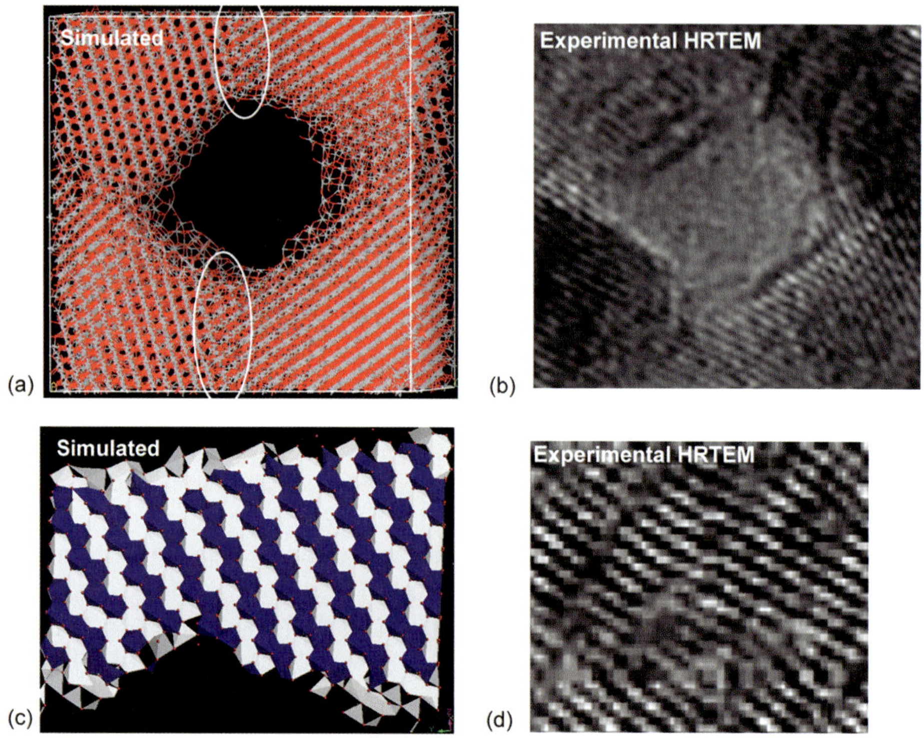

Figure 9.19 *(a) Simulated nanoporous structure of TiO_2, featuring crystallized patterns. Tunnels running into and parallel to the page, together with grain boundaries (depicted by ovals), are observed. Reprinted from M.G. Matshaba © 2012 University of Limpopo. The channel is shown at the center, together with (b), an image from HRTEM. (c) The microstructure can be noted from a cut slice. Ti^{4+} is colored blue (upper plane) and white (lower plane); and zigzag patterns are observed similar to (d) experimental results.(b and d) Reprinted with permission from Y. Ren, L. J. Hardwick and P.G. Bruce., Angew. Chem. Int. Ed., vol. 49, pp. 2570–2574 © 2010 Wiley-VCH Verlag GmbH & Co. KgaA. Reproduced with permission.*

grain boundaries and grain junctions that separate the individual grains, as shown in Figure 9.20b, with an enlarged portion viewed along one of the channels depicted in Figure 9.20c. The crystallization is similar to that observed by TEM reported by Lu *et al.* for metal oxide-based mesoporous materials [86]. In the study by Sayle *et al.* [66] movies were provided depicting such crystallization (with atomistic detail).

The ZnS nanoporous frameworks finally crystallize into various parts (2D system) shown in Figure 9.21. The atomistic structure of a region, which conforms to the wurtzite structure, is given in Figure 9.21a, and a region, comprising four- and eight-membered rings, in Figure 9.21b. In Figure 9.21c, the wurtzite structure can be adjudged to comprise hexagons accommodating a chair conformation, whereas conversely, in the four- and eight-membered ring structure (Figure 9.21d) the hexagons conform to a "boat-like" structure. A region that comprises both types of structure is shown in Figure 9.21e.

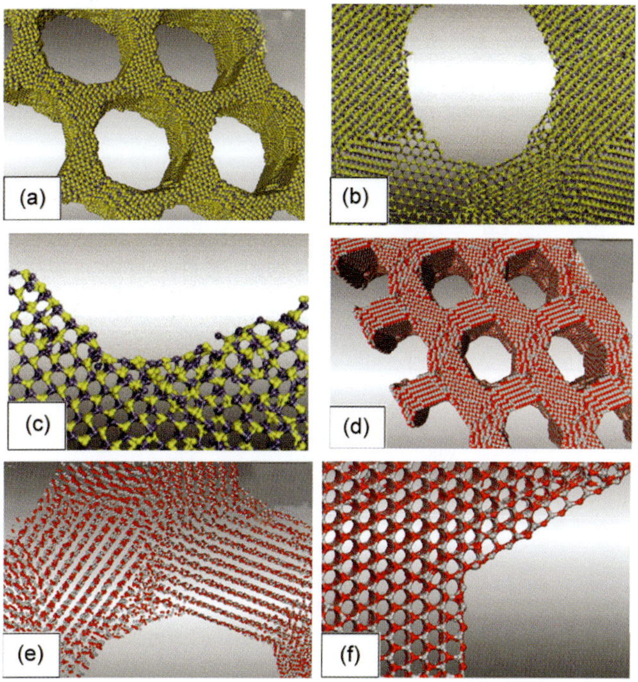

Figure 9.20 *Atomistic models of nanoporous ZnS (a–c) and ZnO (d–f). (a) A view looking along the hexagonal array of channels in ZnS. The channel cross-section appears more octagonal than hexagonal. (b) A view looking through one of the channels in the 2D system revealing the high curvature of the elliptical channel; various grain boundaries and grain junctions are evident in the ball-and-stick model representation. (c) An enlarged ball-and-stick model representation of the 1D system, viewing along one of the channels. (d) A view looking along the hexagonal array of channels in ZnO. (e) The enlarged segment of the ZnO framework where the atom positions are represented by small spheres to reveal that the hexagonal profiles of the channels are facilitated by misoriented grains. The curved structure of the triple junction is also evident from the figure. (f) Here, in contrast to (e), the 120° angle of the hexagonal channel is facilitated by the crystal structure of the ZnO and the figure shows a single crystal. Reproduced with permission from D.C. Sayle, B.C. Mangili, J. Klinowski and T.X.T. Sayle., J. Am. Chem. Soc., vol. 128, pp. 15283–15291 © 2006 American Chemical Society., and S. Seal et al., ACS Nano, vol. 2, pp. 1237–1251, © 2008 American Chemical Society.*

The "four- to eight-membered ring" structure, which is part of the ZnS-based nanoporous material, has the same crystal structure as zeolite BCT, as reported by Hamad and Catlow [51]. ZnS has two structural polymorphs—wurtzite (hexagonal) and sphalerite (cubic)—and therefore two surprising issues emerge from the simulation results: first, no sphalerite has evolved and, second, the evolution of extended regions that conform to the BCT phase. The presence of the four to eight-membered ring structure alludes to similarity of structures whether spontaneously built bottom up from small clusters with a few atoms by GM or top down using A + R, which involves tens of thousands of atoms; this demonstrates agreement on predictive capabilities of evolutionary simulation methods at the nanoscale. The

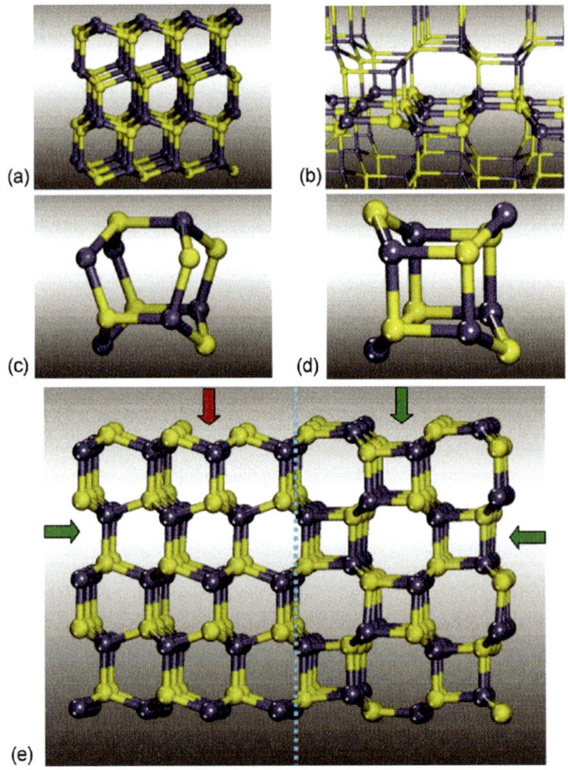

Figure 9.21 Ball-and-stick model representation of the Zn (blue) and S (yellow) atom positions comprising the 2D system. (a) Region conforming to the wurtzite structure; (b) region conforming to the four- to eight-membered ring structure; (c) segment of the wurtzite structure showing the hexagons in a chair conformation; (d) segment of the four- to eight-membered ring structure where the hexagons accommodate a boat conformation; (e) region comprising both the wurtzite and four- to eight-membered ring structures, with the green arrows indicating type I surfaces which have no dipole and the red arrow indicating a type III surface, which is associated with a surface dipole. Note that this segment has been cleaved from the 2D system, and the surfaces indicated by the arrows are hypothetical; they are not exposed at the surfaces of the channels or cavities. Reprinted with permission from D.C. Sayle, B.C. Mangili, J. Klinowski and T.X.T. Sayle., J. Am. Chem. Soc., vol. 128, pp. 15283–15291 © 2006 American Chemical Society.

authors noted that the BCT polymorph was not associated with a surface dipole in contrast to the wurtzite polymorph. Accordingly, at the nanoscale, the authors argue that the quenching of surface dipoles drives the evolution of particular polymorphs. Similar arguments may be applied to the observation of the BCT phase for the ZnS nanoclusters reported by Hamad and Catlow [51].

The final crystallized framework structures for ZnO are presented in Figures 9.20d–f. The ZnO structure comprises a hexagonal array of channels with hexagonal profile. This structure might have been anticipated because these structures can be "cut" from the (single crystal) parent material. However, inspection of the hexagonal channel profile (Figure 9.20e) reveals that the framework ZnO is not a single crystal; rather it comprises several misoriented grains as with ZnS. In addition, Figure 9.20c confirms that the hexagonal profile is facilitated not by the (hexagonal) crystal structure, but by grain boundaries. On the other hand, 120° angles required to facilitate a hexagonal channel were observed in some of the ZnO model structures; an enlarged segment of such a channel depicting the crystal structure is shown in Figure 9.20f. Similarly to ZnS, the polycrystalline framework and associated grain boundaries are attributed to impingement of crystallization fronts, emanating from various seeds, during nucleation and crystallization.

Atomistic models, such as those presented here, are valuable as starting points to calculate, for example, physical and mechanical properties, sensitivity to temperature and thermoelectric properties, catalytic activity at the surface of the channels or cavities (using quantum mechanical calculations), and electronic and optical properties.

9.4 Summary and Conclusion

In conclusion, the use of evolutionary methods has opened new possibilities for studying clusters of different sizes and a variety of nanoarchitectures. GM methods have generated small to large clusters that have similar configurations for both ZnO and ZnS, commencing with the planar, spherical bubbles, onion-like to bulk phases. In the case of ZnS, double bubbles consisting of inner and outer shells were observed; and a BCT structure appeared as an internal structure and had four and six rings. Similar BCT structure with such rings occurred in the A + R-generated nanoporous ZnS, hence indicating that such evolutionary methods yield similar structures at an appropriate scale—bottom up or top down.

Furthermore, smaller clusters produced by GM approach are similar for MnO_2 and TiO_2 at least in the studied range of $n = 1–4$. In both cases the smallest cluster determined by IP methods is linear, whereas the DFT yields a boomerang shape. A key point is that the influence of the d-electrons with regard to bond angles is not usually accounted for in models based on IPs, that is, when searching for plausible structures. Typically, this influence has a greater effect on the bond angles of undercoordinated atoms on the surface, which can always be corrected at a later stage, for example, further refinements of the plausible structures using DFT approaches.

The smallest nucleation seed, which was noted in MnO_2, occurred at 750 ps and no seed was observed before that time. However, that smallest seed consisted of several atoms where the pyrolusite/rutile-type pattern was noted. We assume that in the earlier stages of nucleation smaller clusters similar to those of the 1 : 2 compounds are formed and contribute toward nucleation and crystal growth of the necessary seed until full crystallization is realised. It is proposed that a similar process occurs in various nanoarchitectures and their lithiated forms. In most other cases, such as TiO_2, ZnO and ZnS, multiple seed crystallization fronts impinge on each other resulting in a polycrystalline framework consisting of many grain boundaries.

Crucially, it is the evolution of nucleating seeds within the structure that drives the microstructure. In particular, if only one nucleating seed evolves within the time it takes to facilitate full crystallization, a single crystal will result. Conversely, if multiple seeds evolve prior to full crystallization, a complex microstructure will result. Simulation therefore indicates that if experiment can control such seed evolution, then this will enable control over the microstructure.

Having explored growth mechanisms of clusters and seeds by evolutionary methods, once the lowest energy varying sized clusters up to complex structures have been generated, many interesting properties of various energy materials, such as quantum size effects on photophysical properties, adsorption on rather small nanostructures, sintering, and nucleation, transport, voltage profiles and Ragone plots, can be predicted by different modeling methods. In addition, various complex microstructures for different materials and nanoarchitectures are generated and provide very valuable insights. One of the novel examples of the current simulations has been to predict pores, channels, tunnels and paths that can be followed by Li ions in nanoporous structures. In addition, it has elucidated how symmetric distribution of pores and wall thicknesses can contribute toward enhanced capacity, rate capability and long life cycles before degradation of electrodes can set in. Such properties can inform the design of energy materials for a variety of applications. It is envisaged that with increasing computational power, simulation studies using evolutionary methods will soon be extended to and beyond quarternary compounds and ably facilitate prediction of their properties.

Acknowledgments

We acknowledge the financial support from the National Research Foundation, Pretoria, and Kenneth Kgatwane for technical assistance.

References

1. Cao, G. and Wang, Y. (2011) *Nanostructures and Nanomaterials, Synthesis, Properties and Applications*, Vol. 2, *World Scientific Series in Nanoscience and Nanotechnology*, 2nd edn, World Scientific Publishing, London, UK.
2. Zheng, J.W., Nai, S.M.L., Ng, M.-F. *et al.* (2009) DFT study on nano structures of Sn/CNT complex for potential Li-ion battery application, *J. Phys. Chem. C* **113**, 14015–14019.
3. Jellinek, J., Johnston, R.L. and Ferrando, R. (2008) Nanoalloys: From theory to applications of alloy clusters and nanoparticles, *Chem. Rev.* **108**, 845–910.
4. Landi, B.J., Ganter, M.J., Schauerman, C.M. *et al.* (2008) Lithium ion capacity of single wall carbon nanotube paper electrodes, *J. Phys. Chem. C* **112**, 7509–7515.
5. Samia, A.C., Dayal, S. and Burda, C. (2006) Quantum dot-based energy transfer: perspectives and potential for applications in photodynamic therapy, *Photochem. Photobiol.* **82**, 617–625.
6. Gao, G.Z. (2008) Popcorn-style solar cells, *Photon. Spectra* **42**, 60–61.

7. Zhang, Q.F., Fryxell, G.E., Cao, G.Z. and Chou, T.P. (2007) Hierarchically structured ZnO film for dye-sensitized solar cells with enhanced energy conversion efficiency, *Adv. Mater.* **19**, 2588–2592.
8. Zhang, Q.F., Chou, T.P., Russo, B. *et al.* (2008) Polydisperse aggregates of ZnO nanocrystallites: A method for energy-conversion-efficiency enhancement in dye-sensitized solar cells, *Adv. Funct. Mater.* **18**, 1654–1660.
9. Zhang, Q.F., Chou, T.R., Russo, B. *et al.* (2008) Aggregation of ZnO nanocrystallites for high conversion efficiency in dye-sensitized solar cells, *Angew. Chem. Int. Ed.* **47**, 2402–2406.
10. Zhang, Q., Park, K. and Cao, G. (2010) Synthesis of ZnO aggregates and their application in dye-sensitized solar cells, *Mater. Matt.* **5.2**, 32.
11. Feldmann, C., Jüstel, T., Ronda, C.R. and Schmidt, P.J. (2003) Inorganic luminescent materials: 100 years of research and application, *Adv. Funct. Mater.* **13**, 511–516.
12. Zhao, Q., Xie, Y., Zhang, Z. and Bai, X. (2007) Size-selective synthesis of zinc sulfide hierarchical structures and their photocatalytic activity, *Cryst. Growth Des.* **7**, 153–158.
13. Gilbert, B., Frazer, B.H., Zhang, H. *et al.* (2002) X-ray absorption spectroscopy of the cubic and hexagonal polytypes of zinc sulfide, *Phys. Rev. B Condens. Matter. Mater. Phys.* **66**, 245205.
14. Zhang, H.Z., Huang, F., Gilbert, B. and Banfield, J.F. (2003) Molecular dynamics simulations, thermodynamic analysis and experimental study of phase stability of zinc sulfide nanoparticles, *J. Phys. Chem. B* **107**, 13051–13060.
15. Cheng, F., Zhao, J., Song, W. *et al.* (2006) Facile controlled synthesis of MnO_2 nanostructures of novel shapes and their application in batteries, *Inorg. Chem.* **45**, 2038–2044.
16. Sayle, T.X.T., Maphanga, R.R., Ngoepe, P.E. and Sayle, D.C. (2009) Predicting the electrochemical properties of MnO_2 nanomaterials used in rechargeable Li batteries: Simulating nanostructure at the atomistic level, *J. Am. Chem. Soc.* **131**(17), 6161–6173.
17. Bruce, P.G., Scrosati, B. and Tarascon, J.-M. (2008) Nanomaterials for rechargeable lithium batteries, *Angew. Chem. Int. Ed.* **47**, 2930–2946.
18. Thompson, T.L. and Yates, J.T. (2006) Surface science studies of the photoactivation of TiO_2 – new photochemical processes, *Chem. Rev.* **106**, 4428–4453.
19. Chen, X. and Mao, S.S. (2007) Titanium dioxide nanomaterials: synthesis, properties, modifications and applications, *Chem. Rev.* **107**, 2891–2959.
20. Asahi, R., Ohwaki, T., Morikawa, T., Aoki, K. and Taga, Y. (2001) Visible-light photocatalysis in nitrogen-doped titanium oxides, *Science* **293**, 269–271.
21. Gratzel, M. (2001) Photoelectrochemical cells, *Nature* **414**, 338–344.
22. Khan, S.U.M., Al-Shahry, M. and Ingler, W.B., Jr. (2002) Efficient photochemical water splitting by a chemically modified n-TiO_2, *Science* **297**, 2243–2245.
23. Jiang, C.H., Honma, I., Kudo, T. and Zhou, H.S. (2007) Particles size dependence of the lithium storage capacity and high rate performance of nanocrystalline anatase TiO_2 electrode, *J. Power Sources.* **166**, 239–243.
24. Li, J.R., Tang, Z.L. and Zhang, Z.T. (2005) Preparation and novel lithium intercalation properties of titanium oxide nanotubes, *Electrochem. Solid State Lett.* **8**, A316–A319.
25. Wang, Q., Wen, Z.H. and Li, J.H. (2006) A hybrid supercapacitor fabricated with a carbon nanotube cathode and a TiO_2-B nanowire anode, *Adv. Funct. Mater.* **16**, 2141–2146.

26. Wang, Z.Y., Liu, S.Z., Chen, G. and Xia, D.G. (2007) Preparation and Li-intercalation properties of mesoporous anatase-TiO_2 spheres, *Electrochem. Solid State Lett.* **10**, A77–A80.
27. Kwon, S., Fan, M., Cooper, A.T. and Yang, H.Q. (2008) Photocatalytic applications of micro- and nano-TiO_2 in environmental engineering, *Crit. Rev. Environ. Sci. Technol.* **38**, 197–226.
28. Sayle, T.X.T., Catlow, C.R.A., Maphanga, R.R. *et al.* (2006) Evolving microstructure in MnO_2 using amorphisation and recrystallization, *J. Crystal Growth* **294**, 118–129.
29. Catlow, C.R.A., Bromley, S.T., Hamad, S. *et al.* (2010) Modelling nano-clusters and nucleation, *Phys. Chem. Chem. Phys.* **12**, 786–811.
30. Sayle, D.C. and Johnston, R.L. (2003) Evolutionary techniques in atomistic simulation: thin films and nanoparticles, *Curr. Opin. Solid State Mater. Sci.* **7**, 3–12.
31. Smith, W. and Forester, T.R. (1996) DL_POLY_2.0: A general-purpose parallel molecular dynamics simulation package, *J. Mol. Graph.* **14**, 136–141.
32. Catlow, C.R.A. (1994) Computational solid state chemistry, *Comp. Mater. Sci.* **2**, 6–18.
33. Gale, J.D. (1997) GULP, *J. Chem. Soc. Faradays Trans.* **92**, 629.
34. Koch, W. and Holthausen, M.C. (2001) *A Chemist's Guide to Density Functional Theory*, Wiley-VCH Verlag GmbH, Weinheim.
35. Martin, R.M. (2004) *Electronic Structure. Basic Theory and Practical Methods*, Cambridge University Press, Cambridge, UK.
36. Dovesi, R., Civalleri, B., Orlando, R. *et al.* (2005) Ab initio quantum simulation in solid state chemistry, *Rev. Comput. Chem.* **21**, 1–125.
37. Yip, S. (2005) *Handbook of Materials Modeling*, Springer, Berlin.
38. Jensen, F. (2006) *Introduction to Computational Chemistry*, John Wiley & Sons, Ltd, New York.
39. Behrman, E.C., Foehrweiser, R.K., Myers, J.R. *et al.* (1994) Possibility of stable spheroid molecules of ZnO, *Phys. Rev. A At. Mol. Opt. Phys.* **49**, R1543–R1546.
40. Woodley, S.M. and Catlow, C.R.A. (2008) Crystal structure prediction from first principles, *Nat. Mater.* **7**, 937–946.
41. Catlow, C.R.A., Guo, Z.X., Miskufova, M. *et al.* (2010) Advances in computational studies of energy materials, *Phil. Trans. R. Soc. A* **368**, 3379–3456.
42. Sayle, D.C. (1999) The predicted 3-D atomistic structure of an interfacial screw-edge dislocation, *J. Mater. Chem.* **9**, 2961–2964.
43. Sayle, D.C. and Parker, S.C. (2003) Encapsulated oxide nanoparticles: The influence of the microstructure on associated impurities within a material, *J. Am. Chem. Soc.* **125**, 8581–8588.
44. Seal, S., Wang, Z., Mangili, B.C. *et al.* (2008) Mapping nanostructure: a systematic enumeration of nanomaterials by assembling nanobuilding blocks at crystallographic po, *ACS Nano.* **2**, 1237–1251.
45. Al-Sunaidi, A.A., Sokol, A.A., Catlow, C.R.A. and Woodley, S.M. (2008) Structures of zinc oxide nanoclusters: as found by revolutionary algorithm techniques, *J. Phys. Chem. C.* **112**, 18860–18875.
46. Shevlin, S.A., Guo, Z.X., van Dam, H.J.J. *et al.* (2008) Structure, optical properties and defects in nitride (III–V) nanoscale cage clusters, *Phys. Chem. Chem. Phys.* **10**, 1944–1959.
47. Whetten, R.L. (1993) Processing of single nanocrystals, *Mater. Sci. Eng.* **B19**, 8–13.

48. Becke, A.D. (1993) Density-functional thermochemistry: III. The role of exact exchange, *J. Chem. Phys.* **98**, 5648–5652.
49. Matxain, J.M., Fowler, J.E. and Ugalde, J.M. (2000) Small clusters of II-VI materials: Zn_iO_i, i=1-9, *Phys. Rev. A At. Mol. Opt. Phys.* **61**, 053201–053208.
50. Burnin, A., Sanville, E. and BelBruno, J.J. (2005) Experimental and computational study of the Zn(n)S(n) and $Zn(n)S(n)^+$ clusters, *J. Phys. Chem. A* **109**, 5026–5034.
51. Hamad, S. and Catlow, C.R.A. (2006) Computational study of the relative stabilities of ZnS clusters, for sizes between 1 and 4 nm, *J. Cryst. Growth* **294**, 2–8.
52. Pal, S., Sharma, R., Goswami, B. and Sarkar, P. (2009) Theoretical prediction of ring structures for ZnS quantum dots, *Chem. Phys. Lett.* **467**, 365–368.
53. Spano, E., Hamad, S. and Catlow, C.R.A. (2004) ZnS bubble clusters with onion-like structures, *Chem. Commun.* 864–865.
54. Hamad, S., Catlow, C.R.A., Spano, E. *et al.* (2005) Structure and properties of ZnS nanoclusters, *J. Phys. Chem. B* **109**, 2703–2709.
55. Hu, J., Bando, Y., Zhan, J. and Golberg, D. (2004) Sn-filled single-crystalline wurtzite-type ZnS nanotubes, *Angew. Chem. Int. Ed.* **43**, 4606–4609.
56. Wang, Z., Qian, X.-F., Li, Y. *et al.* (2005) Large-scale synthesis of tube-like ZnS and cable-like ZnS ZnO arrays: preparation through the sulfuration conversion from ZnO arrays via a simple chemical solution route, *J. Solid State Chem.* **178**, 1589–1594.
57. Zhu, Y.-C., Bando, Y. and Uemura, Y. (2003) ZnS–Zn nanocables and ZnS nanotubes, *Chem. Commun.* 836–837.
58. Shao, H., Qian, X. and Huang, B. (2007) Fabrication of single-crystal ZnO nanorods and ZnS nanotubes through a simple ultrasonic chemical solution method, *Mater. Lett.* **61**, 3639–3643.
59. Yin, L.W., Bando, Y., Zhan, J.H. *et al.* (2005) Self-assembled highly faceted wutzite-type ZnS single-crystalline nanotubes with hexagonal cross-sections, *Adv. Mater.* **17**, 1972–1977.
60. Zhang, H., Zhang, S., Pan, S. *et al.* (2004) A simple solution route to ZnS nanotubes and hollow nanospheres and their optical properties, *Nanotechnology* **15**, 945–948.
61. Pal, S., Goswami, B. and Sarkar, P. (2007) Theoretical study on the structural, energetic, and optical properties of ZnS nanotube, *J. Phys. Chem. C* **111**, 1556–1559.
62. Li, L., Zhao, M., Zhang, X. *et al.* (2008) Theoretical insight into faceted ZnS nanowires and nanotubes from interatomic potential and first-principles calculations, *J. Phys. Chem. C* **112**, 3509–3514.
63. Zhang, X., Zhao, M., Yan, S. *et al.* (2008) First-principles study of ZnS nanostructures: nanotubes, nanowires and nanosheets, *Nanotechnology* **19**, 305708.
64. Zhao, M., He, T., Li, W. *et al.* (2008) Theoretical models of ZnS nanoclusters and nanotubes: First-principles calculations, *Solid State Commun.* **147**, 165–168.
65. Hamad, S., Cristol, S. and Catlow, C.R.A. (2002) Surface structures and crystal morphology of ZnS: computational study, *J. Phys. Chem. B* **106**, 11002.
66. Sayle, D.C., Mangili, B.C., Klinowski, J. and Sayle, T.X.T. (2006) Simulating self-assembly of ZnS nanoparticles into mesoporous materials, *J. Am. Chem. Soc.* **128**, 15283–15291.
67. Hamad, S., Catlow, C.R.A., Woodley, S.M. *et al.* (2005) Structure and stability of small TiO_2 nanoparticles, *J. Phys. Chem. B* **109**, 15741–15748.
68. Francisco, E., Martín Pendás, A. and Blanco, M.A. (2005) Global optimization of ionic Mg_nF2_n (n=1–30) clusters, *J. Chem. Phys.* **123**, 234305–234308.

69. Hill, J.R., Freeman, C.M. and Rossouw, M.H. (2004) Understanding γ-MnO_2 by molecular modeling, *J. Solid State Chem.* **177**, 165–175.
70. Maphanga, R.R., Parker, S.C. and Ngoepe, P.E. (2009) Atomistic simulation of the surface structure of electrolytic manganese dioxide, *Surf. Sci.* **603**, 3184–3190.
71. Sayle, T.X.T., Catlow, C.R.A., Maphanga, R.R. *et al.* (2005) Generating MnO_2 nanoparticles using simulated amorphization and recrystallization, *J. Am. Chem. Soc.* **127**, 12828–12837.
72. Zwijnenburg, M.A., Sokol, A.A., Sousa, C. and Bromley, S.T. (2009) The effect of local environment on photoluminescence: a time-dependent density functional theory study of silanone groups on the surface of silica nanostructures, *J. Chem. Phys.* **131**, 034705.
73. Kobayashi, S., Kottegoda, I.R.M., Uchimoto, Y. and Wakihara, M. (2004) XANES and EXAFS analysis of nano-size manganese dioxide as a cathode material for lithium-ion batteries, *J. Mater. Chem.* **14**. 1843–1848.
74. Balachandran, D., Morgan, D., Ceder, G. and van de Walle, A. (2003) First-principles study of the structure of stoichiometric and Mn-deficient MnO_2, *J. Solid State Chem.* **173**, 262–275.
75. Bowden, W., Grey, C.P., Hackney, S. *et al.* (2006) Lithiation of ramsdellite–pyrolusite MnO_2; NMR, XRD, TEM and electrochemical investigation of the discharge mechanism, *J. Power Sources* **153**, 265–273.
76. Maphanga, R.R., Sayle, D.C., Sayle, T.X.T. and Ngoepe, P.E. (2011) Amorphization and recrystallization study of lithium insertion into manganese dioxide, *Phys. Chem. Chem. Phys.* **13**, 1307–1313.
77. Whittingham, M.S. and Zavalij, P.Y. (2000) Manganese dioxides as cathodes for lithium rechargeable cells: the stability challenge, *Solid State Ion.* **131**, 109–115.
78. Zhou, Q., Li, X., Li, Y.G. *et al.* (2006) Synthesis and electrochemical properties of semicrystalline gyroidal mesoporous MnO_2, *Chin. J. Chem.* **24**, 835–839.
79. Jiao, F. and Bruce, P.G. (2007) Mesoporous crystalline β-MnO_2 – a reversible positive electrode for rechargeable lithium batteries, *Adv. Mater.* **19**, 657–660.
80. Lai, M., Kulak, A.N., Law, D. *et al.* (2007) Profiting from nature: macroporous copper with superior mechanical properties, *Chem. Commun.* **34**, 3547–3549.
81. Sayle, T.X.T., Ngoepe, P.E. and Sayle, D.C. (2009) Simulating mechanical deformation in nanomaterials with application for energy storage in nanoporous architectures, *ACS Nano.* **3**, 3308–3314.
82. Chabre, Y. and Pannetier, J. (1995) Structural and electrochemical properties of the proton gamma-mno_2 system, *Prog. Solid State Chem.* **23**, 1–130.
83. Sayle, D.C. and Sayle, T.X.T. (2007) High-pressure crystallisation of TiO_2 nanoparticles, *J. Compt. Theoret. Nanosci.* **4**, 299–308.
84. Matshaba, M.G. (2012) Atomistic Simulation Studies of Pure and Lithiated Nano-Structural Titanium Dioxide, PhD thesis, University of Limpopo.
85. Ren, Y., Hardwick, L.J. and Bruce, P.G. (2010) Lithium Intercalation into mesoporous anatase with an ordered 3d pore structure, *Angew. Chem. Int. Ed.* **49**, 2570–2574.
86. Lu, D., Katou, T., Uchida, M. *et al.* (2005) In situ TEM observation of crystallization of amorphous ordered mesoporous Nb–Ta and Mg–Ta mixed oxides, *Chem. Mater.* **17**, 632–637.

Further Reading

A Chemist's Guide to Density Functional Theory—W. Koch and M.C. Holthausen (John Wiley & Sons, Ltd, 2001)

Applications of Evolutionary Computation in Chemistry—R.L. Johnston in Structure and Bonding Vol. 110 (Springer, 2004)

Band Theory and Electronic Properties of Solids—J. Singleton (Oxford University Press, 2001)

Computational Materials Science: An Introduction—J.G. Lee (CRC Press, 2011)

Computer Modeling in Inorganic Crystallography—C.R.A. Catlow (Academic Press, 1997)

Computer Simulation of Liquids—M.P. Allen and D.J. Tildesley (Oxford University Press, 1989)

Computer Simulation of Solids—C.R.A. Catlow and W.C. Mackrodt (Springer, 1982)

Density-functional Theory of Atoms and Molecules—R.G. Parr and W. Yang (Oxford University Press, 1994)

Dynamical Theory of Crystal Lattices—M. Born and H. Huang (Oxford University Press, 1954)

Electronic Structure—R.M. Martin (Cambridge University Press, 2008)

Elementary Electronic Structure—W.A. Harrison (World Scientific, 1999)

Energy Materials to Combat Climate Change—P. Bruce, C.R.A. Catlow and P.P. Edwards in Philosophical Transactions of the Royal Society A, vol. 368 (2010).

Interatomic Forces in Condensed Matter —M.W. Finnis (Oxford University Press, 2003)

Introduction to Computational Chemistry—F. Jensen (John Wiley & Sons, Ltd, 2006)

Molecular Modelling: Principles and Application—A. Leach (Prentice Hall, 2001)

Quantum Theory of the Solid State: An Introduction—L.N. Kantorovich (Kluwer Academic Publishers, 2004)

Solid State Physics—N.W. Ashcroft and N.D. Mermin (Brooks/Cole, 1976)

Statistical Mechanics: Theory and Molecular Simulation—M. Tuckerman (Oxford University Press, 2010)

Structure and Dynamics—M.T. Dove (Oxford University Press, 2003)

Theory of Defects in Solids—A.M. Stoneham (Oxford University Press, 1979)

Understanding Molecular Simulation: From Algorithms to Applications—D. Frenkel and B. Smit (Academic Press, 2001)

Computational Approaches to Energy Materials, First Edition. Edited by Aron Walsh, Alexey A. Sokol and C. Richard A. Catlow.

Index

Computational Approaches to Energy Materials, First Edition. Edited by Aron Walsh, Alexey A. Sokol and C. Richard A. Catlow.